Seminars in Motor Control

Seminars in Motor Control

MARK L. LATASH

Distinguished Professor, Department of Kinesiology, The Pennsylvania State University, University Park, PA

OXFORD
UNIVERSITY PRESS

Oxford University Press is a department of the University of Oxford.
It furthers the University's objective of excellence in research, scholarship,
and education by publishing worldwide. Oxford is a registered trade mark of
Oxford University Press in the UK and certain other countries.

Published in the United States of America by Oxford University Press
198 Madison Avenue, New York, NY 10016, United States of America.

© Oxford University Press 2025

All rights reserved. No part of this publication may be reproduced, stored in a retrieval system, transmitted, used for text and data mining, or used for training artificial intelligence, in any form or by any means, without the prior permission in writing of Oxford University Press, or as expressly permitted by law, by license or under terms agreed with the appropriate reprographics rights organization. Inquiries concerning reproduction outside the scope of the above should be sent to the Rights Department, Oxford University Press, at the address above.

You must not circulate this work in any other form
and you must impose this same condition on any acquirer

Library of Congress Cataloging-in-Publication Data
Names: Latash, Mark L., 1953- author.
Title: Seminars in motor control / [Mark L. Latash].
Description: New York, NY : Oxford University Press, [2025] |
Includes bibliographical references and index.
Identifiers: LCCN 2024044866 | ISBN 9780197794340 (hardback) | ISBN 9780197794357 (epub) |
ISBN 9780197794364 | ISBN 9780197794371 | ISBN 9780197794388
Subjects: LCSH: Motor ability. | Neurophysiology.
Classification: LCC BF295 .L325 2025 |
DDC 152.3/34–dc23/eng/20241203
LC record available at https://lccn.loc.gov/2024044866

This material is not intended to be, and should not be considered, a substitute for medical or other professional advice. Treatment for the conditions described in this material is highly dependent on the individual circumstances. And, while this material is designed to offer accurate information with respect to the subject matter covered and to be current as of the time it was written, research and knowledge about medical and health issues is constantly evolving and dose schedules for medications are being revised continually, with new side effects recognized and accounted for regularly. Readers must therefore always check the product information and clinical procedures with the most up-to-date published product information and data sheets provided by the manufacturers and the most recent codes of conduct and safety regulation. The publisher and the authors make no representations or warranties to readers, express or implied, as to the accuracy or completeness of this material. Without limiting the foregoing, the publisher and the authors make no representations or warranties as to the accuracy or efficacy of the drug dosages mentioned in the material. The authors and the publisher do not accept, and expressly disclaim, any responsibility for any liability, loss or risk that may be claimed or incurred as a consequence of the use and/or application of any of the contents of this material.

DOI: 10.1093/med/9780197794340.001.0001

Printed by Marquis Book Printing, Canada

Contents

Preface   xi

PART I: BASIC CONCEPTS

1. Philosophy   3
   - 1.1. Laws of nature in the inanimate world   4
   - 1.2. Specificity of living objects   5
   - 1.3. Stretch reflex as a biological law of nature   6
   - 1.4. How many sets of laws of nature are there?   9
   - 1.5. Missing pieces of the mosaic   14

2. Bernstein's Construction of Movements   15
   - 2.1. Nikolai Bernstein: Philosopher and experimentalist   15
   - 2.2. The evolutionary approach to movement construction   17
   - 2.3. Problem of motor redundancy   22
   - 2.4. Sharing and optimality   23
   - 2.5. Missing pieces of the mosaic   25

3. Equilibrium-Point Hypothesis   26
   - 3.1. Roots of the equilibrium-point hypothesis   27
   - 3.2. The equilibrium-point hypothesis: Single-muscle control   30
   - 3.3. The equilibrium-point hypothesis: Single-joint control   35
   - 3.4. Sources of misunderstanding: The alpha-model   39
   - 3.5. Missing pieces of the mosaic   43

4. Motor Programming   44
   - 4.1. Engrams and the generalized motor program   45
   - 4.2. Control with patterns of muscle activation   47
   - 4.3. Internal models   52
   - 4.4. Missing pieces of the mosaic   57

5. Principle of Abundance and the Uncontrolled Manifold Hypothesis   58
   - 5.1. The principle of abundance   59
   - 5.2. The uncontrolled manifold hypothesis   61
     - 5.2.1. Analysis in kinematic spaces   63
     - 5.2.2. Analysis in kinetic spaces   64
     - 5.2.3. Analysis in muscle activation spaces   67
   - 5.3. Dealing with nonlinear systems   69
   - 5.4. Motor equivalence   71
   - 5.5. Missing pieces of the mosaic   74

PART II: CURRENT UNDERSTANDING

6. Synergies   79
    - 6.1. Bernstein's understanding of synergies and its development   80
    - 6.2. Intra-muscle and multi-muscle synergies   83
    - 6.3. Synergies in kinematic and kinetic spaces   86
    - 6.4. Synergies in spaces of control variables   93
    - 6.5. Possible neurophysiological mechanisms   95
    - 6.6. Missing pieces of the mosaic   96

7. Control With Spatial Referent Coordinates   98
    - 7.1. Referent coordinate as generalization of lambda   99
    - 7.2. Hierarchical control with referent coordinates   101
    - 7.3. Synergies in spaces of referent coordinates: Analysis of mechanics   102
    - 7.4. Synergies in spaces of referent coordinates: Analysis of muscle activations   105
    - 7.5. Synergies stabilizing referent coordinates   108
    - 7.6. Missing pieces of the mosaic   109

8. Anticipatory Control of Action   111
    - 8.1. Anticipatory postural adjustments   112
    - 8.2. Early postural adjustments   115
    - 8.3. Grip adjustments to planned actions   117
    - 8.4. Anticipatory synergy adjustments   120
    - 8.5. Distinguishing APAs from ASAs   123
    - 8.6. Missing pieces of the mosaic   125

9. Stability, Agility, and Optimality   126
    - 9.1. Definitions and metrics   126
    - 9.2. Optimization in human movements   130
    - 9.3. Inverse optimization   132
    - 9.4. Optimality-stability trade-off   133
    - 9.5. Agility-stability trade-off   135
    - 9.6. Missing pieces of the mosaic   137

10. Brain Circuitry   139
    - 10.1. What variables are encoded by brain signals?   140
    - 10.2. What is encoded by neuronal populations?   142
    - 10.3. Relations between brain structures and functions   144
    - 10.4. The role of spinal circuitry   147
    - 10.5. Effects of dominance   148
    - 10.6. Missing pieces of the mosaic   151

PART III: EFFECTORS AND BEHAVIORS

11. Synergic Control of a Muscle   155
    - 11.1. Steps and challenges in analysis of motor unit–based synergies   156
    - 11.2. Agonist-antagonist interactions at the motor unit level   159

11.3.	Stabilization of reflex-induced force changes	163
11.4.	Spinal vs. supraspinal synergies	165
11.5.	Missing pieces of the mosaic	168

12. The Hand — 170
12.1.	Muscle organization of the hand	170
12.2.	Indices of finger interaction	172
12.3.	Finger modes	173
12.4.	Grip force	175
12.5.	Prehension synergies	177
12.6.	Principle of superposition	181
12.7.	Force- and moment-stabilizing synergies	182
12.8.	Missing pieces of the mosaic	185

13. Reaching Movement — 187
13.1.	Spinal coordination of multi-joint movements	188
13.2.	Control of reaching with spatial referent coordinates	190
13.3.	Multi-joint synergies	191
13.4.	Equifinality of reaching movements and its violations	194
13.5.	Reach-to-grasp	195
13.6.	Reaching with the dominant and non-dominant arms	197
13.7.	Missing pieces of the mosaic	199

14. Posture and Whole-Body Actions — 201
14.1.	Postural sway and its components	202
14.2.	Posture-stabilizing mechanisms	205
14.3.	Whole-body voluntary movements	207
14.4.	Whole-body synergies	209
14.5.	Locomotion and central pattern generators	212
14.6.	Missing pieces of the mosaic	215

15. Kinesthetic Perception — 217
15.1.	Ambiguity of sensory information	218
15.2.	Perception of muscle length and force	219
15.3.	Stability of percepts: Iso-perceptual manifold	222
15.4.	Vibration-induced illusions	224
15.5.	Interpreting impossible sensory signals	226
15.6.	Missing pieces of the mosaic	227

PART IV: SURPRISING PHENOMENA

16. Drifts in Action — 231
16.1.	Spontaneous force drifts	232
16.2.	Drifts in neural control variables	235
16.3.	Faster drifts triggered by quick force changes	239
16.4.	Unintentional kinematic drifts	241

16.5.	Drifts in whole-body tasks	242
16.6.	Drifts in indices of finger interaction	243
16.7.	Classification of movements	247
16.8.	Missing pieces of the mosaic	248

17. Efference Copy — 249

17.1.	Von Holst's concept of efference copy	250
17.2.	Efference copy as a referent coordinate	252
17.3.	Is efference copy a copy of efference?	253
17.4.	Perception and production of force	256
17.5.	Muscle vibration and stability of percepts	259
17.6.	The place for sense of effort	261
17.7.	Missing pieces of the mosaic	262

18. Equifinality and Motor Equivalence — 264

18.1.	Examples of equifinality and motor equivalence	265
18.2.	Equifinality and the equilibrium-point hypothesis	267
18.3.	Violations of equifinality in different spaces	269
18.4.	Motor equivalence and the uncontrolled manifold hypothesis	271
18.5.	Motor equivalence as a promising clinical index	275
18.6.	Missing pieces of the mosaic	277

19. Muscle Coactivation — 278

19.1.	Surprising behavior of antagonist muscles	279
19.2.	Features of the coactivation command	284
19.3.	Does negative coactivation exist?	287
19.4.	Changes in the C-command and their (mis)perception	289
19.5.	Consequences of increased coactivation	291
19.6.	Missing pieces of the mosaic	292

PART V: IMPROVEMENTS AND IMPAIRMENTS

20. Improvements in Motor Performance — 297

20.1.	Bernstein's three stages	298
20.2.	Can the way our brain controls movements be changed?	300
20.3.	Changes in motor synergies with practice	302
20.4.	Variability vs. stereotypy	304
20.5.	Developmental changes	305
20.6.	Motor rehabilitation: From magic to theory-based approaches	306
20.7.	Missing pieces of the mosaic	309

21. Decline in Motor Performance — 310

21.1.	Fatigue: Peripheral and central effects	311
21.2.	Changes in synergies under fatigue	315
21.3.	Aging: Effects on muscles, neurons, and performance	317
21.4.	Changes in synergies with age	320

21.5.	Adaptive and maladaptive changes	321
21.6.	Missing pieces of the mosaic	322

22. Motor Disorders in Neurological Patients — 324
22.1.	Large-fiber peripheral neuropathy	325
22.2.	Spinal cord injury and spasticity	327
22.3.	Parkinson disease	331
22.4.	Other subcortical disorders	335
22.5.	Stroke	338
22.6.	Missing pieces of the mosaic	340

PART VI: METHODOLOGY

23. Types of Studies and Hypothesis Testing — 343
23.1.	Types of studies	344
23.2.	Is a hypothesis worth testing?	345
23.3.	Can a hypothesis account for the existing knowledge?	347
23.4.	Can a hypothesis be used to make new testable predictions?	349
23.5.	Exploration and development of a hypothesis	353
23.6.	Missing pieces of the mosaic	356

24. Measuring Hidden Variables — 358
24.1.	Measuring "biomechanical" variables	359
24.2.	Measuring lambda	362
24.3.	Measuring lambda in clinical studies	367
24.4.	Missing pieces of the mosaic	369

25. Writing Papers — 371
25.1.	Inviting coauthors and selecting a journal	372
25.2.	Introduction	373
25.3.	Methods	375
25.4.	Results	378
25.5.	Tables and figures	379
25.6.	Discussion	381
25.7.	Responding to reviews	382
25.8.	Rejection: What to do next?	383

References — 385
Index — 431

Preface

This book is a result of my 25+ years of experience running weekly meetings of the Motor Control Laboratory at Penn State University. The laboratory was created in 1995, and its weekly meetings have been crucial to maintaining the continuity and coherent focus of its research program given the changes in the personnel inevitable over such a long time. Numerous graduate students, postdoctoral fellows, and visiting scholars from all over the world trained and worked in the laboratory. They all contributed to the development of the research program and its modifications, expansion, and refinement. Every year, at the beginning of the fall semester, our meetings started with an overview of the main ideas, theories, hypotheses, and toolboxes to facilitate a smooth transition of the newcomers. Then we discussed current and emerging projects and how they could contribute to our understanding of motor control. New projects emerged and were sometimes written on the whiteboard (and then wiped off to make space for newer ones!).

Until a few years ago, lab members would assemble for the meetings around the table in the lab, and the discussion would flow informally and unpredictably. We used loose pieces of paper to draw graphs and diagrams, write equations, and so forth. These pieces were most frequently trashed, although a few of the better-organized lab members collected some of them, organized them, and copied them for others. Still, no formal trace of the meetings existed, and I never thought that such a trace would be necessary.

Everything changed with the COVID pandemic. The lab was closed, we could not meet face to face, and I had to invent an alternative method of running the weekly meetings. Fortunately, Zoom offered a wonderful platform. However, drawing and writing on loose pieces of paper became impractical, and I started to prepare PowerPoint files for each week to guide the discussion. This alternative method of running lab meetings unexpectedly offered a major benefit: Former lab members who had moved to other universities and countries could still join the meetings, frequently at the least convenient local time, and contribute to the discussions. They also encouraged colleagues to join, and a few young and established researchers from a number of countries, who had never visited the laboratory, became regular contributors to the meetings. After the lab and the university reopened, we continued to run these meetings in the mixed format: face to face around the table in the lab and over Zoom. After a few seasons, I realized that the meetings produced a large number of PowerPoint slides organized around specific topics. These collections formed the core of this book.

The style of the book matches the style of the lab meetings as closely as possible. In particular, with all due respect to former generations of great researchers, we assumed no absolute authorities and questioned "established truth," even when it came from such giants as Charles Sherrington and Nikolai Bernstein. We tried to be as exact as possible in using words, and when it was impossible to avoid jargon, we tried to define every meaningful word as precisely as we could, given the existing knowledge. Such commonly used words as *synergy*, *motor program*, *internal model*, *muscle tone*, *motor command*, *efference copy*, and so forth were revisited and defined to the best of our abilities before we accepted some of them for the discussions and banned others as useless and misleading. Overall, we did not reject the everyday jargon developed in the field of movement studies altogether but instead agreed on mutually acceptable definitions for some of the expressions of the jargon and continued to use them. If important pieces of information were unavailable, we tried to devise a mental experiment able to supply these pieces. If we discovered contradictory bits of information, discussions were directed at resolving the controversies theoretically or, more commonly, experimentally. All these features are represented in the book to different extents.

Motor control is a relatively young field of research. It can be compared to pre-Galilean physics: Much information has been accumulated, but there is no universal agreement on a unifying methodology and theory. However, being part of biology, motor control is a branch of the natural science (see the definition of motor control on the webpage of the International Society of Motor Control, www.i-s-m-c.org). At least, this is how we approach this field in the Motor Control Laboratory. So for us, motor control is not a subfield of control theory, engineering, robotics, and many other wonderful areas of basic and applied research. It has to be addressed using the language of the laws of nature, with sets of adequate concepts specific for biological movement, and explored using the scientific method developed in natural science (physics).

Voluntary movements have to possess two basic features. First, they have to reach desired targets in the spaces where tasks are formulated within permissible error margins. Second, these solutions have to be dynamically stable because natural movements take place in unpredictable environments characterized by imperfectly predictable external force fields and are implemented by the body with unpredictably varying internal states. The book focuses on two developments in the field of motor control addressing these two basic features. First, it accepts the theory of the neural control of movements with time profiles of spatial referent coordinates for the effectors, which is a development of the classical equilibrium-point hypothesis (of course, the original lambda-version of the hypothesis!). I would recommend a book by Anatol Feldman (2015) as a valuable resource for those who would like to learn more about this theory. Second, it accepts the principle of abundance and the uncontrolled manifold hypothesis to explore how action stability is ensured and controlled. These two theoretical frameworks are introduced and then used to explore various effectors and phenomena.

The style of the writing and the contents assume more than basic knowledge of neurophysiology, muscle physiology, biomechanics, and other subfields of movement studies as these are taught in typical undergraduate courses in the United States. A perfect reader (participant of the seminar) would also be fluent in statistics and calculus with elements of linear algebra. Information that may be lacking, given the knowledge of the audience, should be supplied by the leader of the discussion (instructor) as needed. I tried to keep the style and level of the seminars somewhat challenging for beginning graduate students and still not boring for more advanced participants, including postdoctoral fellows and visiting scholars.

The structure of the book is rather straightforward. The first part starts with philosophical issues, introduces the views of Nikolai Bernstein, and describes some of the basic concepts. The second part builds on the first one to develop theoretical views in more detail and review related experimental material. The third part applies the introduced concepts and tools to the control of specific effectors and behaviors. There is also a chapter on kinesthetic perception and its relations to neural processes involved in the control of movements. The fourth part is my favorite: It addresses a number of surprising and controversial phenomena and offers their interpretation within the introduced theoretical frameworks. In the fifth part, cases of movement deterioration and improvement are discussed. The final, sixth, part addresses issues that are rarely discussed in formal courses, those of methodology and presentation of studies (writing papers). Similarly to all other parts of the book, these reflect the subjective experience of the author, which does not have to be shared by his respected colleagues.

The order of parts and chapters may be viewed as somewhat arbitrary, a recommended order but not something set in stone. Each chapter was written to stand alone. As a result, some contents overlap and figures are repeated in different chapters, frequently with modifications adapting them to the specific context. However, discussions in individual labs may warrant a different order that could be a better match to the level of preparation and focus of research interests. Some chapters may be skipped, and others extended to more than one session. New topics can and should be added. Each chapter is designed as a guide to lab-specific discussions and expected to take about 1 hour or a bit more. There are between 6 and 14 figures in each chapter. These are mostly schematic drawings and illustrations from published studies. More slides should be added to match the material to the audience.

The main contributors to this book are the former and current members of the Motor Control Laboratory at Penn State. They are too many to be listed by name. I would like, however, to mention those who were particularly active in our lab discussions: Valters Abolins, Satyajit Ambike, Cristian Cuadra, Jaebum Park, Joseph Ricotta, Jae-kun Shim, Stanislaw Solnik, and Tarkeshwar Singh. I am also very grateful to colleagues who had never worked in the laboratory but were very active during the lab meetings, offering interesting comments and unexpected insights. These include, first and foremost, Maria Falikman, Alexei Popov, Jose Praia, Vera Talis, and Serena Woolf. I also want to thank my friends-colleagues who were instrumental in shaping the Motor Control Laboratory and its research program, Alexander Aruin, Anatol Feldman, and Vladimir Zatsiorsky.

Good luck!

Mark Latash

PART I
BASIC CONCEPTS

1
Philosophy

There are two approaches to the neural control of movement dominating the field. One of them assumes that the brain (and the rest of the central nervous system) is a natural object emerged in the process of evolution and, as is true for any natural object, its behavior is defined by laws of nature (reviewed in Latash 2019, 2021a). In particular, the brain of animals does not perform operations with numbers and symbols, or in other words, it does not compute. Of course, its behavior can be described with computational means by researchers similarly to how this is done in classical physics with respect to inanimate objects. However, no sane physicist would assume, for example, that objects with electrical charge compute forces to be applied to other charged objects, as described by the Coulomb law, although the forces between charged objects can be described with high precision with the corresponding equation. Methods of study within the first approach are based on the known laws of classical physics and tools from physiology and behavioral studies in an attempt to discover new laws of nature specific for biological objects.

The alternative approach assumes that the brain is a computational device, which contains and generates *internal models*, that is, computational tools reflecting interactions among parts of the body and between the body and the environment (reviewed in Wolpert et al. 1998; Shadmehr and Wise 2005). Such internal models are used to compute descending neural commands to produce desired movements and to predict future changes in the current state of the body and environment. This approach commonly uses the computational apparatus of the control theory and engineering developed for the control of man-made objects, such as ballistic missiles and robots. It presumes that, to perform tasks typically formulated in terms of peripheral mechanics of involved effectors up to the whole body, the brain can prescribe future mechanical states of these effectors based on the classical Newtonian mechanics by prescribing signals to muscles that should generate requisite force time patterns. This approach has dominated the field for literally hundreds of years.

The latter approach will be discussed in more detail in Chapter 4. Its basic assumption that the goal of motor control is to discover software residing in the brain and defining the presumed computational processes associated with movements looks unappealing. It artificially separates the central nervous system from the rest of the body and endows it with properties that look supernatural; that is, there are no attempts to link them to laws of nature that define processes within the central nervous system and the rest of the body. Of course, humans can learn how to compute, but this ability is a result of education. Other mammals, who are sometimes superior to humans in large classes of movements, are unable to compute even (3×4) no matter how much effort is put into training them. So, assuming that their brains perform very complicated computations within milliseconds seems far-fetched. The heart of the author belongs to the former approach, and much of the material will be discussed within the basic idea that the *goal of motor control is to discover laws of nature specific for the production of purposeful biological actions.*

1.1. Laws of nature in the inanimate world

Classical physics explores regularities in the behavior of objects and tries to describe them in a compact way, commonly in the form of equations. Such descriptions (equations) are addressed as *laws of nature*. Laws of nature are applicable to specific groups of objects. For example, the classical Newton's second law is applicable to objects with mass, not to massless objects, such as, for example, a reflection of a beam of light on a surface. Hooke's law is applicable to objects that can deform under the application of force, accumulate potential energy, and release it after the force has been removed. Such objects are commonly addressed as springs. It is not applicable to nondeformable objects or playdough, which deforms under external force but does not accumulate potential energy. And so on.

Laws of nature involve variables and parameters. For example, consider the aforementioned two laws:

$$F = m \bullet a \qquad (1.1)$$

and

$$\Delta F = -k \bullet \Delta x, \qquad (1.2)$$

where F is force, m is mass, a is acceleration, ΔF is change in force along a coordinate x, k is stiffness, and Δx is deformation (change in coordinate). Here and further in the text, variables are shown in italics, and vectors in bold. These two laws of nature are illustrated schematically in Fig. 1.1.

From the point of view of mathematics, the pairs of symbols on the right sides of both equations are equivalent: You can write $F = m \bullet a$ or $F = a \bullet m$; $\Delta F = -k \bullet \Delta x$ or $\Delta F = -\Delta x \bullet k$. This is not the case in physics. Indeed, if you change force acting on an object with mass, its acceleration will change, not mass. If you compress a spring, its coordinate will change, not stiffness. In other words, some of the symbols in the two equations reflect characteristics of the object that are constrained by the

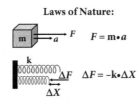

Figure 1.1 An illustration of two classical laws of nature. A: The second Newton law is applicable to objects with inertial mass. It links two variables, force (F) and acceleration (a), with the help of a single parameter, mass (m). B: Hooke's law is applicable to objects that deform under the action of force and accumulate potential energy (springs). It links changes in force (ΔF) and deformation (Δx) with the help of a parameter, stiffness (k).

corresponding laws of nature, while other characteristics are not. The former are addressed as *variables*, and the latter as *parameters*. Commonly, it is assumed that variables can change relatively quickly, and parameters stay constant or change slowly. This does not have to be the case. Indeed, when a rocket flies and burns fuel, its mass can change rather quickly. However, mass still remains a parameter within the second Newton law. There are springs that change their stiffness with deformation—nonlinear springs. However, stiffness remains a parameter within Hooke's law.

Where have the laws of nature come from? Are they set in stone or do they change as the Universe changes? These are good philosophical questions. Human civilization is too young to provide experiment-based answers to the second question. The great physicist Albert Einstein thought that the fundamental laws of nature were indeed unchanging with time. The great philosopher Henri Bergson disagreed: Bergson thought that time was a special coordinate in the four-dimensional space, along which laws of nature could change (reviewed in Canales 2015). They did not come to an agreement, but in this chapter, I will try to argue that, for all practical reasons, laws of nature in the inanimate world may be indeed viewed as eternal, but laws of nature in the world of living systems are indeed pliable and time dependent.

1.2. Specificity of living objects

Let us start with trying to use our own experience and consider the following question: "How do we distinguish a living object from an inanimate one?"—for example, a snake from a twig, a crab from a seashell, or a groundhog from a stone. Movement of inanimate objects is typically predictable. This is not surprising: They obey a relatively small number of established laws of nature. Movement of living objects is much less predictable. This does not mean, of course, that they can violate fundamental laws of nature (at least, we have no evidence for such a supernatural ability). But they seem to be only *constrained by those laws, not driven by them.*

Consider the following examples.

> *Example #1*: You place a cat and a toy cat on the table. Then you go and make yourself coffee. When you return, the toy is still on the table, but the live cat is gone. Why is the toy behaving according to laws of nature, which allow predicting motion of an object if its parameters, initial conditions, and external forces are known, and the cat seemingly ignores them?
> *Example #2*: A toy from the previous example is released at a certain height and falls under the law of gravity. A cat released from the same height also falls down. However, the cat can jump up, and the toy cannot. Why?
> *Example #3*: A toy fish made of a piece of wood and a live fish can both swim with the current, but the live fish can also swim against the current, while the wooden toy fish cannot. Why?

These are all very simple everyday observations. They all suggest that the fundamental laws of nature, as we know them, are insufficient to describe the behavior of biological objects. To make the next step, we must accept a few assumptions that go beyond the physics of the inanimate world. These assumptions are based on solid experimental material, which will be discussed in other chapters.

> *Assumption 1*: Living systems can unite fundamental laws of nature into complex chains and clusters leading to new robust relations among salient variables and involving new parameters. In other words, *living systems can create new laws of nature!* In a sense, this assumption represents a definition of a living system, a definition different from a host of definitions one can find in various textbooks and encyclopedias, which typically do not define living objects but instead enumerate their features.
> *Assumption 2*: To perform actions, living systems change parameters in the new, biology-specific, laws of nature. This assumption is sometimes addressed as *parametric control*. Note that, in the inanimate world, the only way to change movement of an object with mass (and all biological objects have mass!) is to apply force to that object. Note that force is a variable (see the earlier examples of the second Newton law and Hooke's law). The assumption of parametric control suggests that movements of living objects are performed in a qualitatively different way. Of course, time-varying forces emerge during parametric control, but they are not prescribed by the living systems.

These two assumptions deserve detailed discussion, and we will use a specific example, well known in the field of movement science, to illustrate them in the next section.

1.3. Stretch reflex as a biological law of nature

Consider a typical skeletal muscle. Imagine that it has all its connections to the central nervous system intact and that you can record its length (L) and force (F); for example, its distal tendon is separated from the natural insertion point and connected to a motor with coordinate and force sensors. Imagine also that there are no changes in descending signals from the brain; for example, the experiment is performed on an animal preparation, and the experimenter controls descending signals from the brain to spinal structures. If the muscle is stretched slowly, it shows an increase in its force due to properties of the tendon and other connective tissues. This relatively small force increase is shown in Fig. 1.2 with the dashed line. At some value of muscle length, the muscle shows first signs of activation. Further stretch is accompanied by a progressive increase in the activation level, and the slope of the force-length characteristic increases (the solid curve in Fig. 1.2). This behavior is a result of an involuntary mechanism addressed as the stretch reflex (Liddell and Sherrington 1924). Muscle length when the first signs of activation are seen is addressed as the *threshold* of the stretch reflex (λ in Fig. 1.2).

If, for simplicity, we consider only the active force (F_{ACT}) produced by the muscle, this behavior can be described with two equations:

$$F_{ACT} = 0, \text{ when } L < \lambda \tag{1.3}$$

and

$$F_{ACT} = f(L - \lambda) \text{ when } L \geq \lambda. \tag{1.4}$$

1.3. STRETCH REFLEX AS A BIOLOGICAL LAW OF NATURE   7

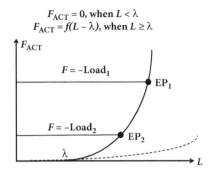

Figure 1.2 Active muscle force (F) as a function of muscle length (L) originates from a particular length value—the stretch reflex threshold (λ). When muscle force equals in magnitude the external load, the whole system is in a state of equilibrium—the equilibrium point (EP, filled circle). A change in the external load leads to another EP along the $F(L)$ characteristic—an involuntary movement. Muscle force changes with length in the absence of muscle activation—the dashed line.

In the last equation, f stands for an unknown smooth function. This equation resembles Hooke's law. The only difference is that Hooke's law assumes a linear relation between changes in force and length with the coefficient called stiffness. To avoid confusion, we are going to use a slightly different term for the slope of the $F(L)$ characteristic, *apparent stiffness* (cf. Latash and Zatsiorsky 1993). This term implies that the behavior of our object of interest looks like that of a nonlinear spring, but the object is not a spring. Note that these equations involve two variables relevant to muscle action, F and L, and one parameter, λ (for now, we consider only very slow processes with velocity and acceleration assumed close to zero).

Equations (1.3) and (1.4) describe a large class of objects, namely numerous skeletal muscles across individual animals and species. They result in behavior that looks rather simple but, in fact, reflects a combination of elemental phenomena, with each of those phenomena being a complex constellation of even more elemental phenomena, all the way down to fundamental laws of physics (Fig. 1.3).

Indeed, when a muscle is stretched by an external force, it shows complex patterns of tissue deformation, which are transmitted to muscle spindles—small structures scattered across the muscle parallel to the power-producing muscle fibers. Deformation of the spindles leads to deformation of small muscle fibers inside the spindles—*intrafusal fibers*—leading to deformation of *sensory endings* located on the intrafusal fibers. This results in changes in the membrane properties of the sensory endings, leading to changes in the permeability of the membrane to several ions including Na⁺. As a result, some of the endings start generating *action potentials*, standard time profiles of changes in the membrane potential, which may be seen as units of information in the nervous system. The action potentials are transmitted along the afferent axons toward the spinal ganglia and then from the ganglia into the spinal cord. Afferent fibers make both direct (monosynaptic) and indirect (oligosynaptic and polysynaptic) projections on the alpha-motoneurons innervating the muscle. The very complex

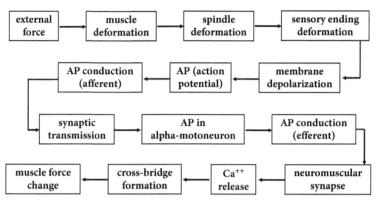

Figure 1.3 A sequence of elemental phenomena leading to the dependence of active muscle force on muscle length. Each elemental phenomenon is based on a number of more basic elemental phenomena.

processes involved in the synaptic transmission lead to changes in the membrane potential of the target alpha-motoneurons. Some of the motoneurons are brought to the threshold for action potential generation. These action potentials are transmitted along the axons (efferent fibers) toward the muscle. Complex processes within the neuromuscular synapses lead to the generation of action potentials on the muscle membrane. These action potentials enter the sarcoplasmic reticulum and change the properties of the cisternae with Ca^{++} ions. The ions are released and trigger a sequence of events within the muscle resulting in the formation of cross-bridges between the actin and myosin molecules. This excitation-contraction mechanism leads to active force generation by the muscle.

In fact, this brief description is grossly incomplete. Each of the mentioned steps is far from being truly elemental; that is, it can be split into many more steps. Some of these steps are relatively well understood and can be described as consequences of multiple physical and chemical processes (e.g., diffusion, chemical reactions, electrical events, etc.). Other steps are known only as input-output regularities. Most of the mentioned steps require energy, and getting the energy may include more complex intracellular processes, in particular those in the mitochondria.

This example satisfies the first part of the definition of a living system. Indeed, the evolution has led to the emergence of a behavior illustrated by the relatively simple and reproducible $F(L)$ characteristic, which represents a combination of chains and clusters of elemental steps ultimately obeying the fundamental laws of nature.

When the external force (load) applied to the muscle is constant, the system "muscle + reflexes + load" reaches an equilibrium state, which is shown in Fig. 1.2 as an equilibrium point (EP). Changes in the external force lead to movement of the system to another point along the same $F(L)$ characteristic. Such movements can be addressed as involuntary. Note that they are associated with changes in the muscle activation level. As described later, in Chapter 3, voluntary movements are produced by changing values of λ, that is, by changing the parameter of the law of nature illustrated in Fig. 1.2. In other words, the central nervous system uses parametric control

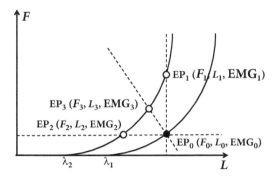

Figure 1.4 Active movements within the control with changes in the stretch reflex threshold, λ. A change in λ by the central nervous system from λ_1 to λ_2 can lead to different peripheral consequences depending on the external load characteristic—see the three possible final equilibrium points (EPs, open circles) for three load characteristics (thin dashed lines). Note that muscle activation levels (EMG) also vary across the three final EPs. The font size reflects the magnitude of the corresponding variable.

to produce movements in compliance with the second part of the introduced definition of living systems.

An important feature of parametric control is that it is *indirect*; that is, it does not prescribe changes in any of the salient variables such as forces, displacements, and their derivatives. Of course, movement of material objects is always associated with changes in forces, but these force changes emerge just like any other variables. Indeed, one and the same change in λ (illustrated in Fig. 1.4) can lead to very different changes in force and coordinate (and muscle activation level!) depending on the external load acting on the muscle. Clearly, producing accurate movements with such a system requires a combination of more or less accurate prediction of external forces and more or less quick correction of ongoing movements if they happen to be inaccurate due to unexpected changes in external forces.

1.4. How many sets of laws of nature are there?

The example of parametric control described in the previous section for a single muscle has been generalized for both multi-muscle systems and intra-muscle elements, such as motor units. These topics are going to be discussed in detail later, in Chapters 7 and 10. For now, let us accept this basic idea and ask the following question: "What kind of laws of nature define time patterns of the parameters, which produce task-specific movements?" This question was asked by several scientists including Nikolai Bernstein, who considered biological actions as emerging primarily from within the body (in contrast to the then-dominant theory of conditioned reflexes introduced by Pavlov and developed by his followers). Bernstein (1966) addressed this idea as the principle of activity and, over the past years of his life, tried to develop a new field of physiology, which he called *physiology of activity*.

This question was also considered by the French philosopher Maurice Merleau-Ponty (1942/1963; reviewed in Kretchmar and Latash 2022), who suggested the existence of three main levels participating in biological actions (Fig. 1.5). The bottom level is the level of inanimate nature where basic laws of nature, as described in physics textbooks, define the behavior of objects. Note that such fields as biophysics, biomechanics, and biochemistry do not imply invoking new laws of nature but rather the direct application of basic laws of nature to biological objects. Hence, these fields of research belong to the bottom part of Merleau-Ponty's pyramid.

The next level describes the behavior of living systems. Processes at that level cannot violate the laws of physics of inanimate nature, but they can configure these basic laws to produce meaningful actions, particularly those giving evolutionary advantage. The definition of life suggested earlier may be seen as a more specific reformulation of this principle based on parametric control. Within the Merleau-Ponty pyramid, the emergence of parametric control in living systems signifies a qualitatively new method of producing movements.

The highest level, according to Merleau-Ponty, is the level of consciousness, which was presumed to be human specific. Processes at that level could not violate laws of nature at the two lower levels, and they defined inputs into those levels ultimately leading to actions.

An important feature of the scheme illustrated in Fig. 1.5 is that processes at a higher level cannot violate laws of nature at lower levels, but they are not driven by those laws. For example, one cannot derive rules of cognitive processes in the brain from physics of the inanimate world, but on the other hand, these processes require energy and are based on elemental phenomena, such as the generation of action potentials, which are constrained by those lower-level laws.

Recently, the scheme in Fig. 1.5 has been reconsidered and modified based on contemporary knowledge (Kretchmar and Latash 2022). In particular, the top level of the main hierarchy has sublevels that may be based on different laws of nature. In addition, hierarchies have been introduced nested in the main one reflecting the complexity of the involved effectors and physiological circuitry (Fig. 1.6). One of those

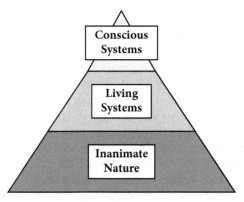

Figure 1.5 Three levels characterized by separate sets of laws of nature after Merleau-Ponty. Modified by permission from Kretchmar and Latash 2022.

1.4. HOW MANY SETS OF LAWS OF NATURE ARE THERE?

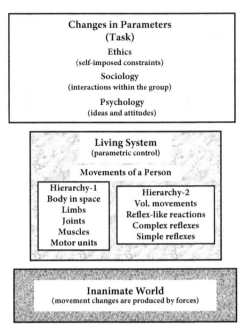

Figure 1.6 A recent development of the Merleau-Ponty scheme shown in Fig. 1.5. Note the embedded hierarchies within each level corresponding to the complexity of effectors and neurophysiological circuits involved in behaviors. Modified by permission from Kretchmar and Latash 2022.

hierarchies considers the involvement of elements at different anatomical levels into coherent functional tasks. Natural movements can involve one joint only, a single extremity, or the whole body. At the extreme of the spectrum is serving in tennis, which is a whole-body action with the task of producing certain kinematic characteristics of the ball. Obviously, to perform the action, the player's brain cannot send signals directly to the ball and must involve different body parts, such as limbs. Each limb action involves movement of individual kinematic degrees of freedom. The kinematics must be produced by muscles, which are the output elements of the neuromotor system. This idea of parametric control has been generalized for movements of any effectors, up to the whole body, using the concept of spatial referent coordinate (RC) for any effector as the parameter in respective laws of nature (Chapter 7; Latash 2010a, 2017).

The other hierarchy (Hierarchy-1 in Fig. 1.6) considers the complexity of neurophysiological circuits involved in the production of motor actions, from reflexes to volitional movements. The term *reflex* is not exactly defined, and there has been major discussion on whether this term is at all useful (Prochazka et al. 2000). Actions differ in such characteristics as the number of synapses in the involved neural circuits, the stimulus-action time delay, stereotypy of the actions, and more. None of them can distinguish reflexes from nonreflexes unambiguously. Nevertheless, we are going to assume that there are circuits (e.g., those involving a single neuro-neural synapse, as is the case for monosynaptic reflexes) that deserve to be called reflexes. Some of the reflexes are rather complex and involve multiple connections within the central nervous system (e.g., the pupillary reflex and the Magnus-De Kleijn reflexes). There are also reflex-like responses to external stimuli that come at delays shorter than that of the simple reaction time but show context-specific and instruction-specific patterns even if the initial state of the body (as far as it can be controlled in an experiment!) and the stimulus remain unchanged (Prochazka et al. 2000; Safavynia and Ting 2013; Pruszynski et al. 2016). They have been addressed with various names such as long-loop reflexes, preprogrammed reactions, and triggered reactions. At least some of them involve transcortical loops. Typical examples include the grasping response when an external force acts unexpectedly on a hand-held object and postural responses when the supporting surface starts to move (e.g., seen in a person standing in a bus when the bus starts to move). At the top of this hierarchy is volitional movement, which does not require an explicit external stimulus to be generated.

Note that the second hierarchy is applicable to actions at all levels in the first one. In particular, reflexes and reflex-like reactions can involve individual muscles, joints, and limbs, and even the whole body. Across all levels and for both hierarchies, action is viewed as produced by time variations of RC. This is true for reflex-like actions and sophisticated voluntary actions, for those involving a small group of muscles (e.g., such as during eye movements), and for those involving the whole body (as in the aforementioned example of a tennis serve).

The second-from-the-bottom level of the main hierarchy in Fig. 1.6 requires an input specifying time changes in task-related parameters such as $RC(t)$. Obviously, these patterns are built on one's understanding of the task, for example, locating an object to be grasped in the external space, selecting a hand to perform the grasping action, and time characteristics of this action. We do not know what structures and circuits within the central nervous system are involved in these processes. So, the top level of the main hierarchy in Fig. 1.6 remains undefined in physiological terms. Most likely, $RC(t)$ patterns are produced by an interaction of broadly defined goals, memory, and sensory signals informing the current state of the environment and the body. So far, the only theory that tries to use concepts from physics and neurophysiology to address the emergence of targets and respective $RC(t)$ patterns is the dynamic field theory (Erlhagen and Schöner 2002; Schöner and Thelen 2006; Martin et al. 2009, 2019), which is based on dynamical interactions within neuronal fields with memory-related and sensory-related signals.

Most frequently, however, processes, such as selection of spatial goals, specification of effectors, action initiation time, and desired duration, have been discussed in the field of psychology. This field, as of now, forms the top level in the hierarchy illustrated in Fig. 1.6. As a null hypothesis, we assume that psychological processes emerge based on parametric control similar in principle to the one introduced for the production of movements. It is possible, however, that psychology is qualitatively different from biology in the involved laws of nature (rules) and modes of control. Unlike the scheme of Merleau-Ponty, we do not see the top level as purely human. Indeed, actions by most animals are defined by numerous processes that are memory based and task and intention specific; these processes are commonly addressed as perceptual and cognitive.

The main scheme in Fig. 1.6 can be viewed as built along an axis corresponding to difficulty in violating laws of nature specific to each of the levels. Laws of inanimate nature seem to be highly resistant to change. Indeed, most textbooks claim that these laws are eternal (or, at least, date back to the Big Bang). On the other hand, Bergson claimed that time is not simply a coordinate in the time-space continuum but a very special coordinate along which laws of nature change (see Canales 2015). If laws of inanimate nature do change, they do so extremely slowly, and, for the goals of this review, we can view them as set in stone and impossible to modify.

Biology-specific laws of nature are much more pliable. Even the simplest, monosynaptic reflexes show plastic changes with extensive practice, which may require hundreds of repetitions over multiple days in an operant conditioning paradigm (reviewed in Wolpaw and Carp 1993; Wolpaw 2007). They also change under certain pathological conditions, in particular spasticity (e.g., Thompson and Wolpaw 2015; Turpin et al. 2017). Since these reflexes form part of the stretch reflex circuitry, this means that the $F(L)$ dependence illustrated in Fig. 1.2 can also change with specialized training. As mentioned earlier, reflex-like preprogrammed responses show strong instruction and context specificity; however, these changes do not have to involve changes in the underlying biology-specific laws of nature, only a change in the rules that define specification of control parameters, $RC(t)$. Overall, the well-documented phenomenon of neural plasticity forms the basis for relatively quick changes in the transformations between levels within the nested hierarchies (Hierarchy-1 and -2 in Fig. 1.6). Practice over a relatively short time (a single 40-minute session) can lead to a change in the sharing of higher-level control parameters among lower-level control parameters and/or in the stability of such a mapping (reviewed in Wu and Latash 2014).

We do not know biological laws that define processes at the level of psychology. However, regularities of experimental data in typical psychological studies are relatively easy to change, suggesting that the respective rules are relatively fragile. Imagine, for example, that a subject in a typical simple reaction time experiment (commonly used in psychological studies of action) decides to ignore the imperative stimulus to initiate action or purposefully delays the reaction. This would immediately lead to a major change in performance, even under perfectly reproducible conditions. Another example is the Fitts' law, which describes changes in movement time with changes in distance and size of a visual target (Fitts 1954; Fitts and Peterson 1964). This law can also be relatively easily violated if the subject elects to do so.

Ethics combines self- and society-imposed rules, which define types and properties of actions a person is more or less likely to perform. These rules may change over one's lifetime, based on education, moving to another country, changing religious views, and many other factors, and can be violated without much practice under certain conditions. Consider, for example, the Old Testament story about Lot and his daughters and multiple examples of cannibalism in famished societies. In these (and many other) examples, neither physics of inanimate nature nor biological laws of nature at the lower level of the hierarchy predict the observed behaviors, which violate the ethical rules accepted in contemporary societies. Indeed, by themselves, laws at the lower levels of the hierarchy in Fig. 1.6 allow a variety of behaviors across individuals under such stressful situations, whether these are ethical or not. Moreover, in the absence of specific ethical constraints, some of such behaviors could be seen as acceptable, as was the case, for example, in societies where cannibalism or incest was acceptable practice.

The hierarchy illustrated in Fig. 1.6 can be viewed as built along the axis corresponding to different sets of rules (laws of nature). The difference between the lower and intermediate levels (inanimate world and biological world) is qualitative, at least with respect to how motor actions are performed. Whether there is a comparable qualitative difference between the intermediate and top levels remains to be seen.

1.5. Missing pieces of the mosaic

The approach to biological movements based on laws of nature seems to have no viable alternative unless, of course, one assumes that the brain was indeed designed by a Supreme Programmer. This book is written based on the evolutionary approach, not the creationist one, which remains a major assumption. Those who accept the idea of creation may stop reading now and start developing an alternative theory of motor control. Within the evolutionary approach, the issue of one (classical physics) vs. two vs. many sets of laws of nature remains open. The views of Merleau-Ponty and their recent developments suggest only one possible scenario, which may happen to be wrong. As we will see in future chapters, very few studies address possible origins of control variables and focus on processes of execution of movements and ensuring their desired properties with respect to salient performance variables. This makes the upper level(s) in the Merleau-Ponty pyramid a true *terra incognita*.

A very important development in the field was associated with stepping outside the "box" of Newtonian mechanics and accepting the idea of parametric control (reviewed in detail in Chapter 3 and the rest of the book). The idea that one can perform accurate movements without prescribing force patterns was truly revolutionary and formed the foundation for the existence of a separate set of laws of nature relevant to the production of biological actions. This idea has been developed over the past 60 years or so (since Feldman 1966), overcame lots of criticisms and claims of refuting this approach, and is now a respected theory of motor control. The problem is that the idea of parametric control is so unusual and unlike all the approaches in the fields of engineering and robotics that few researchers understand it adequately. One of the main goals of this book is to present this approach in its entirety, hoping that a new generation of researchers will understand its main aspects and unique features.

This new "box" of parametric control is still a "box" with its natural limitations (for some of the possible developments of this line of thinking to phenomena outside of motor control see Chapter 15 and Latash 2019). Soon, it may be time to step outside this new "box" and try to discover laws of nature relevant to cognitive processes, particularly those involved in deciding what motor actions would be adequate in each particular situation. Steps in this direction have been made within the field of ecological psychology, which follows the traditions of Gibson (1979), in particular, by the group of Michael Turvey (Kugler and Turvey 1987; Schmidt et al. 1990; Turvey 2007) in their studies of movements performed by individuals and by pairs of individuals. The situation, however, is still far from formulating a viable set of laws of nature.

2
Bernstein's Construction of Movements

The name Nikolai Alexandrovich Bernstein (1897–1967) is well known to every researcher and many practitioners in different branches of movement science (for reviews see Turvey 1990; Bongaardt 2001; Feigenberg 2014; Latash 2020a; Talis 2022). His name is associated with the famous *problem of motor redundancy*, sometimes addressed as *Bernstein's problem*. He introduced the classical three-stage scheme of skill acquisition and explored it across areas of athletics and in the field of motor rehabilitation. He reintroduced the concept of *synergy* into movement science and defined it as exactly as was possible at his time. He developed many methods that were well ahead of their time and used them to study a range of movements, from relatively stereotypical hammering, to running by the world record holder of the time J. Ladoumegue, to playing piano by top-level professional performers. Bernstein himself was a trained musician with perfect pitch, and he used this ability to estimate the frequency of his data acquisition systems by listening to the noise of the rotating shutter placed on the lens of the camera. He was fascinated with bridges and steam engines and thought and wrote about a range of topics outside the field of movement science, such as mathematical analysis of electrophysiological signals, time travel, and linguistics (e.g., see Latash and Talis 2021).

In this chapter, we will touch upon some of those issues. However, we will mostly focus on the famous hierarchical multi-level scheme of the construction of movement developed by Bernstein in the 1930 to 1940s. Bernstein described this scheme in detail in his most comprehensive book published in Russian in 1947. This scheme was not described in sufficient detail in his other books and, as a result, was poorly known to Western colleagues until 2020, when the English translation of the 1947 book became available (Latash 2020a). As a result, misunderstandings of some of Bernstein's central ideas emerged and spread among researchers, publications, and textbooks. Here and further in this book, we will try to clarify Bernstein's views. In this chapter, we will focus on the levels for movement construction within his control hierarchy. Further chapters will discuss in more detail Bernstein's views on redundancy, synergies, muscle tone, and other central concepts.

2.1. Nikolai Bernstein: Philosopher and experimentalist

Nikolai Bernstein is commonly seen as the father of contemporary motor control. However, his contributions to movement science are not limited to the neural control of movements and include such related areas as biomechanics, sport science, and physiology of activity. He was a truly universal scientist able to design and build unique pieces of equipment, develop methods of data processing that were tens of years ahead of his time, combine the existing knowledge into a general theory of movement production, and write philosophical essays on a range of topics including philosophy of time.

Bernstein was trained as a medical doctor and worked as a military physician during World War I. Following a few years of turmoil, which included the end of World War I, the revolution in Russia, and the Civil War, Bernstein returned to Moscow in the early 1920s ready to contribute to the development of the new society. He embraced the main ideas of the new communist government and, during his whole life, had always been its staunch supporter. Bernstein joined the new Institute of Labor in Moscow where he was given the task of inventing new types of labor movements during typical daily activities of factory workers that would be less fatiguing, less boring, and more efficient. In our days, these lines of research would probably be united under the label of ergonomics.

Very early in his career as a researcher, Bernstein realized the importance of precise and objective measurement of salient variables during behaviors. He designed and later improved methods of motion analysis reaching amazing frequencies of data sampling—up to 500 Hz!—in the 1930s (Bernstein and Popova 1930; Kay et al. 2003). In the precomputer age, this was nothing short of a miracle given that all the joint angles in all the individual frames were measured by hand. At about the same time, he also pioneered the computation of joint torque time profiles based on movement kinematics and estimated mass-inertial properties of body segments (using methods of inverse dynamics) and combining kinematic and kinetic variables into the mechanical analysis of very fast movements. He was not afraid of using sophisticated mathematical methods, and at Bernstein's funeral, a great mathematician of the 20th century, Israel Gelfand, addressed Bernstein as "an outstanding mathematician."

Bernstein's philosophical views were strongly materialistic. He viewed human bodies, including the central nervous system, as natural objects that had emerged in the process of evolution and had to obey laws of nature. In the late 1940s, Bernstein suggested the following definition of life: "Living objects and groups are those that can be characterized with aspiration (or display of aspiration) towards their conservation and prolongation" (see Latash and Talis 2021). In other words, if, in the presence of a randomly directed perturbing influence, reactions by a system show preference for certain directions that help maintain homeostasis and evolutionary success, you deal with a living system. Bernstein himself admitted that this definition was deficient because of its obvious teleological underpinnings. However, it resonated well with other definitions offered at about the same time (e.g., by Schrödinger 1944/2012).

Bernstein thought a lot about the nature of time and, at some point, wrote a diary addressing a range of issues including time perception, time travel, and reflections of time in different languages (Latash and Talis 2021). He accepted time as the fourth coordinate in the space we happen to live in but emphasized its difference from other, spatial, coordinates. In particular, he was fascinated by the fact that past events, known to us in minute detail, could not be changed (except in newspapers published in authoritarian regimes, such as in Orwell's 1984, Stalin's USSR, Hitler's Germany, and Putin's Russia), whereas future events, which are never 100% predictable, can be at least partly modified and even controlled by our actions.

2.2. The evolutionary approach to movement construction

When Bernstein wrote his most comprehensive book, *On the Construction of Movements* (1947), he had to draw his conclusions based on the general philosophy and limited existing knowledge about the physiology of skeletal muscles and the central nervous system. Moreover, some of the accepted "knowledge" was incorrect. It is amazing how many of his guesses were true! At the same time, it is no surprise that, sometimes, he drew wrong conclusions.

In that book, Bernstein shows that he was, first and foremost, an evolutionary biologist accepting Darwin's theory of evolution without reservations. He added a new important assumption to the evolutionary theory: At new evolutionary stages, old physiological mechanisms are not replaced by new mechanisms but used as scaffolding for the construction of new mechanisms, resulting in a multi-level hierarchy. This assumption, in combination with the existing information from the field of comparative biology, led Bernstein to postulate two major types of physiological processes in excitable tissues, which he addressed as *paleokinetic* and *neokinetic*. The evolutionarily older, paleokinetic, mechanisms were linked to relatively slow processes acting at relatively short distances, such as currents associated with ion and molecular diffusion. Neokinetic mechanisms were associated with transmission of action potentials over much larger distances and at much higher speeds. In particular, Bernstein viewed intramuscle and synaptic transmission processes as paleokinetic. In our days, the paleokinetic-neokinetic dichotomy would be seen as obsolete, in particular because muscle processes are known to involve action potential generations and transmission.

Bernstein viewed the evolutionary emergence of new levels of control in his multi-level hierarchy as driven by new classes of emerging motor tasks. Overall, his scheme involved five major levels (Fig. 2.1), and, as we will see later, only the bottom two are currently studied experimentally within the field of motor control with sufficient rigor. This means that, for the bottom two levels, there are theoretical frameworks based on

Bernstein's Levels of Movement Construction

Level	Name	Structures
A	The level of tone (tonus)	Red nucleus Spinal cord
B	The level of synergies and patterns	Thalamus Globus pallidus
C C1 C2	The level of the spatial field	 Striatum Extrapyramidal Pyramidal
D	The level of actions	Parietal areas Premotor areas
E	Symbolic, highly coordinated actions	"Highest cortical"

Figure 2.1 The levels in the hierarchical system of the construction of movement according to Bernstein.

laws of nature that allow consistent experimental exploration. As of now, the upper levels are associated with tasks more typically explored in the field of psychology.

The lowest level in Bernstein's hierarchy—level A—was termed the *level of tone* (or *tonus*), or the rubrospinal level. According to Bernstein, level A dealt with the control of relatively slow muscle actions typical of postural tasks or task components and was based primarily on paleokinetic physiological mechanisms. Bernstein defined tone as muscle state, reflecting its preparation for future action. Note that this definition differs significantly from the current practice of quantifying tone in neurological patients when the patient is typically asked to relax and do nothing and an effector (e.g., a joint) is moved over its range of motion. This instruction does not imply preparing for any action and, hence, one has to be careful equating the level of tone with what is measured in a typical clinical examination and also addressed as tone (more on the notion of tone can be found in Chapter 22).

Bernstein realized that the central nervous system was unable to prescribe muscle force (F) or length (L) because the two variables are linked to each other. He thought that the central nervous system specified the muscle activation level, which then translated into a specific $F(L)$ dependence. In other words, he ignored the contributions of reflex loops to muscle activation level (stretch reflex; e.g., Liddell and Sherrington 1924), which lead to changes in muscle activation with length. This makes Bernstein's views close to the so-called alpha-model of the equilibrium-point hypothesis (see Chapter 3; Polit and Bizzi 1978; Bizzi et al. 1982), which was developed based on studies of deafferented animals when reflex loops could not contribute to muscle activation.

In his illustration, however (presented in a more concise, schematic way in Fig. 2.2), Bernstein drew a family of $F(L)$ lines that differed by the threshold when force started to increase with length. This makes the illustration intrinsically contradictory because, if the central nervous system could fix a level of muscle activation independently of possible changes in muscle length, this should have led to active force production within the whole range of muscle length changes. The contradiction is resolved if the lines correspond not to fixed levels of muscle activation but to fixed values of

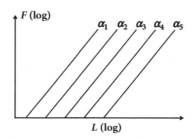

Figure 2.2 A schematic illustration of the dependence of muscle force (F) on muscle length (L) according to Bernstein. Note that the force-length characteristics (they are shown as straight lines in the log-log plot) differ by the neural command associated with the level of muscle activation (shown as α). Controversially, the curves differ by the length at which the muscle starts generating active force.

the stretch reflex threshold as suggested about 20 years later by Anatol Feldman in his classical papers introducing the equilibrium-point hypothesis (λ-model; Feldman 1966; see Chapter 3).

Addressing level A as the rubrospinal level reflected Bernstein's understanding of the central role of the cerebellum and its output to the spinal cord mediated, in particular, by the red nucleus in the control of muscle state. Currently, one group of the common consequences of cerebellar disorders is addressed as disorders of muscle tone (Thach et al. 1992; Rothwell 1994; Bastian et al. 1999), in line with Bernstein's guess.

The next level—level B or the *level of synergies and patterns*—was supposed to serve two main functions. One of them is linked to the famous *problem of motor redundancy* covered in detail in Chapter 6: At any level of description, the number of elements involved in typical motor tasks is higher than the number of task-related constraints, resulting in an infinite number of possible solutions. How does the brain select specific solutions? The level of synergies was presumed to organize elements into a relatively small number of groups, with each group controlled with the help of a single control variable. The number of groups was still redundant, but this step was supposed to mitigate the problem of motor redundancy (Fig. 2.3).

The other function of level B was to ensure *dynamical stability* of movements. Bernstein realized that all movements were always performed in conditions of imperfectly predictable external forces. For example, during walking, we frequently step on small objects (e.g., pebbles) or on uneven surfaces. Such external perturbations propagate through the mechanically coupled effectors and change their kinematics. Since muscle forces are length and velocity dependent, they also change. According to Bernstein, these factors make it impossible for the brain to prescribe mechanics of individual effectors, and dynamical stability of movements becomes functionally paramount. Bernstein understood under dynamical stability the natural ability of a moving system to return toward a state or trajectory in cases of relatively small transient perturbations. He viewed the organization of dynamically stable movement patterns as a central step in the development of a motor skill.

To account for dynamical stability of natural movements, Bernstein introduced the principle of sensory corrections. Any deviations of salient performance variables from their desired trajectories were assumed to be sensed by sensory organs, which led to

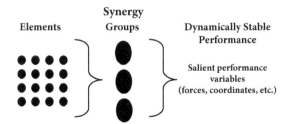

Figure 2.3 Two main functions of the level of synergies (level B). The first function unites elements (Bernstein associated them with individual muscles) into robust groups. The second function ensures dynamical stability of salient performance variables using feedback loops (addressed as the *principle of sensory corrections*).

adjustments in the outputs of all elements contributing to those variables. Bernstein emphasized that the organism never responded to a local deviation by a local correction. Instead, corrections were expected to involve all the relevant elements. These insights were later confirmed by studies of a variety of actions including speech and prehension (Abbs and Gracco 1984; Cole and Abbs 1987; Latash et al. 1998).

While describing different levels of movement construction, Bernstein had a special subsection for each level dedicated to the clinical consequences of malfunctioning of the level. In the description of level B, Bernstein described numerous cases of neurological patients with lost dynamical stability of movements, but not patients with too many variables to control. In other words, he illustrated and emphasized the latter function of level B, not the former one. In contrast, over recent years, the former feature of synergies—grouping of elements—has been studied extensively using various methods of matrix factorization applied to correlation or covariation matrices (reviewed in Tresch et al. 2006; Ting and McKay 2007; Santello et al. 2016; see Chapter 6). The latter feature—ensuring dynamical stability of actions—attracted little attention until the end of the 20th century. It has mostly been studied within the framework of the uncontrolled manifold (UCM) hypothesis (Schöner 1995; Scholz and Schöner 1999; see Chapters 5 and 6).

Bernstein addressed level B as the thalamopallidar level; that is, he linked it to one of the structures within the basal ganglia (the pallidum) and the thalamus, a constellation of nuclei involved in two major brain loops, through the basal ganglia and through the cerebellum. This guess has received support in recent studies documenting major problems with the control of dynamical action stability in patients with basal ganglia disorders (such as Parkinson disease) and cerebellar disorders (such as olivopontocerebellar atrophy) (reviewed in Latash and Huang 2015; see Chapter 22).

The next level within the hierarchy—level C or the *level of spatial field*—was subdivided by Bernstein into two sublevels, C1 and C2. Both were linked to the concept of *spatial field*—a novel concept introduced by Bernstein and developed later by prominent psychologists, including John Gibson and Michael Turvey, within the field of ecological psychology (Gibson 1979; Kugler and Turvey 1987). Bernstein realized that objective classical mechanics might not be adequate to describe actions by animals with external objects in the natural environment. Animal actions depend not on actual distance to an object but on whether the object is reachable or requires making a few steps to reach it. They depend not on the actual weight of the object but on whether the object can be lifted and manipulated. The spatial field reflects properties of the nearby space via its *affordances*—a term introduced later in ecological psychology to describe opportunity for action. The spatial field is characterized by its geometry and metrics, but these differ from the geometry and metrics in the objective Euclidian space.

The two sublevels differ primarily in their sharing of responsibilities with respect to typical tasks. Level C2 is concerned with getting to a target, while level C1 deals with selection of specific trajectories leading to the target, for example, circumventing obstacles on the way to the target. Bernstein emphasized the importance of *error compensation* among involved effectors (e.g., individual joint trajectories during reaching) with respect to reaching the desired target or following the desired trajectory. This term means that if, during an action, one of the elements (e.g., a joint) shows an unexpected deviation from its desired trajectory, spontaneously or because of an unexpected change in external forces, other joints would also deviate to minimize the effects of the original error on task-specific salient variables (cf. Abbs and Gracco 1984; Cole and Abbs 1987; Jaric and Latash 1998; Latash et al. 1998). This feature makes level C intimately linked to dynamical stability of movements, that is, to a feature assigned to level B. Bernstein emphasized that performing a task multiple times is never associated with perfect reproduction of the contributions of the involved elements. He coined a special term for this phenomenon, *repetition without repetition*: When people repeat solving one and the same motor problem, they never repeat details of trajectories. This insight formed the core of one of the methods developed to study movement stability within the framework of the UCM hypothesis (see Chapter 5).

Level C was also addressed by Bernstein as the pyramidal-striatal level. This name emphasizes the importance of the loop cortex–basal ganglia–thalamus–cortex in the functioning of this level. Indeed, disorders of the basal ganglia commonly lead to problems with object manipulation and weakening of dynamical stability of movements (cf. Latash and Huang 2015 and Chapter 22). Bernstein did not explicitly invoke the cerebellum in the functioning of level C but mentioned that typical disorders of that level led to ataxias and dysmetrias, typical signs of cerebellar disorders. Currently, intimate links between the subcortical loops involving the basal ganglia and cerebellum have been assumed based on observations of changes in the cerebellar loop in patients with Parkinson disease, which is the most common disorder of the basal ganglia (Lewis et al. 2007; Yu et al. 2007).

The next level—level D—was termed the *level of actions* and involved movements with a purpose, for example, not only to reach for a tea kettle but also to do so to pour water into a cup. These actions are common among humans but can also be observed in some animals, in particular those that can use objects as tools. This level includes motor skills. Movements at level D may be counterintuitive. For example, to open a can one has to turn the lid, although the ultimate goal is to move it away from the can.

The highest level in Bernstein's hierarchy is level E, the *level of symbolic actions*, which can be seen only in humans. These include writing, playing musical instruments, and similar actions when the goal cannot be seen in the motor part of the action but transcends it. Both levels D and E were presumed to be cortical in nature, and injuries to the involved structures were associated with different types of apraxia.

According to Bernstein, any natural movement is built on several levels, with one of them playing the role of the *leading level* and others providing necessary background corrections. Bernstein emphasized the role of predominant sensory modalities used in the functioning of each level. According to his views, assigning primary control over a group of tasks was defined by the sensory modality best suited for these tasks. In particular, actions such as standing and walking were linked to vestibular and somatosensory signals, actions related to object manipulation were linked to vision, and so forth. Finding the right leading level for any new skill was a crucial step in skill development, and skill refinement was associated with assigning corrections to possible complicating factors (such as, e.g., typical changes in the environment) to appropriate lower levels.

2.3. Problem of motor redundancy

Among Bernstein's numerous contributions to motor control, the formulation of the *problem of motor redundancy* (a.k.a. the *Bernstein problem*; Turvey 1990) arguably has had the strongest influence on the field. There are two parts to this problem. The first is related to the aforementioned excess of elements at each level of analysis of systems involved in typical motor tasks. Indeed, we have more digits than strictly necessary to perform common object manipulation tasks; more axes of joint rotation than task constraints associated with reaching and pointing tasks; multiple muscles crossing individual joints, affording numerous solutions for the problem of producing a certain magnitude of the joint moment of force; and numerous motor units within each muscle that allow reaching a desired level of muscle activation with an infinite number of recruitment patterns. In each example, the situation resembles solving a system of n equations with m unknowns, where $m > n$. Such systems generally have an infinite number of solutions. However, during each specific movement, a single solution is observed.

There is, however, a second part to the problem. Even if only one element is involved in a task of changing its output from a certain initial value to a certain desired final value, this task can be solved with an infinite number of trajectories leading to the desired outcome. The former problem has been addressed as *state redundancy* and the latter one as *trajectory redundancy* (Latash 2012a, 2021b). In terms of Bernstein's hierarchical system, the former problem emerges at level C1 and the latter at level C2. Both problems of redundancy are illustrated in Fig. 2.4.

As discussed in more detail in Chapters 5 and 6, the excess of elemental variables (those produced by elements at a selected method of analysis) is intimately related to dynamical stability of movements. This point escaped Bernstein's attention and he assigned ensuring dynamical stability to level B (the level of synergies), while the problem of motor redundancy was associated with level C.

Many of my colleagues would probably charge me with blasphemy, but I am going to suggest that the problem of state motor redundancy is ill-formulated. This

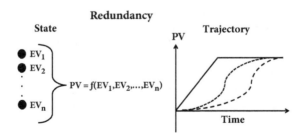

Figure 2.4 Two problems of motor redundancy. The first (state redundancy) is associated with an infinite number of solutions to produce a certain magnitude of a performance variable (PV), which gets contributions from many elements producing elemental variables (EVs). The second (trajectory redundancy) is associated with an infinite number of trajectories leading from a value of a performance variable to another value.

formulation misled generations of researchers and forced them to search for different computational methods and criteria to solve this nonexistent problem. Later, in Chapters 5 and 6, we will discuss in more detail why this problem is ill-formulated and what would be a better attitude to the apparent excess of elements at different levels of analysis. At this time, let me only ask the reader to imagine that a person coactivates strongly muscles of a limb performing a reaching task. This would lead to a drop in the excursion of some of the joints moving under the action of joint coupling forces. Some of those joints would move very little such that their motion could be considered nonexistent. So, at the level of joint kinematics, this strategy would reduce the number of variables and contribute to solving the problem of redundancy. This argument has been used in a number of studies addressing the effects of coactivation as *freezing of degrees of freedom* (Newell 1991; Vereijken et al. 1992).

Now, consider the same example at the levels of muscle activation. It is obvious that coactivating multiple muscles potentially increases the number of involved elements and makes the problem of motor redundancy worse. Coactivating muscles definitely increases the number of recruited motor units, thus making the problem worse at that level as well. So, depending on the selected level of analysis, one and the same action may be viewed as helping to mitigate the problem of motor redundancy or exacerbating it. Clearly, something is wrong with the way the problem has been formulated.

The problem of trajectory redundancy seems to be less problematic. Indeed, there are preferred and less preferred trajectories between the initial and final values of salient variables. Although many studies have provided evidence for preference for straight trajectories with more or less bell-shaped velocity profiles (e.g., Morasso 1981, 1983), the possibility of analysis of trajectories in different spaces makes this observation somewhat unclear. Indeed, a straight trajectory during reaching in the external Euclidian space is not the same as a straight trajectory in the space of joint rotations or in the space of muscle activations. We should also keep in mind that the level of spatial field is associated with non-Euclidian systems of coordinates and metrics. Trajectories associated with affordances are probably more salient for movements controlled at level C. Properties of such trajectories remain mostly unexplored. These examples suggest that the most crucial step in analysis of any action is selection of a proper analysis space. We will return to this issue after introducing a language for motor control processes (Chapters 3 and 7) rather than using languages for motor performance such as those of kinematics, kinetics, and muscle activation.

2.4. Sharing and optimality

Based on our everyday experience, it is obvious that typical motor tasks can be shared differently among the involved elements. Reaching with a hand for a target placed at the edge of reach can be done with or without major trunk involvement, with or without making a step. When we grasp an object, the gripping force can be shared differently among the four fingers opposing the thumb. In particular, we can easily lift one of the fingers without losing the grasp, although this action obviously reduces to zero the contribution of the lifted finger to the gripping force. Nevertheless, when a task is performed without additional constraints, the pattern of sharing salient task-specific variables across elements is relatively consistent both across subjects and across variations of the task parameters.

Imagine, for example, that you are asked to press with the four fingers of a hand and produce a slow ramp of the total force. If individual finger forces are recorded, their relative contribution to total force stays about the same over a wide range of total force magnitudes (Li et al. 1998; Fig. 2.5). So, there is a consistent preferred sharing pattern that is not imposed by the task and may reflect self-imposed constraints. In particular, in the described example, the pattern of force sharing across the fingers has been linked to trying to minimize the resultant moment of force with respect to the longitudinal axis of the hand and forearm. Note that this requirement adds one more constraint on possible finger forces. In other words, it adds one more equation to the one for total force:

$$F_{TOT} = F_I + F_M + F_R + F_L \tag{2.1}$$

$$M_{TOT} = d_I \bullet F_I + d_M \bullet F_M + d_R \bullet F_R + d_L \bullet F_L = 0,$$

where F and M stand for force and moment of force; d stands for level arms; and the subscripts I, M, R, and L stand for the index, middle, ring, and little finger, respectively.

The system of two equations (2.1) with four unknowns still has an infinite number of solutions. This meets the "minimal moment" criterion constraining the solution space but still is insufficient to account for the observed consistent sharing patterns across the four fingers.

Bernstein thought that particular sharing patterns were selected based on some optimality criterion (although he did not use this terminology). This idea has been developed over years leading to a variety of optimization criteria associated with different methods of computing costs of performing actions (reviewed in Nelson 1983; Seif-Naraghi and Winters 1990; Prilutsky and Zatsiorsky 2002; Diedrichsen et al. 2010; see Chapter 9). It has always been assumed that the central nervous system tries

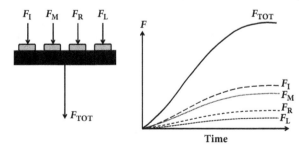

Figure 2.5 Left: A schematic illustration of four-finger total force (F_{TOT}) production. Right: During a slow increase in F_{TOT} produced by the four fingers pressing in parallel, the individual finger forces show a relatively stable sharing pattern; that is, they produce about the same percentages of F_{TOT} over the range of magnitudes. I, index; M, middle; R, ring; L, little.

to share the task among sets of effectors in a way that reduces the overall cost of action. In general, this idea looks appealing. Indeed, one can offer justifications for various costs, starting from required energy—the "universal currency" of the body (Yufik and Friston 2016)—to accumulated fatigue, discomfort, possibility of injury, and many others. It is difficult, however, to accept that a single cost defines all the variety of actions. For example, minimizing fatigue or energy expenditure makes sense while running a marathon, but not when performing a high-jump or writing a short letter.

One of the main problems with the idea of movement optimality is that specific optimality criteria are selected by individual researchers rather arbitrarily, sometimes based on intuitive considerations and reflecting their personal scientific views. There have been a few attempts to address this problem and introduce methods of computing costs (cost functions or objective functions) based on observed behavior of the system of interest. These approaches have been addressed as inverse optimization (Bottasso et al. 2006; Siemienski 2006; Terekhov et al. 2010). So far, however, such approaches have been relatively impractical, applicable only to some types of actions (see Chapter 9 for more details).

2.5. Missing pieces of the mosaic

It is a pity that we cannot ask Bernstein to explain in more detail some of his concepts, conclusions, and illustrations. This may be seen as an illustration of Bernstein's aforementioned fascination with time: We know a lot about the past but cannot change a single aspect of it! The missing pieces of the puzzle of his heritage seem to be beyond our reach, and we have to compare his works written at different times in attempts to understand what he meant under such crucial concepts as tone, synergy, and spatial field. Bernstein always tried to be very exact in his definitions and usage of the main terms, but the paucity of factual information at his time led sometimes to contradictory and confusing statements. Some of them, in particular interpreting intramuscular processes in terms of paleokinetics, the confusing illustration of the muscle force-length relationship, and the association of synergies with dynamical stability at level B but separating dynamical stability from excess of elements assigned to level C, all beg for explanations. Nevertheless, even when Bernstein accepted questionable axioms, his further logics have always been impeccable.

Some of the missing pieces now can be refined and brought together into a logical system based on the accumulated knowledge and theoretical advances. This will be a major goal of the rest of this book. We remain in relative darkness with respect to other issues. For example, the current knowledge of processes related to the upper levels within the Bernstein scheme is still far from sufficient for formulating viable hypotheses within the field of natural science. Most of the processes explored at levels D and E are studied using black box approaches, as input-output relations. Their links to neurophysiological processes have been explored at the level of correlations with electrophysiological and brain imaging techniques, which remain relatively crude. One should also keep in mind that correlations, even strong ones, do not mean causation.

3
Equilibrium-Point Hypothesis

Most hypotheses in the field of natural science either get accepted or disproven within a few years or even faster. The equilibrium-point (EP) hypothesis is unique. It was introduced by Anatol Feldman nearly 60 years ago (Feldman 1966), and it remains a controversial issue now. It has not been disproven in spite of many attempts to falsify it in experiments (see later in this chapter), but neither it has been accepted by most researchers. I would even make a stronger statement: It has not been understood by most researchers. In the author's humble opinion, the EP-hypothesis remains unique in the field because of its combination of a solid physiological foundation with the elegance of a physical theory.

The main problem seems to be the counterintuitive mode of control suggested within the EP-hypothesis. The impressive success of the physics of inanimate nature in the past 200 years and the ensuing progress in engineering biased researchers across various fields including biology. Indeed, movements of objects of moderate size at moderate velocities have to obey the laws on Newtonian mechanics. This means that, to ensure a desired movement of an object with inertial mass (e.g., a limb of the human body) to a new location, requisite time patterns of force have to be applied to that object. To do this, any neural controller has to produce neurophysiological control signals that, after being processed by the involved intermediate neural structures and muscles, generate those forces. This logic seems to be impeccable. But the EP-hypothesis effectively rejected this conclusion and offered an alternative, which has been a tough pill to swallow for many researchers in the field of motor control.

The main flaw in the mechanics-based thinking resides in the underlying assumption that muscles are force-generating structures within the body, just like actuators in robots. They are not. Muscles generate forces that are functions of their length and velocity. So, to ensure that a muscle generates a certain force, one has to predict perfectly (!) its length and velocity at the time that excitation signals from the central nervous system reach the muscle and induce its contraction. And this is impossible—a conclusion reached by Bernstein many years ago (1947; see Chapter 2). So, the idea of control with prescribed force patterns is incompatible with the design of the human body. What could be an alternative? The EP-hypothesis suggested one.

In this chapter, we will try to understand the origins of the EP-hypothesis, its core, origins of misunderstanding of this hypothesis, attempts to falsify it, and why these have failed. We will limit this chapter to analysis of the neural control of a single muscle and a pair of muscles acting against each other, agonist and antagonist, and controlling a simple joint with one kinematic degree of freedom. Generalization of the EP-hypothesis to submuscle and multi-muscle systems will be covered in Chapters 7 and 10. Later, we will also discuss implications of this hypothesis for kinesthetic perception (Chapter 15) and for some pathological conditions (Chapter 22).

3.1. Roots of the equilibrium-point hypothesis

The EP-hypothesis rests on a number of pillars provided by earlier research. One of its central concepts in that of *stretch reflex*, which can be traced back to classical studies by Sir Charles Sherrington and his colleagues (e.g., Liddell and Sherrington 1924). Sherrington considered stretch reflex as a major mechanism involved in both voluntary and involuntary movements. He also viewed voluntary movements as consequences of modulation of reflex parameters. These views have been all but rejected by later studies providing evidence for central pattern generators (CPGs) within the central nervous system for a variety of rhythmical movements including locomotion, breathing, chewing, scratching, and so forth (Graham Brown 1914; reviewed in Stein 1984; Grillner and Wallen 1985; Orlovsky et al. 1999). Within the idea of CPG, the main pattern of muscle activation is provided by structures within the central nervous system, and the role of reflexes is reduced to tuning the muscle activations and adjusting them as necessary in cases of changes in the external force field. It took quite a few years to combine the two views and suggest that the output of a typical CPG defines not patterns of muscle activation but time profiles of reflex parameters, which then translate into patterns of muscle activation given the actual movement kinematics (Feldman et al. 2021).

One of the pioneers of electromyography (EMG), Kurt Wachhölder, together with his younger colleague, Hans Altenburger, performed a series of studies from 1925 to 1927, and in particular, they asked a seemingly naïve question: Can humans relax agonist-antagonist muscle pairs at different joint positions? Indeed, everyday experience suggests that this is easy to do. On the other hand, the spring-like properties of muscles and tendons are well known. When a joint rotates to a new position, one of the opposing muscles (and its tendons) gets stretched and the other shortened (Fig. 3.1). This means that the balance of their moments of force is expected to be changed, and if the net moment of force was zero in the initial position, it cannot be

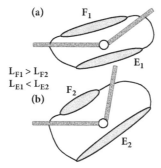

Figure 3.1 If relaxed agonist and antagonist muscles (flexor [F] and extensor [E]) balance their moments of force at a specific joint position (A), this balance has to be violated at another position (B) due to the changes in muscle length (L): shortening of one of the muscles and stretching the other one and the dependence of passive muscle force on muscle length.

zero in the final position and requires nonzero muscle activation to keep the final position. Wachhölder and Altenburger (see translation in Sternad 2002) recorded EMG signals and concluded that the muscles were indeed quiescent in both joint positions. Of course, their equipment was noisy and provided information on only a portion of the muscle fibers, and drawing such a conclusion based on the available data was far-fetched. Nevertheless, they concluded that, during voluntary movements, parameters of the muscle spring-like behavior were reset. As we will see, the ability of the central nervous system to change parameters of the force-length (spring-like) muscle behavior forms one of the important components of the EP-hypothesis.

As mentioned in Chapter 2, Nikolai Bernstein emphasized the importance of the length dependence of muscle force and viewed this dependence as one of the main reasons for the inability of the central nervous system to prescribe muscle force patterns during movements. He, however, did not consider the contribution of muscle reflexes to muscle activation and illustrated muscle force-length characteristics assuming that each characteristic corresponded to a particular fixed level of muscle activation (see Fig. 2.1 in Chapter 2). As a result, this illustration in Bernstein's main book (1947) is intrinsically contradictory, as discussed earlier. Bernstein did not make the natural next step to assuming that parameters of the stretch reflex could be modified by the central nervous system, leading to families of force-length characteristics defined by both peripheral muscle properties and reflex-mediated changes in its activation.

Another important step was made by von Holst and Mittelstaedt (1950/1973), who introduced the so-called *posture-movement paradox* (Fig. 3.2). These researchers were puzzled by the fact that an external perturbation applied to an effector involved in a postural task (e.g., to the trunk of a standing person) produces posture-stabilizing responses that could be seen at latencies shorter than the typical simple reaction time delay (these have later been addressed with various names including long-loop reflexes and preprogrammed reactions; reviewed in Nashner and Cordo 1981; Chan and Kearney 1982; Reschechtko and Pruszynski 2020), while quick voluntary

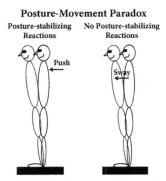

Figure 3.2 An illustration of the posture-movement paradox. A deviation of the body induced by an external force pulse produces short-latency posture-stabilizing reactions (A). A similar deviation induced by voluntary body sway does not produce short-latency posture-stabilizing reactions (B).

movements leading to a similar effector deviation (e.g., quick voluntary sway) did not produce such apparently involuntary responses. How can animals produce voluntary movements without triggering those involuntary posture-stabilizing reactions? This question has been a major litmus test for hypotheses in the field of motor control. As we will see later, the EP-hypothesis provides a natural explanation for this observation.

Arguably, the closest precursor of the EP-hypothesis was the servo-model introduced by Merton (1953). This hypothesis assumed that signals from the brain acted on gamma-motoneurons and set a value of muscle length, which was then established with the help of the stretch reflex (Fig. 3.3A). Signals from gamma-motoneurons changed the activity of muscle spindle endings, which led to reflex-mediated changes in the activity of alpha-motoneurons, leading to changes in muscle contraction state. The reflex was assumed to act as a perfect servo; that is, the force-length characteristics were assumed to be nearly vertical (Fig. 3.3B) such that the prescribed muscle length was achieved independently of the external load.

The servo-model was indeed very elegant, and it may be viewed as the first truly motor control hypothesis because it specified explicitly a physiological parameter modulated by the central nervous system to produce movements, as well as the involved neurophysiological circuitry. This hypothesis, however, suffered from a few major flaws. First, one could theoretically control movements with the help of nearly vertical force-length characteristics. However, using such characteristics to produce accurate muscle force changes in isometric conditions was next to impossible, because a tiny change in the control signal could produce a disproportionally huge force change (Fig. 3.3). Second, the assumed neurophysiological scheme posited that any movement started with a change in signals to gamma-motoneurons, and alpha-motoneurons changed their activity after a substantial time delay defined, in particular, by the relatively slow condition speed along the relatively thin and long axons

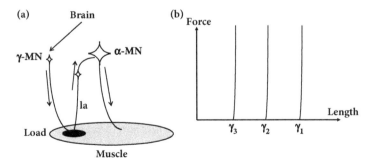

Figure 3.3 Schematics of the servo-model. A: Voluntary movements are initiated by signals to gamma-motoneurons, which change the set point of muscle spindle endings and encode a new value of muscle length. B: Force-length characteristics within the servo-model are nearly vertical to ensure that any changes in the external load are readily compensated, resulting in the same muscle length value. Different neural commands (γ_1, γ_2, and γ_3) encode different locations of the force-length characteristic. Reproduced by permission from Latash 2019.

of gamma-motoneurons. Experiments have shown, however, that this was not the case: Changes in activation of alpha- and gamma-motoneurons happen nearly simultaneously, a phenomenon addressed as alpha-gamma coactivation (see Matthews 1966, 1972). Finally, Ake Vallbo (1971, 1974) measured the gain of the stretch reflex and showed that it was modest; that is, the muscle force-length characteristics were very far from vertical.

Very important experimental data for the formulation of the EP-hypothesis were provided by the experiments of Peter Matthews (1959) on decerebrate cat preparations. In particular, those studies have documented force-length muscle characteristics with modest slopes, and moreover, they showed nearly parallel displacements of those characteristics along the length axis with minimal changes in their shape under a variety of manipulations including, in particular, anesthesia of the efferent fibers from gamma-motoneurons. In other words, those experiments have suggested that changes within the central nervous system were associated with changes in the spatial threshold of the stretch reflex.

All the aforementioned studies created the background for the next important step—the introduction of a general scheme for the neural control of a muscle that would unite them all into a noncontradictory neurophysiological hypothesis. A young physicist, Anatol Feldman, performed a series of experiments on human subjects looking for *invariants of behavior* under the instruction "not to interfere voluntarily" with movements induced by changes in the external load. The logic was that if a parameter, behavioral or neurophysiological, stays unchanged under such manipulations, it may be viewed as a candidate for the neural control parameter specified to perform the original task.

3.2. The equilibrium-point hypothesis: Single-muscle control

The original experiment by Feldman (1966) explored elbow joint movements induced by changes in the external load placed on the hand. In the initial state, the subject was asked to keep a certain elbow joint angle while holding a load on the hand (point EP_0 in Fig. 3.4). Then, part of the load was removed smoothly. This led to joint motion to a new position and corresponding muscle length changes where the movement stopped (point EP_1 in Fig. 3.4). Then another portion of the load was removed, leading to a new position (point EP_2). Such a sequence of unloading procedures led to a sequence of points on the force-length plane that were interpolated to discover the point of intersection between the interpolated line and the characteristic of the joint when muscles were relaxed (not shown in Fig. 3.4; see Fig. 1.2).

Note that, under the main assumption that the subjects were following the instruction "not to interfere voluntarily," the steady states shown by EP_0, EP_1, ..., EP_n corresponded to a fixed voluntary command but differed by the magnitudes of muscle length and force, as well as levels of muscle activation. Hence, the observed changes in muscle activation level, length, and force resulted from an involuntary, reflex mechanism associated with the classical stretch reflex. The point of deviation from the characteristic of the relaxed muscle was, therefore, accepted as the threshold of the stretch reflex (λ), which gave name to the term *λ-model*.

3.2. THE EQUILIBRIUM-POINT HYPOTHESIS: SINGLE-MUSCLE CONTROL   31

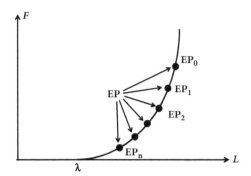

Figure 3.4 When a person does not intentionally "interfere voluntarily" with deviations of an effector induced by a change in the external load, steady states (filled circles) cling to a smooth force-length characteristic of the muscle with an intercept with the length axis, λ.

Then, the subject was asked to occupy a different joint position against the same external load (compare points a_0 and b_0 in Fig. 3.5), the sequence of unloading manipulations was repeated, and the observed equilibrium points were interpolated, resulting in a new value of λ (compare λ_a and λ_b in Fig. 3.5). The observations summarized in these two figures were interpreted in the following way. When a person keeps a neural command to a muscle constant, its equilibrium states are constrained to a single force-length (moment-angle) characteristic, which is associated with a fixed value of λ. When a person changes the neural command to the muscle voluntarily, a new value of λ is set, and this value defines a new force-length characteristic. So, the neural control variable manipulated by the central nervous system is λ!

Note that shifts in λ do not define any of the variables measured in typical experiments (kinetic, kinematic, and EMG) because all of them depend also on the external

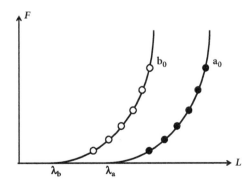

Figure 3.5 When a person intentionally occupies a different initial position against the same load (compare a_0 and b_0), the same sequence of changes in the external load leads to steady states that cling to another force-length muscle characteristic with a different threshold value (compare λ_a and λ_b).

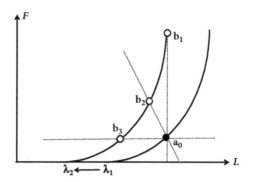

Figure 3.6 A change in λ (e.g., from λ_1 to λ_2) can lead to different peripheral consequences depending on the external load characteristic (from the initial state shown as point a_0 to various final states shown as points b): a change in muscle force in isometric conditions (vertical load characteristic), a change in muscle length in isotonic conditions (horizontal load characteristic), and a change in both in more natural conditions (see the slanted load characteristic). The final steady states differ in muscle force, length, and activation level values.

load, which may be time-varying and unpredictable or, at the very least, imperfectly predictable. In particular, one and the same shift in λ can lead to a change in muscle force, length, or both depending on the external load characteristic as illustrated in Fig. 3.6. Since no physiological processes happen instantaneously, the rate of λ shift is limited. These shifts can happen within a range of speeds. Changing the speed of λ shift defines, indirectly, the speed of the action.

To discuss the neurophysiological mechanisms of muscle control within the EP-hypothesis, let us start with some of the basic properties of alpha-motoneurons. An alpha-motoneuron generates an action potential to the target muscle when the neuron's membrane is depolarized, and its membrane potential reaches the threshold value. For simplicity, let us consider three groups of excitatory inputs to an alpha-motoneuron: those from descending pathways (I_{DES} in Fig. 3.7), those along reflex pathways from proprioceptors within the target muscle (I_{REF}), and those from other sources including receptors in other parts of the body (I_{OTHER}). Signals from I_{REF} define the shape of the force-length characteristic described earlier. I_{DES} and I_{OTHER} can be viewed as contributing to the effective threshold for muscle activation, λ^*. One should also consider the well-known velocity sensitivity of primary spindle endings, which makes the stretch reflex threshold velocity sensitive. This can be described as another contributor to λ^*, making it different from λ in pseudo-static conditions (as in the earlier Fig. 3.4). It has also been known that properties of neuronal membranes are history dependent. So, in a linear approximation and assuming additive effects from all the mentioned sources, the effective threshold of the stretch reflex can be represented as

$$\lambda^*(t) = \lambda_{DES}(t) + \lambda_{OTHER}(t) + \mu V(t) + \rho(t), \tag{3.1}$$

3.2. THE EQUILIBRIUM-POINT HYPOTHESIS: SINGLE-MUSCLE CONTROL

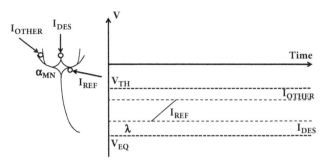

Figure 3.7 A schematic illustration of three inputs into a pool of alpha-motoneurons, from descending pathways (I_{DES}), from peripheral sensory endings (I_{REF}), and other inputs (I_{OTHER}). I_{DES} produces subthreshold depolarization of the membrane, which leads to a change in the effectiveness of the I_{REF} input translating into a change in the stretch reflex threshold (λ). The third input can involve, in particular, effects of persistent inward currents (PICs), effectively reducing the distance to the threshold of action potential generation. V_{EQ}, equilibrium membrane potential; V_{TH}, threshold membrane potential for action potential generation.

where μ is a coefficient reflecting the velocity sensitivity of spindle primary endings, and ρ is a term reflecting the history dependence of the membrane properties.

Generally speaking, descending signals (I_{DES}) can bring the target alpha-motoneuron to the threshold (we will call such inputs suprathreshold) or depolarize the membrane below the threshold (subthreshold). If structures in the brain send a suprathreshold signal to a neuron, this signal forces the neuron to generate an action potential independently of other possible inputs because of the "all or none" rule for the action potential generation. If such a signal were kept for some time at a suprathreshold level, the neuron would continue generating action potentials at the highest possible frequency, also independently of other possible inputs. Hence, this type of control makes alpha-motoneurons insensitive to other inputs, in particular to inputs from peripheral sensory endings. Effectively, this would eliminate the natural adaptive properties of alpha-motoneurons that make muscles adjust to the actual conditions of performance. This mode of control is possible and may be used in deafferented animal preparations as well as in patients with large-fiber peripheral neuropathy—a rare disorder eliminating both reflexes and kinesthetic perception (see Chapter 22).

In healthy animals, however, the level of muscle activation shows strong sensitivity to changes in muscle kinematics, reflecting the important role of reflex loops in defining muscle activation patterns. This is possible if the central input produces only subthreshold depolarization of the target alpha-motoneurons, which require additional excitation via reflex inputs to generate action potentials. Respectively, at the level of neurophysiology, λ has been associated with subthreshold depolarization of the membrane of alpha-motoneurons controlling the target muscle.

In Fig. 3.7, I_{DES} reflects descending inputs. Its changes lead to subthreshold membrane depolarization and reduce the distance between the actual membrane potential and the threshold for action potential generation. I_{REF} produces a steady depolarizing input, reflecting the muscle state, which leads to increasing depolarization of the membrane. At some time, the membrane potential reaches the threshold and leads to action potential generation. The neuron resets, and the process continues. Within this scheme, the frequency of firing of a neuron is defined by two main factors: (1) the effective distance between the steady potential on the membrane defined by I_{DES} and the threshold value and (2) the rate of depolarization produced by I_{REF}, which increases with muscle length.

Figure 3.7 and Eq. (3.1) are of course major simplifications. In particular, recent series of studies have brought attention to the phenomenon of persistent inward currents (PICs; reviewed in Heckman et al. 2005, 2008a,b), which can contribute significantly to decreasing the distance between the membrane potential and threshold for action potential generation. As the name suggests, PICs are long-lasting processes, which depend on ion motion through channels without the phenomenon of inactivation, typical of Na^+ channels. Originally, PICs were assumed to reflect primarily motion of Ca^{++} ions, although recently Na^+ channels without inactivation have been reported.

Note that in two of the figures (Figs. 3.4 and 3.7), the same parameter (λ) is expressed in units of muscle length and in units of membrane depolarization, respectively. The former is salient for movement mechanics, and the latter for neurophysiological processes. So, we can conclude that the mechanism of stretch reflex plays the role of a synchronous interpreter between the worlds of mechanics and neurophysiology without assuming any computational mechanisms that could be involved in communication between these two worlds.

It is very useful to distinguish among three types of trajectories during voluntary movements (Fig. 3.8). The first is the time profile of the control variable, $\lambda(t)$, which we assume, for simplicity, to be under brain control. This variable can change at a very fast rate, although not instantaneously. Existing estimates put the highest rate of change of $\lambda(t)$ during arm joint movements performed "as fast as possible" at about 800°/s (e.g., Latash et al. 1991). A change in λ leads to translation of the force-length characteristic and of a comparably fast shift of the instantaneous EP, the point of intersection of the force-length characteristic with the external load characteristic (which is assumed to be spring-like in Fig. 3.8). Note that EP shift is characterized by time changes in two mechanical variables, length and force, $L_{EP}(t)$ and $F_{EP}(t)$. Unlike λ, EP shifts are not under exclusive neural control because they depend on the actual external load, which may be unpredictable. In typical experiments, observed variables are those of actual muscle length and force, $L(t)$ and $F(t)$. These may change at a considerably slower rate, which depends, in particular, on the inertia of the moving effector. The former two trajectories, the control trajectory $\lambda(t)$ and the EP trajectory $\{L_{EP}(t); F_{EP}(t)\}$, are not directly observable. To compute them, one has to use observable trajectories, such as $\{L(t); F(t)\}$, and compute the hidden ones assuming a model of the moving effector. As we will see later (see Chapter 24), such models can turn into major sources of controversy.

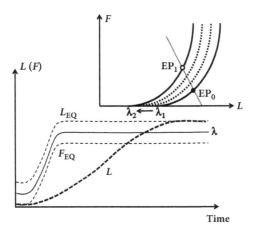

Figure 3.8 Three types of trajectories associated with a fast movement. The control trajectory is λ(t) changing from $λ_1$ to $λ_2$ (see the insert). The equilibrium trajectory has two components, $L_{EQ}(t)$ and $F_{EQ}(t)$, depending on the external load characteristic (shown as a slanted dotted line). The actual trajectory, L(t), is slower and is defined by movement mechanics, including inertia. Force trajectory, F(t), is not shown so as not to clutter the drawing.

3.3. The equilibrium-point hypothesis: Single-joint control

All effectors in the body are controlled by at least two muscles opposing each other. The presence of such agonist-antagonist (flexor-extensor, adductor-abductor, etc.) muscle pairs is a mechanical necessity because muscles can produce active force in one direction only. Consider a simple joint with one kinematic degree of freedom spanned by only two muscles with similar force-length characteristics. This is, of course, a cartoon system (Fig. 3.9): Most joints in the human body have more than one kinematic degree of freedom and are spanned by more than two muscles. This cartoon, however, is useful to illustrate the first step in the generalization of the EP-hypothesis from the neural control of a single muscle to the control of more complex systems (considered later in Chapter 7).

Figure 3.9 illustrates this system using muscle moment-angle characteristics. We switch here from linear variables (force and length) to rotational ones (moment of force and angle) to describe the control of rotational joint action. Assuming that each muscle is controlled by setting its λ, the system of two muscles has two control variables, $λ_{AG}$ and $λ_{ANT}$, for the agonist and antagonist muscle, respectively. Setting values of $λ_{AG}$ and $λ_{ANT}$ defines the spatial positions of the muscle characteristics (thin curves in Fig. 3.9). The antagonist characteristic is drawn "upside down" because the antagonist muscles are assumed to generate negative moment values. Mechanical behavior of the joint is defined by the algebraic sum of the opposing muscle characteristics (the thick line in Fig. 3.9). Given a particular magnitude of the external moment of force (load), this system would settle in an EP where the net active moment of force balances the load.

3. EQUILIBRIUM-POINT HYPOTHESIS

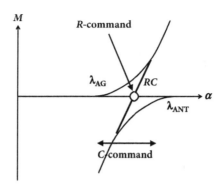

Figure 3.9 The control of a simple joint with one rotational degree of freedom can be described as setting the time function of $\lambda_{AG}(t)$ and $\lambda_{ANT}(t)$ for the agonist and antagonist muscles, respectively. It can also be described as changes in two basic commands, reciprocal and coactivation (R and C). The R-command defines the referent coordinate (RC) for the joint—a position it occupies in the absence of external forces. The C-command defines the spatial range of simultaneous activation of the opposing muscles, that is, the spatial interval between $\lambda_{AG}(t)$ and $\lambda_{ANT}(t)$.

Note that the scheme in Fig. 3.9 assumes that two variables at the neural control level can be manipulated to perform movements of this system with only one kinematic degree of freedom. This abundance at the level of neural control is a very important feature that can be used to ensure the stability of salient task-specific performance variables, such as net moment of force produced on the environment or joint position, as discussed in a future chapter (Chapter 7).

Another pair of control variables was introduced to describe the system in Fig. 3.9. One of them defines joint angle where the net moment of force is zero. In other words, this variable defines joint position that would be occupied if the external load were zero. This position can be addressed as the referent coordinate (RC) for the joint (see Feldman 2015 for review and Chapter 7). Note that, if the external load is removed, a single muscle would contract and reach λ. So, for a single muscle, λ = RC. The notion of RC is central within the current generalization of the EP-hypothesis, and we will return to it in Chapter 7.

The control variable that defines RC for a joint has been addressed as the reciprocal command or the R-command. This name reflects the fact that a change in the R-command shifts both muscle characteristics in the same direction, corresponding to better conditions for activation of one of the muscles and worse conditions for the opposing muscle (Fig. 3.10A). This results in a shift of the joint torque-angle characteristic along the angle axis without a major change in the shape of this characteristic. For a system with two perfectly symmetrical muscles (which is unrealistic!), $RC = (\lambda_{AG} + \lambda_{ANT})/2$.

The other control variable reflects the distance between λ_{AG} and λ_{ANT}. It has been addressed as the coactivation command or the C-command. If the external load is zero, changing the C-command leads to parallel changes in the activation level of the

3.3. THE EQUILIBRIUM-POINT HYPOTHESIS: SINGLE-JOINT CONTROL 37

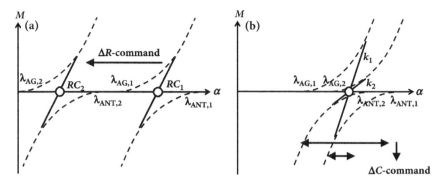

Figure 3.10 An illustration of the effects of changes in the *R*-command (A) and *C*-command (B) on the torque-angle joint characteristic. Note that a change in the *R*-command (ΔR) leads to a shift in the referent coordinate (RC) of the joint. A change in the *C*-command (ΔC) leads to a change in the slope of the moment-angle characteristic (its apparent stiffness, *k*).

opposing muscles, that is, a decrease or an increase in the amount of their coactivation (Fig. 3.10B). At the level of joint mechanics, the *C*-command defines the slope of the torque-angle characteristic without major effects on its spatial location. We will address the slope of this characteristic as its *apparent stiffness* (*k* in Fig. 3.10B; cf. Latash and Zatsiorsky 1993). The word "apparent" implies that we deal here not with a single spring, but with a complex system that, within a certain range of its behavior, shows spring-like properties (i.e., close-to-linear dependence between net torque and angle).

Figure 3.11A illustrates the effects of changing the *R*-command in isotonic and isometric conditions. In isotonic conditions, a change in the *R*-command leads to movement to a new joint position. In isometric conditions, the same change in the *R*-command leads to a change in the active joint torque. So, similarly to λ, the *R*-command does not prescribe changes in joint mechanics, which depend on the external load characteristic.

Figure 3.11B illustrates the effects of the *C*-command. These depend not only on the type of external load but also on its magnitude. In particular, if the external load is isotonic and set at zero level, changes in the *C*-command lead to no joint movement. If the external load is nonzero, the same change in the *C*-command produces joint movement. In isometric conditions, *C*-command changes produce changes in the net torque. Some of the nontrivial effects of changing the *C*-command will be discussed later, in Chapter 19.

Before moving to sources of misunderstanding of the core elements of the EP-hypothesis, let us summarize its main contributions and developments over the past 60+ years.

The EP-hypothesis is deeply rooted in both physics and neurophysiology. It introduces a novel principle of control—parametric control—and specifies explicitly the control parameter manipulated by the central nervous system to produce voluntary

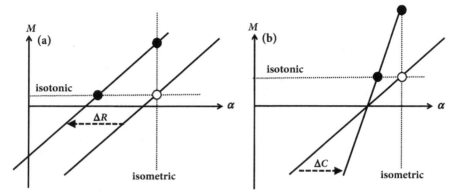

Figure 3.11 An illustration of the mechanical effects of changes in the R-command (ΔR) (A) and C-command (ΔC) (B) under different external load characteristics. In isometric conditions, changes in either command lead to a change in the net joint torque. In isotonic conditions, the effects of the C-command changes depend on the external load magnitude.

movements. It offers both physical and neurophysiological interpretations of this parameter: threshold of the force-length dependence and subthreshold depolarization of the alpha-motoneuronal membrane.

The EP-hypothesis solves the posture-movement paradox without any additional assumptions. Indeed, if a change in the external force moves the muscle away from its EP, posture-stabilizing mechanisms act to stabilize the EP of the system "muscle + reflexes + external load." Voluntary movement does not move the system away from its EP. Instead, it leads to translation of the EP itself, and the previous muscle state becomes a deviation from the newly established EP. The same posture-stabilizing mechanisms move the muscle to the new EP. So, within the EP-hypothesis, posture and movement are different peripheral outcomes of essentially the same control process.

Within the framework of the EP-hypothesis, a range of issues from neurophysiology and motor behavior have been analyzed since the introduction of the hypothesis in the mid-1960s. Here is an incomplete list (reviewed in Feldman 1986, 2015; Latash 1993, 2008, 2019):

- A range of voluntary movements, from eye movements to locomotion
- The origin of postural sway
- Triphasic patterns of muscle activation during fast movement
- Patterns of muscle activation during whole-body movements
- The role of muscle coactivation
- Recruitment of motor units across muscle pairs
- Patterns of control parameters during a variety of motor tasks
- Motor variability and speed-accuracy trade-offs
- Effects of stimulation of different descending systems
- Synergies stabilizing salient performance variables

- Kinesthetic perception and illusions
- The origin of spasticity in neurological patients

Some of the mentioned topics are discussed in detail in the following chapters.

3.4. Sources of misunderstanding: The alpha-model

The first reactions of the research community to the introduction of the EP-hypothesis in the mid-1960s were predominantly negative. The impressive success of studies on the interactions among reflex pathways at the level of spinal interneurons (e.g., Lundberg 1966; Jankowska 1979) made neurophysiologists well aware of the complexity of the spinal reflex machinery. As a result, they met with strong suspicion the EP-hypothesis suggesting that the neural control of a muscle with its apparently very complicated reflex pathways could be described with time changes of a single parameter: threshold, λ, of the stretch reflex. Engineers trained in the control theory expected a hypothesis based on predicting and prescribing time profiles of forces and torques to be produced by the apparent actuators—muscles. The ideas of parametric control looked alien and weird. The progress in EMG made it hard to accept the idea that muscle activation patterns represented another peripheral variable only indirectly related to the neural control process. Overall, on the surface, the main idea of control looked too simple to be true and too disconnected from the mainstream research to be taken seriously.

There were a few more factors that complicated understanding and acceptance of the EP-hypothesis. Some of them originated from a few unfortunate terms, such as "invariant characteristic" for the force-length muscle characteristic. Others originated from examples selected to illustrate some of the main ideas of the hypothesis, in particular the mass-spring analogy, which was offered as a metaphor but taken explicitly by many researchers. This situation was exacerbated by the fact that the original papers were written in Russian, and their published English versions were prepared by translators who were not experts in the field. In the late 1970s, another version of the EP-hypothesis (the alpha-model; Polit and Bizzi 1978; Bizzi et al. 1982) was introduced, thus adding to the overall confusion. Let us consider some of the main sources of misunderstanding in more detail.

Addressing the force-length muscle characteristic as *invariant* was done originally to emphasize the fact that changes in the neural control variables, λ, led to shifts of those characteristics along the length axis without a major change in their shape. Of course, the shape of the force-length characteristic did not remain completely unchanged, as shown in particular in earlier studies by Matthews as well as in Feldman's original experiments. In addition, the effects of hysteresis had been known by the time the EP-hypothesis was introduced (Partridge 1965), leading to different shapes of the force-length characteristics reconstructed with loading and unloading procedures. The term *invariant characteristic* was criticized in a series of papers by Gottlieb and Agarwal (1986, 1988), who confirmed experimentally that the shape of the characteristic differed depending on the loading/unloading history. In addition, a study on the decerebrate cat preparations showed that muscle force-length characteristics could potentially intersect in response to electrical stimulation of the magnocellular nucleus, meaning that their shape was not always kept invariant (Nichols and Steeves 1986). Let us keep in mind, however, that the possibility of changes in the shape of the force-length characteristic is a technical issue, which does not question the core ideas of the EP-hypothesis: The brain can control movements using the thresholds of a family of characteristics with different shapes.

The EP-hypothesis on many occasions has been described as a mass-spring model or, more generally, as the control of a second-order linear system with changes in zero length of the spring element (cf. Schmidt and McGown 1980; Simmons and Richardson 1984; de Lussanet et al. 2002). This is, of course, a major simplification, as emphasized in many publications (Feldman 1986, 2015; see also Ostry and Feldman 2003; Feldman and Latash 2005). Accepting this model implies, in particular, that the system has to demonstrate a feature termed *equifinality*: If a particular time profile of the control variable is defined by the central nervous system, transient perturbations applied in the course of the movement should not violate its final coordinate as long as the load in the final position remains the same. Indeed, a number of studies documented equifinality during quick movements associated with transient perturbations (Schmidt and McGown 1980; Hoffer and Andreassen 1981; Latash and Gottlieb 1990). However, violations of equifinality have also been reported, in particular in experiments with destabilizing transient forces proportional to movement velocity, such as the Coriolis force (Lackner and DiZio 1994) and negative damping force (Hinder and Milner 2003). Note that negative damping does not exist in nature but could be simulated with a robotic system. Forces proportional to movement velocity act only during the movement but not when the effector stops; that is, these forces should not affect the equilibrium state achieved after the movement. Observations of violations of equifinality were claimed to refute the EP-hypothesis.

First, let us state that human active muscles are not springs even though, under certain conditions, they display spring-like behavior. The same can be said about damping properties of intact muscles. Muscle are active elements, and their mechanical behavior is mediated, in particular, by reflex effects on their activation level. Properties of these reflex loops are known poorly. In particular, the reflex-mediated damping has never been measured properly (see Chapter 24). The idea of fractional power damping has been developed to explain how fast movements can be associated with effective stopping at the target (Gielen et al. 1984; Novak et al. 2000). Within this idea, reflex-mediated changes in muscle activation produce force (F) proportional to velocity (V) in a fractional power: $F = b \cdot V^{(1/n)}$, where b is a constant and n is an integer. At high velocities, this equation leads to smaller force values as compared to traditional, linear damping. At low velocities, however, the effects of damping on force are higher than those of linear damping, thus helping to avoid terminal oscillations at movement completion.

Second, the prediction of equifinality assumes not only action of transient forces but at least two more factors. First, there should be no modifications in the control variable (λ) in response to the perturbation. Note that such modifications can happen unintentionally and not be perceived by the subject (see also Chapter 16). Second, peripheral properties of the muscle fibers should stay unchanged; otherwise, the shape of the force-length characteristic can change, leading to shifts of the final EP. Such changes may be expected, for example, due to the so-called catch property of muscle fibers—a change in their force-generating capabilities in response to a standard activation signal following a quick contraction (Burke et al. 1970, 1976). So, equifinality is possible but in no way guaranteed within the control scheme offered by the EP-hypothesis.

Let us emphasize that the tendency to move toward the target was seen in the cited experiments of reaching movements by subjects sitting in a centrifuge. The subjects sat over the axis of rotation, moved in complete darkness, and were unaware of the centrifuge rotation. So, during a quick movement to a target, the Coriolis force acted always in the same direction (Fig. 3.12). The hand trajectory, however, first showed a deviation away from the straight trajectory observed in the absence of rotation but then reversed, moved toward the target (i.e., against the Coriolis force), and stopped shy of the target. So, there was a residual error, but there were also signs of a tendency toward equifinality. This error has been interpreted as a likely consequence of non-perceived corrections by the subject (Feldman and Latash 2005).

A major confusing factor was the introduction of the alpha-model, which became viewed by some researchers as a more advanced version of the EP-hypothesis. The alpha-model was based on a series of ingenious experiments on deafferented monkeys trained to perform upper limb movements to visual targets without seeing the moving limb (Polit and Bizzi 1978, 1979; Bizzi et al. 1982). The deafferentation had two consequences. First, the monkeys could not feel the moving limb and, therefore, were not expected to introduce corrections if a movement was unexpectedly perturbed by a change in the external force. Second, the stretch reflex was eliminated, and the monkeys had to develop a different method to control their muscles. As illustrated earlier (Fig. 3.7), in the absence of reflex inputs to alpha-motoneurons, the only way to control their output is to generate an adequate time profile of their suprathreshold presynaptic input. This input effectively prescribes the level of muscle activation independently of movement kinematics. Of course,

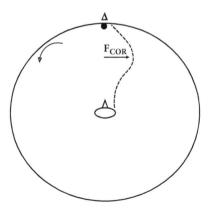

Figure 3.12 An illustration of the arm trajectories during reaching in the rotating centrifuge. Note that the Coriolis force acts always in the same direction. However, the arm tends to return to the target but does so with a consistent undershoot (shown as Δ).

this is a method of control of a pathologically changed system. A somewhat similar condition in humans—the large-fiber peripheral neuropathy (see Chapter 22)—is associated with poor motor coordination even during movements performed under continuous visual control.

Two major results of the experiments on deafferented monkeys should be mentioned. First, when an unexpected transient force was applied to the moving limb, the final position of the limb remained unchanged; that is, these movements showed equifinality, a feature of movements expected within the EP-hypothesis. Second, when the limb was quickly brought into the target by the strong motor and then released, it showed a movement toward the initial position and then reversed and continued back into the target.

The authors correctly interpreted their results as consequences of setting a control trajectory, $\alpha(t)$, resulting in a time profile of the equilibrium states, an equilibrium trajectory, which led to arm movement at a limited speed (cf. Fig. 3.8). Within this model, however, the nature of the control parameter is changed. The force-length muscle characteristics are defined only by the peripheral properties of muscle fibers at a given, prescribed centrally, level of muscle activation. So, within the whole range of muscle length values, active muscle force is always nonzero. This is illustrated in Fig. 3.13 with a family of characteristics corresponding to different levels of muscle activation. Unlike the original version of the EP-hypothesis, the control parameter loses its meaning of a spatial referent coordinate for the muscle, and the mode of control is changed qualitatively. It retains, however, some of the features of EP-control such as the aforementioned equifinality.

Of course, the alpha-model cannot be used to interpret results of studies of movement of intact animals, including humans. Some of its predictions, such as constant muscle activation over the whole range of muscle length, are obviously wrong. But the model was not designed to describe the neural control of movements in the presence of muscle reflexes. Unfortunately, the qualitative difference between the two versions of the EP-hypothesis was missed in some influential publications (e.g., Shadmehr and Wise 2005), leading to a suspicious attitude toward the original version of the hypothesis.

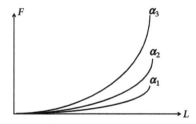

Figure 3.13 An illustration of the α-model with the muscle force-length characteristics corresponding to different levels of muscle activation. Note that, unlike the λ-version of the equilibrium-point hypothesis (cf. Fig. 3.5), the muscle produces active force over the whole range of muscle length values.

3.5. Missing pieces of the mosaic

Although the EP-hypothesis is about 60 years old, it has been developed actively only by a handful of researchers. It is of no surprise that many phenomena have not been explored or discussed, even in the most comprehensive reviews of the hypothesis (such as Feldman 2015; Latash 2019, 2021b). These include, to name only a few, the relative role of spinal and supraspinal structures in defining the values of λ for participating muscles; the similarities and differences in λ(t) patterns for apparent synergists such as the heads of the triceps surae, quadriceps femoris, triceps brachii, and other muscles with multiple heads; changes in the mode of control with various peripheral and central neurological disorders (although see Chapter 22); the role of reflex-mediated damping of movements related to the velocity sensitivity of primary spindle endings; and many others.

Of course, no movements are produced by single muscles or even by single agonist-antagonist pairs. Generalization of the EP-hypothesis to the control of multi-muscle movements is going to be discussed in Chapter 7. It remains mostly unknown what neural circuits contribute to this generalization. Sharing of the higher-level neural control variables among λs to individual muscles has not been studied. As noted by Bernstein (see Chapter 2), the central nervous system never reacts to a local event locally, but does so by changing the contributions of multiple muscles. There have been a number of studies exploring intermuscle and interjoint reflexes in animals (e.g., Nichols 1994, 2002), but such a mapping is all but absent for human muscles. This makes one uncertain with respect to the sources of muscle activation changes in response to a local load change because such a change may involve changes in activation of other muscles, which could also affect the muscle of interest.

To summarize, achieving progress in the exploration and development of the EP-hypothesis requires efforts of numerous researchers, not only of Anatol Feldman and his students. As mentioned in the introductory notes, this is not an easy hypothesis to work with, primarily because of the counterintuitive mode of control it assumes. But if the goal is to discover new important pieces of information (rather than publishing a few more papers), this effort would be well spent.

4
Motor Programming

The idea that the central nervous system programs specific patterns of peripheral variables to satisfy task requirements has been dominating the field of motor control for centuries. It can be traced to ancient Greek philosophers, in particular Aristoteles, who compared the brain to the charioteer controlling the horses (effectors) (see Meijer 2001). Over the centuries, the success of the Newtonian mechanics led to the view that continues to dominate the field of motor control: To produce a desired movement of a material object (of course, all our effectors are material objects), one has to apply proper time functions of forces. It is hard to argue with this statement. However, commonly, the next logical step is made, which sounds natural but in fact is much less obvious and, as we hope to show, misleading: To perform voluntary movement, the brain sends signals to the output neurons (alpha-motoneurons) and, ultimately, to muscles encoding the required force time functions. As described in previous chapters (Chapters 2 and 3), the last statement is incompatible with the design of the body, in particular with the length and velocity dependence of muscle force mediated by both properties of peripheral tissues and spinal reflexes.

Nevertheless, the idea that the brain uses its immense computational power (maybe overvalued—see later) to overcome the dependence of muscle force on movement kinematics and other apparently complicating factors, in particular the unavoidable time delays in neural signal transmission, has been highly influential (e.g., Hollerbach 1982). This view has received a crucial boost from the control theory, a field of mathematics developed to control the motion of man-made objects such as ballistic missiles and robots. The control theory was developed based on the idea that forces acting on objects of control can be programmed by the controller, implemented by nearly perfect force generators (actuators) and corrected, if needed, using accurate sensors and feedback loops with minimal delays. These assumptions are true for torque motors, artificial sensors, and electric feedback loops. The relatively sluggish muscles, fuzzy sensors, and slowly conducting pathways within the body make the powerful apparatus of the control theory awkward and require more and more assumptions to be applied to human movements.

During the 20th century, the development of electrophysiological methods, in particular electromyography, resulted in a group of motor control hypotheses based on the assumption that the brain could guide movements by programming time changes in the outputs of alpha-motoneuronal pools and, therefore, defining muscle activation patterns. This group of hypotheses avoids problems associated with the peripheral length and velocity dependence of muscle force. However, these approaches face another major complicating factor: reflex loops from peripheral sensory endings, which make muscle activation patterns dependent on the actual movement mechanics. Predicting all the details of peripheral movement mechanics is unrealistic, which makes it equally unrealistic to predict the contributions of reflex pathways to muscle activation levels. Moreover, as discussed in future chapters (Chapters 5 and 6), movement mechanics do not reproduce perfectly over repetitive attempts to perform a task, even if the task has been perfectly learned and is performed in perfectly predictable external conditions. This was emphasized by Bernstein (1930, 1947), who coined the expression "repetition without repetition" to reflect the fact that repeating the process of solving a motor task is associated with signals that do not repeat themselves at all levels of the involved structures, neural and muscular.

In spite of all the mentioned factors, the idea of programming peripheral patterns, mechanical or muscle activation ones, has shown striking resilience. This is probably a reflection of human psychology: If one has a hammer, everything looks like a nail. If the only piece of equipment is a force platform, the problem of the neural control of vertical posture is seen as that of prescribing forces measured by that platform. If one has a motion analysis system, all processes are seen as prescribing movement kinematics. If one has a multi-channel electromyographic system, all processes are seen as prescribing muscle activation patterns. To summarize, researchers commonly think along the lines compatible with the available equipment and theories (e.g., the control theory), even if those theories were developed for other objects and processes.

4.1. Engrams and the generalized motor program

The idea that learning a motor skill is associated with the creation and storage of a neural pattern somewhere in the brain encoding salient features of the required action was suggested and developed by Bernstein (1935, 1947, 1996). Experimental studies of a range of skilled movements, from handwriting to playing musical instruments to athletic movements, led Bernstein to a conclusion that neural patterns stored in memory encoded topological, but not metric, properties of the learned action. In particular, this conclusion was based on the ability of humans to preserve individual features of handwriting while performing this action with a novel effector, which has never been used during earlier practice, such as during writing with the non-dominant arm, holding the pencil gripped in one's mouth, or with the implement attached to a foot or an elbow.

Bernstein used the term *engram* (see also Semon 1921) for such hypothetical neural traces of a learned movement, which can be scaled by both magnitude and time. He did not specify what exactly engrams encoded, although this list definitely excluded kinetic variables, such as muscle forces and joint torques, as well as kinematic variables, such as joint angle, velocity, and acceleration. Indeed, one can write on a piece of paper with a pencil or on the blackboard with a piece of chalk. Clearly, joint rotations and moments of force in the dominant arm are very different in these two examples, while salient features of the individual handwriting are preserved. We will return to the issue of variables that may be used for engrams stored in memory in Chapters 10 and 20. Since most skilled actions involve effector or external object movement in the three-dimensional space where we happen to live, likely engrams encode topological features of variables also expressed in spatial units. In the previous chapter, we discussed the basics of the equilibrium-point hypothesis, which assumes control variables expressed in such units—trajectories of the threshold of stretch reflex for a muscle expressed in units of muscle length. More recent developments of the idea that neural control operates with spatial referent variables will be discussed in Chapter 7.

The idea of engrams led to the development of the concept of the generalized motor program and associated schema theory by Richard Schmidt and his colleagues (Schmidt 1975, 1980). The concept of the generalized motor program is similar to engrams in a number of aspects, such as the possibility of scaling learned actions by amplitude and time and transfer effects of practice to untrained effectors. There is, however, a major difference: The generalized motor program associates the learned patterns of neural signals with patterns of peripheral variables such as forces, torques, and muscle activations. As mentioned earlier, this is a very questionable step, which was not made by Bernstein because he understood that neural signals could not possibly prescribe patterns of those variables.

Important support for the concept of the generalized motor program came from studies of relative timing during the performance of complex skilled actions, which showed that the action of elements scaled with a single timing parameter while preserving the relative timing of involvement of the elements. In particular, when professional typists were asked to type the same phrase multiple times, the timing of hitting individual keys varied across trials with the natural variation in the overall speed of typing, but the relative timing of key presses was preserved (Viviani and Terzuolo 1980).

Later studies, however, questioned this crucial finding by showing that changing the speed of action can lead to abrupt, unintentional changes in the relative timing of elements with changes in the overall action speed or frequency. In classical experiments by Scott Kelso, Michael Turvey, and Peter Kugler (Kelso et al. 1980; Kelso and Holt 1980; Kugler et al. 1980), subjects were asked to perform bilateral actions such as moving the two index fingers rhythmically while being paced by a metronome. There were only two stable regimes, in-phase and out-of-phase. An increase in the movement frequency led to spontaneous switch from the out-of-phase regime to the in-phase regime at some critical frequency (shown schematically in Fig. 4.1). No such switch was seen from the in-phase to the out-of-phase regime. Later, these findings were reproduced using a variety of effectors and tasks (Schöner and Kelso 1988; Carson 1995), leading to the development of the so-called dynamical systems approach to motor control (reviewed in Kelso 1995; Schöner 2002). Overall, these studies have provided ample evidence that peripheral movement patterns, even their basic characteristics such as the relative phase of involvement of two effectors, could not be prescribed by the brain but emerge given various action parameters such as action frequency, loading of the effectors, and so forth. More recent studies have also shown that features of such actions and their switches between stable regimes depend crucially on the instruction and type of feedback provided to the actor (Kovacs et al. 2010; Lafe et al. 2016; Reschechtko et al. 2017).

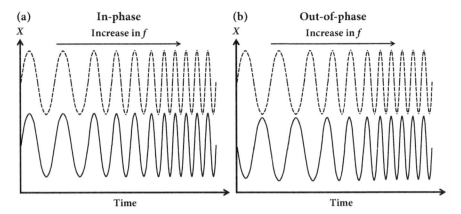

Figure 4.1 When two effectors (e.g., two index fingers) move at a prescribed frequency, only two regimes are stable, in phase (A) and out of phase (B). An increase in the movement frequency does not change the in-phase regime but leads to a spontaneous switch from the out-of-phase to the in-phase movement.

4.2. Control with patterns of muscle activation

This group of hypotheses represents a step from programming mechanical variables to programming patterns of activity of alpha-motoneurons innervating participating muscles, which is synonymous with programming patterns of muscle activation. This step recognizes some of the problems inherent to the force programming approach, such as the dependence of muscle force on its length and velocity for a given level of muscle activation. However, it faces another major problem: dealing with reflex contributions to muscle activation patterns. Reflex loops originate from peripheral sensory endings sensitive to a variety of kinematic and kinetic variables. Hence, if the central nervous system is unable to prescribe movement mechanics, it is equally unable to predict reflex contributions to muscle activation patterns.

Two solutions for the problem of reflex contributions have been discussed within this group of hypotheses. The first assumes that reflex contributions are small and, during voluntary movements, the gains in reflex loops are tuned down to make these contributions even smaller. Experimental studies have shown, however, that both these assumptions are false. For example, if you are holding a very heavy load on the hand and suddenly the load is removed, the unloading reflex is seen in the previously active muscle, which shows a period of nearly complete silence. This example shows that reflex loops are able to cancel out very high levels of muscle activation; that is, effects of these loops are not small. Along similar lines, studies of decerebrate animal preparations have shown that reflexes have significant contributions to muscle activation level and force (Nichols and Houk 1976). Experiments with the application of unexpected changes in the external load during voluntary movements have documented very strong reflex effects, showing that reflex gains are not necessarily reduced during movements (Gottlieb 1996).

The other solution was to assume that reflex contributions to muscle activations were predicted by the brain and descending signals to alpha-motoneurons were corrected by those predicted values. This assumption, however, faces a major problem, previously mentioned. If muscle forces and the ensuing kinematics cannot be perfectly predicted, as emphasized by Bernstein (1947), signals from peripheral sensory endings sensitive to muscle force, length, and velocity also cannot be predicted. This makes effects of reflex loops originating from those sensors (e.g., muscle spindles and Golgi tendon organs) also unpredictable. The problem is exacerbated by the existence of gamma-motoneurons, which show activation simultaneously with alpha-motoneurons and modulate the sensitivity of sensory endings in muscle spindles—a major source of reflex effects on muscle activation. This seems to be a wicked circle, which cannot be broken in principle. In the next section, we will discuss more sophisticated attempts to break this circle with the help of powerful computational devices assumed to exist in the brain—the so-called internal models.

In spite of the aforementioned problems, hypotheses assuming programming of patterns of muscle activation have been very popular. Two of those hypotheses addressed the control of agonist-antagonist muscle pairs acting at individual joints with one kinematic degree of freedom. One of them—the *dual-strategy hypothesis*—was originally based on studies of muscle activation patterns during voluntary movements (Gottlieb et al. 1989a). The other one—the *pulse-step model*—was primarily based on observations of muscle activation patterns during force production in isometric conditions (Ghez and Gordon 1987; Gordon and Ghez 1987; Scheidt and Ghez 2007). Both were successful in accounting for the observed muscle activation patterns during quick actions within the studied groups of tasks. Such actions are associated with the so-called triphasic patterns of muscle activation (Fig. 4.2): An action is initiated with an activation burst in the agonist muscle accompanied by a relatively small level of coactivation of the antagonist, then an antagonist activation burst is

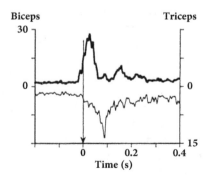

Figure 4.2 An illustration of the triphasic pattern of muscle activation during a fast elbow flexion movement. Note the activation of the antagonist (triceps) during the first agonist (biceps) burst, the nearly complete quiescence of the agonist during the antagonist burst, and the increased level of muscle coactivation following the movement. Time $t = 0$ corresponds to the first detectable joint deviation. Reproduced by permission from Latash et al. 1995b.

4.2. CONTROL WITH PATTERNS OF MUSCLE ACTIVATION

seen while the agonist is relatively quiescent, and then the second, smaller agonist burst is seen at action termination. There is an increased level of agonist-antagonist coactivation in the final state, which declines slowly over a few seconds or even tens of seconds.

Both hypotheses struggled or required major modifications when they were applied to other motor tasks, in particular to more natural movements. And, as already mentioned, both suffered from the common problem of ignoring reflex contributions to the observed muscle activation patterns.

The dual-strategy hypothesis assumes that patterns of activation of alpha-motoneuronal pools can be modeled as rectangular pulses with varying height and duration (Fig. 4.3). One of the two basic strategies assumed within this framework consists of modulating the duration of the excitation pulse without a change in its height. Originally, this strategy of control—termed the *speed-insensitive strategy*—was illustrated with movements performed as fast as possible over varying distances and against varying inertial loads (Gottlieb et al. 1989b). Later, the idea was generalized to movements performed at submaximal speeds when the actor tried not to modulate the movement speed voluntarily (Gottlieb et al. 1990). For example, predictable features of muscle activation were observed under the instruction to move "at the same speed" over different distances. Note that this is a very ambiguous instruction because movement speed is a bell-shaped time function and both average speed and peak speed scale with distance. So, what does it mean "to move at the same speed"? Surprisingly, naïve subjects have no problems interpreting this instruction and do not ask for clarifications. By itself, this is an important observation suggesting that, in spite of the scaling of effector speed with distance, there is a neural variable that is kept constant when humans try to follow this instruction, which has little sense in the world of mechanics.

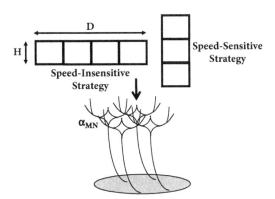

Figure 4.3 The dual-strategy hypothesis assumes that the neural control of a muscle can be modeled as a rectangular presynaptic excitation pulse into the alpha-motoneuronal pools. The two strategies are associated with modulating only the duration (D) of the pulse (speed-insensitive strategy) and with modulating its height (H) with or without changes in the duration (speed-sensitive strategy).

The other strategy—the *speed-sensitive strategy*—was associated with modulation of the excitation pulse height with or without modulation of its duration (Corcos et al. 1989). This strategy was assumed to be used when actors purposefully modulated movement speed (or time) or when it was modulated unintentionally, for example, as a consequence of the classical movement time scaling with the target size, as dictated by Fitts' law (Fitts 1954).

The dual-strategy hypothesis introduced order and classification into the preexistent body of literature describing muscle activation patterns over a variety of joints, instructions, and task parameters. It was successful in accounting for features of the first agonist burst and their modulation with movement distance, speed, and inertial load. This should not be very surprising because reflex effects on muscle activation emerge at time delays comparable with the duration of the first agonist burst during very fast movements. Problems emerged when the same rules were applied to later phenomena such as the antagonist burst. In particular, during fast movements over varying distances, the peak of the first agonist burst scales with movement distance, but the peak of the antagonist burst does not—it increases over a range of distances and then starts to decrease (Gottlieb et al. 1989b). This behavior reflects the effects of damping, in particular reflex-mediated damping, which helps to slow movements down and works synergistically with the antagonist burst (Jaric et al. 1998). The dual-strategy hypothesis did not even try to address later phenomena such as the second agonist burst and terminal coactivation.

The hypothesis was applied to other motor tasks, such as accurate torque production in isometric conditions and arm movements in the three-dimensional space. Those attempts were somewhat successful but required broadening the menu of control functions, in particular by adding excitation steps and ramps to the excitation pulses (e.g., Corcos et al. 1990). In a way, those adjustments of the original formulation looked like trying to fit the data with new assumed ways of modifying the presynaptic input into the agonist and antagonist alpha-motoneuronal pools.

The pulse-step hypothesis shared a few features with the dual-strategy hypothesis. In particular, it also assumed a scalable standardized set of excitation signals (steps and pulses) to the participating muscles during the joint torque production in isometric conditions and voluntary movements in isotonic conditions. The basic idea of this hypothesis is that the control of the transition from the initial state to the final state is separate from the control of the final state (Fig. 4.4). The former was assumed to be implemented with a pulse of excitation, which could be scaled to produce transitions at different rates and over different amplitudes, similarly to the assumptions of the dual-strategy hypothesis. The excitation step was assumed to define the level of agonist-antagonist coactivation after the movement termination and the final level of torque in isometric conditions.

Two types of isometric contractions formed the original basis for the pulse-step hypothesis: step and pulse contractions (Ghez and Gordon 1987; Gordon and Ghez 1987; see also Corcos et al. 1990). Step contractions required the subject to increase joint torque up to a certain level, while pulse contractions required also returning quickly back to the initial, commonly zero, level of net joint torque. Other instructions have been used to specify the time of the torque increase, its magnitude, accuracy constraints, and so forth. Fast isometric contractions are

4.2. CONTROL WITH PATTERNS OF MUSCLE ACTIVATION 51

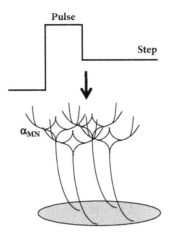

Figure 4.4 The pulse-step control assumes two control processes, those related to transition to a new state (pulse) and those related to keeping this new state (step). Both can be modified to fit the task requirements.

accompanied by triphasic muscle activation patterns similar to those observed during fast movements. Scaling of the bursts with task parameters was used to estimate the parameters of the pulse and step portions of the assumed control process.

The idea of control with muscle activation patterns has been expanded to multi-muscle systems participating in natural movements under the label of multi-muscle synergies. The notion of *synergy* will be discussed in detail in Chapter 6 (also see Chapter 2). Here we only mention a few basic ideas and features of the concept of multi-muscle synergies. This concept is a natural generalization of the reviewed dual-strategy and pulse-step hypotheses to larger sets of muscles.

The main tools in the identification of multi-muscle synergies are the so-called matrix factorization methods applied to sets of pairwise correlation or covariation coefficients computed between indices of integrated activation quantified for individual muscles. Depending on the type of data and objectives of individual studies, different matrix factorization methods have been used, including principal component analysis, independent component analysis, and non-negative matrix factorization. They all produce a set of independent eigenvectors, orthogonal or non-orthogonal, which account for reasonably high amounts of variance in the original data set. These eigenvectors are addressed in different studies as factors, modes, modules, primitives, or synergies (reviewed in Flash and Hochner 2005; Ivanenko et al. 2006; Hogan and Sternad 2012, 2013; Giszter 2015; Overduin et al. 2015; Latash 2020b). They represent linear combinations of individual muscle activation indices, with some muscles represented significantly, according to selected criteria, and others represented nonsignificantly. Such muscle groups have been reflected in neuronal activation patterns at the spinal level. In some studies, specific time profiles of such modules have been viewed as building blocks for movements.

All the approaches mentioned in this section share the same assumptions, same attractive features, and same pitfalls. The main shared assumption is that, somehow, the central nervous system prescribes patterns of muscle activation in task-specific ways. The main attraction is that patterns of muscle activation can be relatively easily and objectively recorded and quantified, thus creating a feeling that one is able to look into brain strategies. The main pitfall is that the main assumption is incompatible with the design of the human body.

4.3. Internal models

The obvious problems facing the idea of programming patterns of peripheral variables, mechanical and electromyographic, led to the development of an influential approach to motor control (and also some other brain functions) under the label of *internal models* (reviewed in Wolpert et al. 1998; Kawato 1999; Shadmehr and Wise 2005). This approach assumes that the brain contains computational devices, sometimes placed into the cerebellum and sometimes distributed across multiple brain areas, which solve the highly nontrivial problems of interaction among parts of the body and between the body and the environment in real time during the performance of movements. So, the task of the researchers is viewed not as understanding laws of nature leading to the observed behaviors but as deciphering the software used by the presumed computational units. From the subjective view of the author, this approach is unacceptable philosophically, contradicts the basics of evolutionary biology, assumes impossible mathematical operations, grossly overestimates the computational speed of the brain, and overall leads into a dead end. But first, let us review the basics of this approach.

Consider a simple action by an effector, for example, moving the hand from an initial to a final position (Fig. 4.5). To move the hand, a number of joint angles have to change. Defining joint rotations based on the task—the problem of inverse kinematics (reviewed in Zatsiorsky 1998)—is not trivial because, typically, the number of joint rotation axes is larger than the number of task constraints. To move the joints, muscles have to change their forces. Since each axis of joint rotation is served by several muscles, this is also a nontrivial problem—the problem of inverse dynamics (reviewed in Zatsiorsky 2002). Muscle forces depend on the level of activation and muscle length and velocity. So, to define signals from alpha-motoneurons innervating the muscles, one has to know the length and velocity of each of the involved muscles at the time that signals from the alpha-motoneurons reach it.

The output of a neuron depends on the combined effect of all the presynaptic inputs. Moreover, neurons are "all or none" threshold elements. Note that it is impossible to define the input into a threshold element based on its output: For example, with what force do you press on the light switch to turn the electric bulb on? This is clearly an unsolvable problem because the force has to be only over the threshold, but its actual value can be any within a large range to produce the same effect. In a very much simplified scenario, the input signals to each of the involved alpha-motoneuronal pools form two large groups, those from descending pathways and those from reflex pathways originating from peripheral sensory endings. Let us

4.3. INTERNAL MODELS

Steps	Complicating Factors
Hand coordinates $\{X_0\} \to \{X_{FIN}\}$	Trajectory redundancy
Joint configurations $\{\Phi_0\} \to \{\Phi_{FIN}\}$	State redundancy
Joint trajectories $\{\Phi(t)\}$	Trajectory redundancy
Joint moments $\{M(t)\}$	
Muscle forces $\{F(t)\}$	State redundancy
Muscle activations $\{EMG(t)\}$	Force-length and force-velocity dependences
Output of motoneurons $\{\alpha(t)\}$	
Input to α-motoneurons	Threshold properties of neurons Enormous state redundancy
Central input to α-motoneurons	Reflex contributions $\{\alpha(t)\}$

Figure 4.5 A sequence of steps involved in moving the hand from an initial (X_0) to a desired final (X_{FIN}) coordinate in external space and some of the associating complicating factors.

assume that the descending signals are under the control of the actor. To define the overall level of activity along those pathways, one has to know how much excitation is expected along the reflex pathways. Predicting the reflex input requires knowing all the mechanical variables that affect the firing of the sensory endings. Assuming that all the aforementioned problems are solved by hypothetical *inverse internal models*, a required descending signal is computed, and the planned action can be initiated. These models are referred to as *inverse* because they compute signals in the opposite order as compared to what happens during actions—from the descending signals to muscle activation and to movement mechanics.

The described sequence of assumed computational steps is already unsolvable and requires precise prediction of many of the involved variables. Most of the mentioned steps also involve solving the problem of redundancy when many unknown variables have to be computed based on a smaller number of known variables. We will discuss these problems in Chapter 5. However, this is only the tip of the iceberg of problems. Now we have to take into consideration that signals travel along neural pathways at a limited speed and the conduction of these signals involves time delays on the order of several tens of milliseconds. In other words, all the crucial information is outdated by the time of its delivery by the involved sensory pathways, and these time delays cannot be ignored because, during fast movements, they are associated with significant changes in all the variables.

The problem of prediction has been recognized and led to the emergence of another group of internal models addressed as *direct models* or *predictors*. These models perform computations based on the current state of the body, as reflected in signals

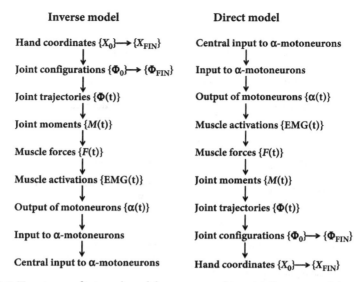

Figure 4.6 Two types of internal models are assumed to exist. Inverse models compute required descending neural inputs into alpha-motoneuronal pools to produce a desired mechanical effect. Direct models compute changes in the state of peripheral structures based on the current sensory signals and efferent output to compensate for the conduction time delays.

from sensory endings (which are outdated by the time they reach brain structures!), and the current descending signals to predict changes in the body state in the future (Fig. 4.6). Their purpose is to estimate all the important mechanical and electrophysiological variables, which are crucial for the computations assumed in the inverse models. Some of the studies within this line of thinking introduced multiple direct and inverse models to improve the prediction of body states and required descending signals to perform movements (Imamizu et al. 2003, 2004). Note that these models assume highly nontrivial computational steps including problems of mechanics, conversion of electrical signals into mechanical ones, optimization, and so forth.

There are no arguments that animals can predict events in the environment. Such predictions vary in their characteristic timing. For example, many actions by humans have a goal far away in the future, for example, attending a university. Predators commonly run not toward the current location of the prey but to its predicted location. Actions that are expected to produce perturbations to posture are preceded by postural adjustments (see Chapters 8 and 14). There are changes in the stability of an ongoing action in preparation of a quick change in a salient variable (see Chapter 8). And so on. Does addressing these phenomena with the expression *direct model* add value and/or explanatory power? The word *model* is loaded with an implied meaning of underlying computational processes. And here we come to a very basic question: Do natural biological systems compute?

Of course, this question requires a definition for the word *computation*. We assume, as it is done in many studies invoking the concept of internal models (e.g., see the equations in the aforementioned papers), that computation involves operations with numbers and symbols. For example, when students use Newton's second law and the law of gravity to predict the movement of a stone rolling down a hill, they perform computations. When the stone actually rolls down the hill, the stone does not perform computations but acts under the mentioned laws of nature, although the outcomes of both processes can be the same. Under this definition, animals do not compute—they behave. There has not been a single piece of evidence that animals can multiply 3 by 4. Hence, the assumption that parts of the brain of animals can perform rather complex computations involved in direct and inverse models seems far-fetched at best.

Consider the following example. You are asked to move an index finger as quickly as possible following a visual signal (e.g., a flash of light). Typical reaction time in this condition will be within the range of 100 to 150 milliseconds. Now, let us make the task a little bit more complex: You have to move the left finger if the light is red and the right finger if the light is green. The reaction time will nearly double. So, it takes the brain an extra 100 milliseconds or so to solve the trivial problem of selecting between moving the right or left finger. These observations suggest that the speed of brain processes involved in movement selection and initiation is far from infinite. However, internal models commonly assume that movement adjustments in cases of unexpected changes in external forces happen nearly instantaneously.

Sometimes the expression "internal model" is used in a more general meaning, as a neural representation of a motor task and involved body parts. The idea that the brain represents tasks and effectors is very old and dates back at least to seminal papers by Hughlings Jackson (1889). This nonspecific understanding of internal models seems more acceptable, but it is not clear why one should introduce a new expression for the phenomena addressed as neural representations.

If one assumes a model, any model, the first natural questions to ask are: "What are the input and output of your model and in what units are they expressed?" I asked these questions to a few champions of this approach on a number of occasions and typically got the answer: "An internal model can link any input to any output." In other words, these models have supernatural abilities, which are not limited by any constraints. Many centuries ago, our ancestors would address entities with such abilities as "Gods." If we want motor control to become an area of natural science, "internal models" have to be replaced with laws of nature.

A major source of experimental support for the concept of internal models comes from studies of movements in artificial force fields (Shadmehr and Mussa Ivaldi 1994; reviewed in Shadmehr and Wise 2005). A typical study involves a person grasping a handle and performing arm reaching movements from a fixed initial location to targets arranged along a circle, the so-called center-out task (Fig. 4.7). The task is limited to a plane and involves rotation of two major joints, the elbow and shoulder. So, it is not redundant in terms of kinematics. Such tasks are typically performed with straight hand trajectories in the external space (cf. Morasso 1981). After the subject practices performing this task, programmable motors are used to create an artificial force field, with the force acting on the handle proportional to speed and acting orthogonal to the trajectory (see the insert in Fig. 4.7). The trajectories become curved in the direction of the force field. After some

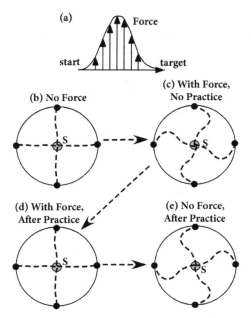

Figure 4.7 An illustration of typical behavior in the center-out arm movement task. Without an external force field, the trajectories are nearly straight (b). Turning on a velocity-dependent field (see a) leads to curved trajectories (c). With practice, trajectories become straight again (d). Turning the field off leads to trajectories curved in the opposite direction (e). Reproduced by permission from Latash 2012b.

practice, the trajectories become straight again. Now, if the force field is turned off, the trajectories become curved in the opposite direction (compare panels C and E in Fig. 4.7). These observations were interpreted as the brain building an internal model of the external field and generating hand force time profiles to compensate for its kinematic effects. In the absence of the field, these force time profiles lead to trajectories curved in the opposite direction.

Numerous findings have questioned this interpretation. In particular, practicing movements in a range of directions did not generalize well to movements outside this range (Malfait et al. 2005). Such generalization can be expected if a model of the force field is built because the rule for the force field generation remains unchanged across all directions. When more joints participated in the action and the limb became kinematically redundant (abundant; see Chapter 5), adaptation to the new force field induced a significant increase in joint rotations, which did not affect the hand movement at all; that is, they were motor equivalent (Yang et al. 2007; see Chapters 5 and 18). Such rotations were apparently irrelevant to task performance and, as such, could not be predicted by any adjusted internal model. An alternative interpretation of these results was offered within a different theoretical framework compatible with the equilibrium-point hypothesis (Gribble and Ostry 2000; Ostry and Feldman 2003).

4.4. Missing pieces of the mosaic

The biggest problem in the motor programming approach seems to be the lack of clear, exact definitions of central notions such as "motor program," "internal model," "neural computation," and so forth. For example, to discuss a model, it is essential to define its input and output. As they are defined now, internal models are magical constructs that convert any signals into any signals, which have infinite power to describe regularities in experimental findings but do not offer much explanatory power. In most experimental studies using this concept, however, the inputs of the models are coordinates of movement targets and their outputs are muscle forces, joint torques, or muscle activations. This justifies describing them under the motor programming umbrella. But this also makes all the problems associated with the force programming approach inherent to the idea of internal models.

The term *internal model* may be useful as a temporary substitute for lacking scientific knowledge. For example, as described in Chapter 1, we do not know the laws of nature that lead to processes commonly referred to as cognitive. Regularities in the input-output mappings observed in experiments suggest certain properties of those processes, which have for many years been addressed as "neural representations"—an imprecise term, synonymous with "internal model," but without the implied computational processes.

Overall, unless the champions of this approach define the central notions clearly and unambiguously, using this approach or arguing with it seems to be a waste of time. To quote Israel Gelfand (from one of our conversations): "The worst method of discussing complex issues is doing this with hints."

5
Principle of Abundance and the Uncontrolled Manifold Hypothesis

The problem of motor redundancy has been at the center of attention in the research community for decades (reviewed in Bernstein 1947; Turvey 1990; Latash 2008). It was emphasized by Bernstein, who stated that the essence of motor control was in the elimination of redundant degrees of freedom. The problem relates to the availability of numerous apparent elements at any level of description of the system for movement production being greater than the number of constraints associated with typical tasks. In addition, there is an infinite number of trajectories able to bring an element from its initial state to a desired final state. As a result, any task is associated with numerous (an infinite number of) solutions. So, the problem is: How does the central nervous system select specific motor solutions observed each time a motor task is performed?

Bernstein illustrated this problem at the level of joint kinematics during reaching. Indeed, the number of joint axes of rotation in the human arm is larger than three, which is the number of spatial coordinates describing a target. The problem also emerges at other levels (illustrated in Fig. 5.1). For example, the number of muscles acting at any joint is larger than the number of axes of joint rotation. As a result, an infinite number of combinations of muscle forces can be produced to match a desired joint torque magnitude. A given level of muscle activation can be produced by an infinite number of combinations of recruited motor units and their frequencies of firing, even if the recruitment order follows the size principle (Henneman et al. 1965). An action potential can be generated as a result of a subset of sodium ions crossing the membrane from the zillions of those ions available in the vicinity. The last example is obviously ridiculous because the central nervous system does not micromanage at that level: It does not care which individual ions cross the membrane. But does it care about specific patterns of motor unit recruitment, patterns of muscle activation, joint rotations, and, in general, selecting specific patterns of solutions in a variety of other apparently redundant problems?

This may sound like blasphemy, but let me suggest that the problem of motor redundancy does not apply to biological movements and, in its original formulation, it has been misleading generations of researchers. Consider, for example, that a person learns an unusual multi-joint movement and tries to minimize peak-to-peak deviation in a specific joint by co-activating the muscles acting at that joint. At the level of joint kinematics, this may be seen as a reduction of the number of degrees of freedom and a step toward solving the problem of motor redundancy. However, at the levels of muscle forces and motor unit recruitment, this strategy would make the problem worse. This example shows that apparent problems of motor redundancy formulated at different levels of analysis are commonly in conflict with each other, and trying

Problems of Motor Redundancy

Joint Angles		Limb Endpoint Coordinates
$\{\alpha_1, \alpha_2, ... \alpha_n\}$	n > 3	$\{x_1, x_2, x_3\}$
Muscle Forces		Joint Torque
$\{F_1, F_2, ... F_i\}$	i > 1	$\{T\}$
Motor Units		Muscle Activation
$\{f_{MU1}, f_{MU2}, ... f_{MUj}\}$	j >> 1	$\{E\}$
Na^+ Ions		Membrane Potential
$\{Na^+_1, Na^+_2, ... Na^+_k\}$	k >>> 1	$\{V\}$

Figure 5.1 Schematic illustrations of the problem of motor redundancy at various levels of analysis. At each level, the number of constraints is smaller than the number of elemental variables, resulting in an infinite number of solutions.

to solve one of them sometimes makes other problems worse. In Chapter 7, we will discuss apparent problems of redundancy at the level of control variables, but even at that level, specific solutions are not computed and implemented by the central nervous system but allowed to emerge given the actual, frequently unpredictable external conditions of movement execution.

In Chapter 9, we will review briefly methods developed to address the problem of motor redundancy, in particular those based on the concept of optimization. These methods may be very useful in applied areas such as robotics and prosthetics, but in motor control they should all be viewed with suspicion and a large grain of salt.

5.1. The principle of abundance

An alternative approach to the apparently redundant design of the body originates from a very early experiment of Bernstein on professional blacksmiths (Bernstein 1930; Fig. 5.2). Bernstein recruited professional blacksmiths with multiple years of experience and asked them to perform their typical labor movement multiple times: hitting the chisel held by the nondominant hand with the hammer moved by the dominant arm. Given their lifetime experience, these were the best-trained subjects to perform the required action involving the kinematically redundant upper extremity. Their movement kinematics were recorded using the motion analysis system designed and built by Bernstein. The system involved placing a set of electric bulbs on strategically selected parts of the body and filming a sequence of hitting actions on the black background using a highly sophisticated camera (described in detail in Bernstein 1930; Bernstein and Popova 1930). Further, the film was developed and analyzed to create a sequence of stick figures representing the moving extremity. Each stick figure was used to measure the joint angles (by hand, of course!), resulting in time profiles for all individual joint trajectories. Bernstein reached

60   5. PRINCIPLE OF ABUNDANCE

Figure 5.2 A schematic illustration of the experiment on blacksmiths (as in Bernstein 1930). Markers are shown with open circles. The trajectory of the tip of the hammer was variable across trials but less variable compared to other markers.

amazing frequencies of analysis, over 500 Hz, which would be very much respected in our times.

The experiment on blacksmiths produced two major results. First, trajectories of all the joints and the hammer showed substantial variability across successive hits, suggesting that there was not such a thing as a single optimal trajectory, even for those perfectly trained individuals. Second, the hammer trajectory showed the smallest inter-trial spatial variability as compared to the trajectories of individual joints. This was a truly spectacular result! Obviously, the brain could not send signals to the hammer, only to muscles acting at individual joints. How could it happen that the signals to the muscles produced highly varying joint trajectories, while the hammer followed a relatively invariant trajectory? Bernstein invoked the concept of multi-joint synergy but did not develop it to address this finding. As we will see, this observation was only a step away from the concept of abundance and the uncontrolled manifold (UCM) hypothesis.

Briefly, the main result of the experiment on blacksmiths can be formulated in the following way: Individual joints compensated for each other's errors (defined as the deviations from the mean-across-trials trajectory) with respect to their effects on the hammer trajectory. Numerous studies provided evidence for such error compensation phenomena across effector sets and tasks (Abbs and Gracco 1984; Cole and Abbs 1987; Jaric and Latash 1998; Kurtzer 2015; Pruszynski et al. 2016). An unexpected perturbation applied to an effector (or the natural unavoidable variability in the initial state and/or external forces) produced very short-latency corrective actions in other effectors involved in a common action, which attenuated the effects of the original perturbation on task-specific salient variables during a variety of actions including uttering sounds, pinching, grasping, pointing, standing, walking, and many others.

Similar corrective reactions in effectors involved in a multi-effector action are seen when a self-initiated action introduces predictable changes in salient performance variables. For example, imagine a person who is asked to produce a constant level of force while pressing with the four fingers of a hand (with the help of visual feedback) and then to tap quickly with one of the fingers (Latash et al. 1998). During the first tap, the tapping finger obviously stops contributing to the total force. Even in the absence of any instruction regarding total force during tapping, subjects show out-of-phase changes in the forces of the other fingers, which compensate for up to 95% of the total force change expected from the instructed tapping action. These phenomena have been collectively addressed as *error compensation*.

Taken together, all the reviewed observations suggest that the apparently redundant degrees of freedom are not eliminated during biological movements but used to ensure accurate task performance in the natural, unpredictable environment. This conclusion forms the core of the *principle of abundance*. The principle states that there are no uniquely optimal solutions to motor tasks and that elements at any level of analysis have the freedom to change their contributions to performance in the actual variable conditions of movement execution.

5.2. The uncontrolled manifold hypothesis

Consistent with the principle of abundance, Gregor Schöner suggested in 1995 that biological movements can use abundant sets of elements to ensure task-specific stability of salient performance variables. This idea led to the formulation of the UCM hypothesis a few years later (Scholz and Schöner 1999). The hypothesis suggests that the neural controller is able to organize large sets of elements in such a way that most inter-trial variance in the space of elemental variables (those produced by elements) is constrained to a subspace where a salient performance variable does not change. Note that, in the presence of unavoidable variability in the initial conditions prior to each trial (Fig. 5.3), trajectories in more stable directions are expected to converge more (diverge less) as compared to trajectories in less stable directions. As a result, quantifying inter-trial variance in different subspaces produces indices of action stability within those subspaces along each of the trials. This is a nontrivial step: using indices of inter-trial variability to characterize stability along individual trials.

Figure 5.4 illustrates the main concepts of the UCM hypothesis using a very simple task to produce a certain magnitude of total force while pressing with two effectors (e.g., the right and left index fingers) on individual force sensors. After a little bit of practice, anybody can perform this task with high accuracy. The slanted line with

62 5. PRINCIPLE OF ABUNDANCE

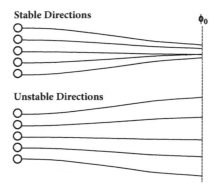

Figure 5.3 An illustration of trajectories in a sequence of trials at the same well-trained task in stable directions (A) and unstable directions (B) assuming some variability in the initial states (shown with open circles). Note that measuring inter-trial variance at a certain phase (ϕ_0) is expected to produce smaller values in stable directions than in unstable directions.

negative slope in the space of elemental variables $\{F_1; F_2\}$ corresponds to perfect performance (i.e., no change in total force). This line is the UCM for this task. Note that the line affords an infinite number of finger force combinations that can solve the task perfectly. If the system resides somewhere on the UCM, there is no reason to modify control signals to the elements, which is the source of the term *uncontrolled*. If the system deviates in directions orthogonal to the UCM (along the line labeled ORT in Fig. 5.4), this leads to errors in performance. Note that deviations within the UCM, by definition, have no effects on performance.

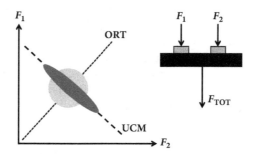

Figure 5.4 An illustration of the task of producing a certain magnitude of total force (F_{TOT}) while pressing with two effectors, for example, two fingers, producing forces F_1 and F_2. The solution space ($F_1 + F_2 = F_{TOT}$ = const.) is shown as the slanted dashed line. This is the uncontrolled manifold (UCM) for the task. The line orthogonal to the UCM (ORT) corresponds to changes in F_{TOT}. Two distributions of data points are shown with ellipses, nearly spherical (light gray, $V_{UCM} \approx V_{ORT}$) and elongated along the UCM (dark gray, $V_{UCM} > V_{ORT}$), where V stands for inter-trial variance.

Experimental studies have shown that, indeed, data clouds collected over repetitive attempts at such a task are commonly elongated along the UCM such that inter-trial variance shows an inequality $V_{UCM} > V_{ORT}$ (Latash et al. 2001; Scholz et al. 2002; reviewed in Latash et al. 2007; Latash 2008, 2019). Of course, comparison of variance between spaces has to be done taking into account their actual dimensionality as required by classical statistics (both UCM and ORT are unidimensional in the example in Fig. 5.4). The two outcome variables, V_{UCM} and V_{ORT}, have been reduced in some studies to a single index reflecting the relative difference between V_{UCM} and V_{ORT}, for example, $\Delta V = (V_{UCM} - V_{ORT})/V_{TOT}$, where V_{TOT} stands for total variance, and each index in the right side of the equation is normalized per dimension. Note that the ΔV index reflects the shape of the datapoint cloud, not its size. So, large values of ΔV do not imply accurate performance, which is defined by V_{ORT} only. There may be very accurate behavior when $V_{UCM} \approx V_{ORT}$ and both are very small—an example of stereotypical behavior. There may also be sloppy behavior when $V_{UCM} > V_{ORT}$ but V_{ORT} is large. Situations when $V_{UCM} > V_{ORT}$ have been addressed as *synergies* stabilizing the performance variable for which the UCM was computed and analysis performed. We will discuss the term *synergy* in more detail in Chapter 6.

Analysis of specific actions within the UCM hypothesis involves several important steps and assumptions. First, one has to commit to a space of elemental variables (those produced by the elements at the selected level of analysis). Second, a potentially important salient variable is selected. Third, the UCM has to be computed, that is, a subspace within the space of the elemental variables where the performance variable does not change. This is commonly done in linear approximation by mapping the relations between small changes in each of the elemental variables and the performance variable or, in other words, partial derivatives of the performance variable with respect to each of the elemental variables. Assume that the space of elemental variables is n-dimensional and the performance variable is m-dimensional, $m < n$. So, the relations between them can be expressed as $\Delta PV = J \cdot \Delta EV^T$, where J is the $m \times n$ matrix of the mentioned partial derivatives, the Jacobian matrix of the system. The UCM can be approximated linearly by setting ΔPV to zero, which defines the null-space of J, null(J).

The first step is particularly important and nontrivial. In particular, elemental variables should not contain task-independent patterns of covariation, which, by pure chance, can lead to $V_{UCM} > V_{ORT}$ or $V_{UCM} < V_{ORT}$, resulting in spurious conclusions regarding the stability of the salient performance variable. Such situations and methods of dealing with them are discussed in more detail later in this section.

5.2.1. Analysis in kinematic spaces

The first two studies using the framework of the UCM hypothesis analyzed synergies within spaces of segmental angles and anatomical joint angles (Scholz and Schöner 1999; Scholz et al. 2000). Selection of elemental variables is important because segmental angles are non-independent assuming anatomical integrity of the moving kinematic chain (e.g., a limb). Indeed, moving actively the most proximal segment leads to motion of all other segments of the limb and changes in their segmental angles.

So, in the absence of any task-specific control (e.g., when only one segment is moved intentionally), changes in segmental angles are expected to co-vary. Moreover, patterns of this covariation depend on the limb configuration and may lead to different relations between V_{UCM} and V_{ORT} with respect to a selected performance variable. This problem disappears when anatomical joint angles are selected as elemental variables: One can rotate one joint about one axis without changing angles in other joints of the limb, at least hypothetically.

The last statement is far from trivial. Indeed, joints are coupled by both mechanical and reflex-mediated effects. The presence of bi-articular muscles allows expecting coupled rotations in the spanned joints. There are also inter-joint reflexes, which can lead to joint coupling. So, can a person rotate one joint at a time? So far, there have been no experiments exploring this question. So, analyses of inter-trial variance in spaces of anatomical joint rotations have been based largely on intuition.

Linking joint rotations to potentially salient performance variables (e.g., coordinates of the limb endpoint) has been typically based on geometric models of the moving kinematic chains (reviewed in Zatsiorsky 1998). Since such mapping is nonlinear, involving trigonometric functions, it requires linearization to make analysis of inter-trial variance possible. Typically, such analysis is performed for each movement phase assuming small deviations of the individual trajectories of each elemental variable from its average-across-trials trajectory. Under this assumption, the J matrix is computed, and its null-space is used as a linear approximation of the nonlinear UCM. In some studies, rotational and translational variables have been united to form the group of elemental variables, for example, joint rotations and scapular translation during arm movements (e.g., Yang et al. 2007). By itself, this does not present a problem, as long as the proper J matrix is used, which makes the effects of small changes in the elemental variables commensurable.

Studies of the stability of performance variables in kinematic spaces resulted in a number of important findings. In particular, indices of stability (ΔV) were different with respect to different potentially important performance variables (Scholz and Schöner 1999; Scholz et al. 2000). This means that this method of analysis allows asking the central nervous system questions: Do you care about this particular performance variable? Do you care about variable A more than about variable B? In these questions, "care" means "stabilize." In addition, the variance components (V_{UCM} and V_{ORT}) and the ΔV index showed nontrivial modulation within the movement time (or phase, for cyclical actions).

5.2.2. Analysis in kinetic spaces

Studies in spaces of kinetic variables can be classified into two groups: those addressing action of serial chains (such as a limb) and parallel chains (such as the four fingers of a hand).

Note that, if the resultant force/moment vector acting at the end of a serial chain (e.g., a human limb) and its geometric configuration are known, each of the individual joint moments is defined unambiguously and there is no abundance at the level

5.2. THE UNCONTROLLED MANIFOLD HYPOTHESIS   65

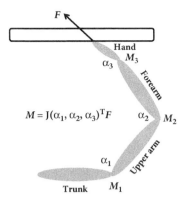

Figure 5.5 During natural movement of a multi-joint kinematic chain, mapping from the joint moment space (*M*) to the endpoint force (*F*) on the environment is configuration dependent. Joint angles are shown as α_1, α_2, and α_3.

of joint moments, no matter how many joints are involved. So, the solution space is a point, the UCM is zero-dimensional, and V_{UCM} is zero. The situation changes qualitatively in two cases: first, if only some of the components of the endpoint force/moment vector are prescribed by the task and others are allowed to vary (e.g., Xu et al. 2012), the UCM in the joint moment space can be computed, and second, if the geometry of the chain is allowed to change (e.g., Auyang et al. 2009; Yen et al. 2009), the UCM can be formally computed, but it combines variance in the joint configuration space and in the joint moments of force.

Overall, analysis of synergies in serial chains is complicated because the mapping of elemental variables, such as individual joint moments, on a performance variable, such as the force produced by the endpoint of the limb on the environment, changes with joint configuration, $M = J^T F$, where *M* is the vector of joint moments, *F* is the vector of endpoint force, J is Jacobian, and T is the sign of transpose (Fig. 5.5). In natural actions (e.g., during stepping or hopping), joint configuration of the limb changes with time and can show substantial variability across repetitive trials (cycles). Of course, for each phase, the average joint configuration can be selected to compute the mapping between the joint torque change and the endpoint force change. However, the structure of the inter-trial variance, that is, the relation between V_{UCM} and V_{ORT}, can depend on covaried adjustments of the joint torques and/or covaried adjustments in the joint angles. So, the outcome of such analysis can reflect adjustments both at the level of joint kinematics and at the level of joint torques. A number of studies have provided evidence for stabilization of components of the force-moment vector produced by the endpoint of a serial chain, possibly by a combination of covaried adjustments in both joint moments and angles.

Within a system of parallel manipulators involved in a one-dimensional pressing task (Fig. 5.6), the mapping between changes in the forces produced by each element and the resultant force remain unchanged, for example, $J = [1\ 1\ 1\ 1]^T$ for four

66 5. PRINCIPLE OF ABUNDANCE

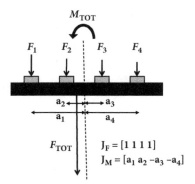

Figure 5.6 During four-finger pressing tasks, the mapping between changes of individual finger forces and changes in total force (F_{TOT}) can be represented by a Jacobian, $J = [1\ 1\ 1\ 1]^T$. The mapping to changes in the total moment of force (M_{TOT}) with respect to the midline between fingers 2 and 3 can be represented by another Jacobian, $J = [a_1\ a_2\ -a_3\ -a_4]^T$, where a stands for the lever arm of each of the fingers.

fingers pressing in parallel. The same is true for the mapping on rotational action, that is, total moment of force with respect to a preselected axis: $J = [a_1\ a_2\ a_3\ a_4]^T$, where a stands for the lever arm of each of the four elements. There is a potential problem, however, related to the fact that forces by the individual elements may be mutually dependent due to factors unrelated to the task. In particular, when a person tries to press with a finger, all four fingers of the hand produce non-zero forces, a phenomenon addressed as "enslaving" (Li et al. 1998; Zatsiorsky et al. 2000; for more detail see Chapter 12). So, individual finger forces are expected to show particular patterns of positive inter-trial covariation in the absence of any task-specific neural control process.

To handle this problem, the notion of finger *modes* has been introduced (Latash et al. 2001; Danion et al. 2003), hypothetical neural commands sent by the actor's brain when the actor tries to change the force of one finger only (Fig. 5.7). Each finger mode leads to proportional force changes by all four fingers with different, person-specific coefficients of proportionality. These effects can be combined into a 4 × 4 matrix, addressed as the enslaving matrix, E, where the diagonal entries reflect desired force production and off-diagonal entries reflect the effects of enslaving. If such a matrix is computed for a person, any set of four finger force values (F) can be converted into a set of four finger mode values (*m*), which can be manipulated by the central nervous system one at a time, at least hypothetically: $m = E^{-1}F^T$. The mapping between finger mode values and resultant force or moment of force can also be computed easily.

This approach, however, assumes that enslaving is a robust phenomenon, which allows applying the E matrix across multiple trials and conditions. A number of recent studies, however, have shown that enslaving tends to increase with time both within a trial and across multiple trials (Abolins et al. 2020; Hirose et al. 2020; for more detail

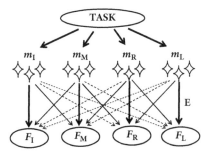

Figure 5.7 An illustration of the control of multi-finger tasks with variables encoding finger modes (*m*), which map on finger forces (*F*) via the enslaving matrix (*E*). The strength of projections from *m* on *F* is shown schematically with line thickness and dashes.

see Chapter 16). Typically, during a single trial, enslaving increases by 15% to 20% over 10 s, possibly due to spread of excitation over neighboring finger representations in the primary motor cortex. Whether this is an acceptable error in estimation of multi-finger interaction remains an open question.

5.2.3. Analysis in muscle activation spaces

Mapping muscle activation on potentially important performance variables (typically mechanical ones) involves two additional steps as compared to the analyses described earlier. In particular, the number of muscles involved in typical tasks is large, and it would be naïve to assume that the central nervous system acts in a space of elemental variables with one dedicated variable per muscle. It is commonly assumed that muscles are grouped, and the neural controller uses one variable per group to define the gain of its involvement. The groups have been addressed with a variety of names including factors, modes, modules, primitives, and synergies (reviewed in Latash 2020b, 2021b). We will use the term *muscle mode* (M-mode) for consistency with finger modes described earlier. The gains of mode involvement are time functions, and sometimes the concept of mode has been applied to a specific basic time function of recruitment of a muscle group—a building block for movement (reviewed in Zelik et al. 2014; d'Avella et al. 2015).

Muscle modes have been defined using a variety of matrix factorization methods applied to correlation or covariation matrices within the space of muscle activation indices integrated over reasonable time intervals. Most commonly used methods have involved non-negative matrix factorization (NNMF) and principal component analysis (PCA); independent component analysis (ICA) has also been used. All of them result in a set of a relatively small number of eigenvectors in the muscle activation space as compared to the number of muscles. Defining how many M-modes to accept is based on criteria that vary across studies, in particular depending on the signal-to-noise ratios, which can be relatively low for studies based on surface electromyography, in particular if muscle activation levels are modest. The composition of M-modes is task specific and can change with relatively small variations in the task as well as with practice (Asaka et al. 2008, 2011). This also applies to the number of M-modes accepted based on specific criteria.

To perform inter-trial analysis of variance in the subspaces of elemental variables, UCM and ORT, which can be of different dimensionality, it is much more convenient to deal with a set of orthogonal eigenvectors. This gives an important advantage to PCA, whereas NNMF and ICA produce sets of independent but not orthogonal eigenvectors.

After a set of modes is defined, the next step is to map small changes in the modes onto changes in a salient performance variable, that is, to identify the J matrix. There is no reliable model that would map small changes in muscle activation indices (changes in M-mode magnitudes, Δm) on changes in mechanical performance variables (ΔPV) explored in typical studies, such as, for example, center of pressure coordinate, limb endpoint coordinate, center of mass coordinate, endpoint force vector, and so forth. This requires assuming that the mapping is relatively robust and linear, at least within a certain range of values. Under this assumption, multiple linear regression methods have been used to discover the relations between Δm and ΔPV, that is, the J matrix (Krishnamoorthy et al. 2003a,b; Danna-dos-Santos et al. 2007; reviewed in Latash 2020b). The null-space of the J matrix has been used as a linear approximation of the UCM.

Recently, this method of analysis has been modified for application to sets of motor units recorded within a muscle or across muscles (Madarshahian et al. 2021; reviewed in Latash et al. 2023). This is not a trivial generalization because activity patterns of motor units are discrete in nature. Nevertheless, the concept of M-modes has been developed for motor unit modes (MU-modes), the method of multiple linear regression has been used to identify the J matrices in mapping changes in MU-mode magnitudes to changes in salient mechanical variables, and inter-trial analysis of variance within the UCM and ORT led to nontrivial findings confirming $V_{UCM} > V_{ORT}$ in some analyses but not in others. This topic will be discussed in more detail in Chapters 6 and 11.

Across studies, the experimentally observed modulation of the variance components (V_{UCM} and V_{ORT}) and ΔV index with movement time suggests that V_{UCM} is strongly dependent on the magnitude of the salient performance variable but not so much on its rate of change. In contrast, V_{ORT} showed strong dependence on the rate of change of the performance variable. Assuming that performance variability is defined by inter-trial variability in setting two parameters, spatial and timing, at a control level (Gutman et al. 1993; Goodman et al. 2005), these findings suggest that covariation of elemental variables is efficient in dealing with spatial deviations but not with timing deviations. This conclusion fits the general scheme of control illustrated in Fig. 5.8, which is a simplified version of the scheme suggested earlier (Martin et al. 2009, 2019). Within this scheme, the neural control of timing is hierarchically higher than the level assumed to implement patterns of covariation within elemental variables. As a result, any timing error is common across all the elemental variables and cannot be compensated for at the lower level.

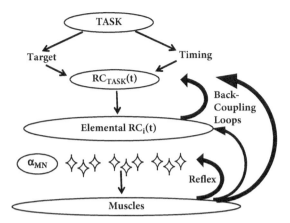

Figure 5.8 A simplified scheme on the control of a system with many elements. Note that action timing is defined at a hierarchically higher level than the level organizing performance-stabilizing synergies with the help of back-coupling loops from the sensory endings and within the central nervous system.

5.3. Dealing with nonlinear systems

In certain situations, the mapping between elemental and performance variables cannot be linearized. This happens, for example, during analysis of multi-joint kinematic synergies when one of the joints approaches its anatomical limit of rotation. This may also happen when the performance variable is a function of elemental variables, **PV** = f(**EV**), that cannot be linearized even locally; for example, it represents a power function of an elemental variable or the product of several elemental variables. If the function f is known and allows running nonlinear regression analysis, running this analysis across observations in individual trials would be the most straightforward method of estimating covariation in the space of elemental variables.

One can, however, ask a more general question: Is there covariation among elemental variables that helps to reduce the inter-trial variability of the performance variable? To answer this question, a method of data randomization has been developed, which creates covariation-free surrogate data sets derived from the actual data (Kudo et al. 2000; Müller and Sternad 2003). The method involves choosing values of individual elemental variables from different trials randomly and computing corresponding values of the performance variable multiple times. Of course, the method requires knowledge of the mapping of the elemental variables on the performance one, that is, the function f in **PV** = f(**EV**). Note that the randomization procedure leads to similar mean values and standard deviations (SDs) of each of the elemental variables in the actual data set and in the surrogate ones. After a number of surrogate data sets have been created and corresponding **PV** values computed, a measure of variability of **PV** across trials can be estimated, for example, its SD, SD(**PV**$_{SUR}$), and compared to the actual variability of **PV**, SD(**PV**$_{ACT}$). If the original data set contained

EV covariation that helped reduce the variability of **PV**, SD(**PV**$_{SUR}$) > SD(**PV**$_{ACT}$) is expected.

This method has been used in a number of studies, in particular those mapping elemental mechanical variables produced by the body on task-specific performance (Kudo et al. 2000; Müller and Sternad 2003; Hasanbarani and Latash 2020). For example, to perform an accurate basketball shot, one has to co-vary the coordinates of basketball release (X_R), speed (V_R), and angle (α_R) at release (Fig. 5.9). This allows computing the basketball trajectory, in particular its coordinate when it passes the plane of the basket on the descending limb of the trajectory (X_B). The mapping function in $X_B = f(X_R, V_R, \alpha_R)$ may be complex and highly nonlinear, but it is important that this function is known from classical mechanics. Then, sets of variables {X_{Ri}, V_{Rj}, α_{Rk}} are selected at random, where i, j, and k correspond to trial numbers, $i \neq j \neq k$. The procedure is repeated multiple times, and the surrogate sets are used to compute $X_{B,SUR}$ values. If, on average, $X_{B,SUR}$ deviates from the center of the basket more than X_B, one can draw a conclusion that the actor covaried X_R, V_R, and α_R in a way that helped achieve higher accuracy of the shots.

This method can be applied to various sets of elemental variables, those describing processes within the body and those describing action on the environment. The UCM hypothesis explores mapping between body-level variables (those produced by body elements at the selected level of analysis) and performance variables in the environment (e.g., on the manipulated object). The randomization method can be used to address similar issues in cases when linearization inherent to the analysis of inter-trial variance is impossible (Ambike et al. 2016a; see Chapter 7). Most studies using the randomization method, however, explored mapping between elemental variables acting on an object and a more general, task-specific performance variable, also describing behavior of the same object.

Another method to explore covariation among elemental variables and performance variables when both are defined at the level of the environment has been developed based on the concept addressed as goal-equivalent manifold (GEM; Cusumano and Cesari 2006). As the name suggests, the concept is applicable to movements with an explicitly defined goal for a performance variable, a function of a set of elemental variables. Since this method does not address processes within the body, it is only indirectly relevant to the field of the neural control of biological movements.

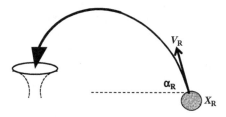

Figure 5.9 An illustration of a task (basketball throw) with a strongly nonlinear dependence between the elemental variables, the coordinate of basketball release (X_R), its speed (V_R) and angle (α_R) at release, and accuracy of the throw.

5.4. Motor equivalence

The term *motor equivalence* has been used to reflect the ability to perform tasks with different effectors or different effector involvement within a set of effectors (Hughes and Abbs 1976; Kelso et al. 1984; Wing 2000; Levin et al. 2003). This ability is rather directly related to the principle of abundance. Indeed, the ability to use effectors with variable involvement requires having more effector-level variables (i.e., abundance of those variables) as compared to the number of task-related salient performance variables. Phenomena of motor equivalence have been described in a broad range of tasks under conditions with unexpected perturbations of ongoing actions as well as with intentional changes in effector involvement.

Arguably, the most famous example of motor equivalence is the preservation of individual features of one's handwriting when using various implements, writing on different surfaces, with different involvement of dominant arm segments (e.g., during writing with a pencil on a piece of paper vs. writing on a blackboard with a piece of chalk), and even with different effectors (Bernstein 1947; Raibert 1977; reviewed in Latash 1993). Bernstein claimed that the individual handwriting features were the same during writing with a pencil held in the dominant and nondominant hands, attached to one of the elbows, attached to a foot, and even gripped by teeth. This example was used as an argument in favor of the existence of *engrams*—hypothetical functions stored in the brain, related to topological features of practiced movements—that can be applied to different effector sets and scaled in both magnitude and time.

Other examples of motor equivalence include the adjustments of ongoing actions in response to unexpected external perturbations. Such adjustments are seen at short time delays, shorter than simple reaction time. For example, a perturbation applied to the lower jaw during speech produces quick adjustments in the action of other articulators, which keep the sound only mildly affected (Abbs and Gracco 1984). When a person performs a gripping action with the thumb and index finger, an unexpected obstacle stopping the movement of one of the fingers leads to a larger motion of the opposing finger (Cole and Abbs 1987).

To link motor equivalence to the UCM hypothesis, consider the following example. Imagine that you hold a small spring (e.g., the spring from a typical pen) between the thumb and index finger. Now imagine that you try to squeeze it very quickly along its main axis. What will happen? As it is easy to predict, the spring will buckle and jump sideways (see Valero-Cuevas et al. 2003). Why does it move not along the desired direction but orthogonal to it? This happens for two reasons: First, the spring is relatively stable along its main axis and relatively unstable in the orthogonal plane; second, your action is never perfect and has components both along the desired direction and orthogonal to it. This example shows that a quick voluntary action can produce movement not necessarily in the desired direction but in less stable directions. As described earlier, the least stable directions during natural voluntary movements commonly span the UCM for the salient performance variable.

Figure 5.10 illustrates the space of two finger forces (elemental variables) involved in the task of producing a certain value of total force (F_{TOT}). Imagine that the subject has been instructed to double F_{TOT} quickly and then to come back to the initial F_{TOT} value. This action involves transition from an original state (black circle) within the

72   5. PRINCIPLE OF ABUNDANCE

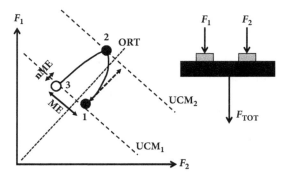

Figure 5.10 Two fingers generate forces F_1 and F_2 and are involved in a task of producing a magnitude of total force (F_{TOT}). If the subject increases the force magnitude quickly, from point 1 to point 2, and then tries to return to the initial force magnitude, the final finger force combination (point 3) will correspond to a larger deviation along the uncontrolled manifold (UCM; motor equivalent deviation [ME]) than orthogonal to it (along ORT, non–motor equivalent deviation [nME]). A typical minimum-norm trajectory is shown as the straight dashed line.

UCM corresponding to the initial F_{TOT} value to some point on another UCM corresponding to the desired higher F_{TOT} value, and back to the original UCM. There is an infinite number of trajectories that can solve the task. As we will discuss in a later chapter (Chapter 9), a robot would likely be programmed to perform this action following an optimal trajectory (e.g., the so-called minimum-norm trajectory), which is shown as a line orthogonal to the UCM, during both force increase and decrease. Humans, however, perform such a task following a trajectory with large components along the UCM, which results in a final state (open circle) displaced significantly along the UCM compared to the initial state.

Note that deviations along the UCM, by definition, have no effect on the magnitude of the salient performance variable, which means that they are motor equivalent. Deviations along ORT are non–motor equivalent since they lead to a change in the performance variable. Since nobody is perfect, in the described example, both initial and final states will deviate from the required force level. This means that they will have non-zero non–motor equivalent deviations, shown as nME in Fig. 5.10. However, these deviations will be smaller than those along the UCM, shown as ME (motor equivalent), ME > nME.

There have been many examples of the inequality ME > nME during voluntary movements (Scholz et al. 2007; Mattos et al. 2011, 2013, 2015a). For example, imagine that a person is asked to perform a series of quick pointing movements to a target with the pointer held by the dominant hand. Unexpectedly, in the initial state, the elbow joint happens to be spanned by a spring-like load (e.g., a rubber band). The subject will correct the induced movement deviation quickly. If one quantifies the corrective movement within the UCM and ORT computed for the pointer tip spatial coordinates, the former component (ME) would be much larger than the latter one (nME). This was shown to be true for analysis in the joint configuration space and in the muscle activation space (Mattos et al. 2011, 2013). In other words, much of the corrective movement did not correct anything because it was directed along the UCM for the salient performance variable (pointer tip coordinates). This strategy looks wasteful because ME motion requires energy, but it is natural if one considers the difference in stability along the UCM and ORT defined by the ongoing neural control process.

Recall now an earlier described experiment with a person asked to press with the four fingers of a hand and produce a constant force level, and then asked to tap quickly with one of the fingers (Latash et al. 1998). Experiments show that, during the tap, forces of the non-tapping fingers increase, and this increase compensates for nearly 100% of the force drop induced by the lost contribution of the tapping finger. In other words, without any special instruction, the deviation in the finger force space is mainly constrained to the UCM for F_{TOT}.

Another example of motor equivalence is illustrated in Fig. 5.11 (Mattos et al. 2015a). In this example, the subject was asked to press with the four fingers of a hand and follow the sine-like template shown on the screen with the cursor corresponding to the current F_{TOT} value. Unexpectedly, one of the fingers was lifted smoothly (over 1 cm in 0.5 s), leading to an increase in its force, and after some time, the finger was lowered to its initial position. Lifting the finger naturally led to an increase of its force, and lowering the finger led to a drop of its force due to both peripheral muscle

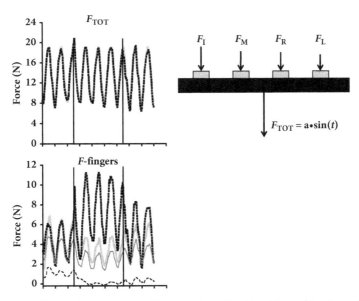

Figure 5.11 An illustration of a task with accurate cyclical total force (F_{TOT}) production by four fingers pressing in parallel. Lifting a finger (first dashed vertical line) leads to a quick F_{TOT} correction (top panel) accompanied by a large change in the sharing pattern of force across the fingers (bottom panel). Lowering the finger back to its initial coordinate (second dashed line) does not restore the original sharing pattern. Reproduced by permission from Mattos et al. 2015a.

properties and the action of stretch reflex. As shown in Fig. 5.11, subjects corrected the F_{TOT} deviations very effectively so that only minimal effects could be seen in the F_{TOT} time series. However, this unchanged performance was associated with major redistribution of F_{TOT} across the four fingers, that is, with large ME motion along the corresponding UCM.

Both pairs of variables, {ME; nME} and {V_{UCM}; V_{ORT}}, reflect stability along the two subspaces, UCM and ORT. Not surprisingly, these indices correlate as expected from classical statistics assuming that all measurements are sampled from the same normal distributions. Indeed, if we assume that each measurement represents a sample from a normal distribution with the SD = σ, the difference between two random samples will be another normal distribution with the mean $\mu_D = 0$ and SD $\sigma_D = \sigma\sqrt{2}$. Since ME and nME are absolute distances between two points, the distribution of those distances is non-negative (folded) with a new mean (μ_X) and new SD (σ_X). From classical statistics (Leone et al. 1961), $\mu_X = \sigma\sqrt{2/\pi}$, $\sigma_X^2 = \sigma^2 - \mu_X^2$. These equations can be applied separately to data in the UCM and ORT. They suggest that ME is expected to be proportional to SD within the UCM (σ_{UCM}), while nME is expected to be proportional to SD within the ORT space (σ_{ORT}). This insight was confirmed experimentally (Falaki et al. 2017a). Note that, if a subject performs a voluntary action, the final state may differ from the initial one in the distribution of datapoints within the space of elemental variables, and then the mentioned correlations may be lost (cf. Cuadra et al. 2018). As we will see later (Chapter 18), analysis of ME and nME deviations can have certain advantages as compared to analysis of the variance components, V_{UCM} and V_{ORT}.

5.5. Missing pieces of the mosaic

The main concepts described in this chapter, such as the principle of abundance, the UCM hypothesis, and motor equivalence, seem to be well established and supported in numerous experimental studies. These concepts will be used as the foundations for future chapters. Limits of their applicability, however, have not been defined. For example, does the principle of abundance apply only to organisms with central nervous systems? Can it be extended to animals such as insects? Can it be extended to single-cell animals and single cells in the body? In other words, is this principle equivalent to the notion of a living system?

The UCM hypothesis has offered a powerful tool to explore the stability of performance variables produced by abundant sets of contributing elements. The associated computational methods, such as analysis of inter-trial variance, by their nature are only applicable to systems that can be linearized, at least locally. This is an obvious limitation that has been acknowledged and alternative methods have been developed. We should also keep in mind that performance of a single element can be stabilized to various degrees using feedback from sensory endings sensitive to this performance variable. How is stability of the overall performance of a task related to stability of the outputs of individual elements and stability organized in the multi-element system? As discussed later (Chapter 6), this is a nontrivial issue because large variance along the UCM for a performance variable requires proportionally large variances of the contributing elemental variables. So, individual elemental variables cannot be stabilized too strongly to allow them to vary sufficiently to produce large variance within the UCM. This trade-off has not been explored in much detail.

Another underexplored issue is the relation between inter-trial variance indices and those of motor equivalence. These variables seem to be related to each other but only under certain assumptions, which have not been identified in sufficient detail.

Arguably, the most obvious limitation is the paucity of information on neurophysiological mechanisms involved in these hypotheses and phenomena. We will return to the available hypothetical schemes later (Chapter 6, 10, and 11), but so far, they remain speculative and rather general.

PART II
CURRENT UNDERSTANDING

6
Synergies

The term *synergy* in motor control is tightly linked to another term, *coordination*. Emphasis on coordination of biological movement as its distinguishing feature from movements in inanimate nature has been emphasized since Aristoteles. In the Middle Ages, the concept of synergy was used by St. Gregory Palamas in a seemingly very remote field, theology. Palamas (1983, 1988) used this term to describe the cooperation of God and Man toward salvation of Man. He emphasized that God gave less help to those with strong faith and more to those with weak faith to bring them all to salvation. This negative covariation of the efforts by two actors involved in a common task is very close to the current understanding of one of the features of synergies described later in this chapter. This understanding of synergy also fits well with the meaning of this word in Greek, "work together," which implies the presence of a number of actors and a clearly identified goal or purpose of a synergy.

Applications of the concept of *synergy* to the field of motor control started at the end of the 19th century. In particular, discoordination observed in patients with injuries to the cerebellum was addressed by Babinski (1899) as *asynergia* or *dyssynergia*. At about the same time, Hughlings Jackson (1889) developed his scheme of multiple representations of the motor system in the primary motor cortex with an implied possibility of mapping high-dimensional muscle representations onto lower-dimensional joint and limb representations.

Currently, the word *synergy* is used in three major meanings. The first reflects clinical observations of typical movement disorders after injuries to certain brain structures (e.g., after cortical stroke) leading to stereotypical patterns of muscle activation that interfere with the ability to perform functional movements. Examples include "flexion synergy" and "extension synergy" when all the major flexors or extensors of an affected limb show large levels of activation resulting in a gross flexion or extension movement leading close to the limits of joint rotation (Bobath 1978; DeWald et al. 1995). Such pathological patterns of muscle activation are clearly dysfunctional, and the goal of rehabilitation is eliminating these patterns with the hope that more functional movements will be facilitated. Addressing these patterns as synergies does not seem proper given that the elements (muscles) have no clearly defined goal; they do not "work together" but are only activated together.

Most commonly, the word *synergy* is used to address sets of elemental variables (e.g., muscle activations, joint angles, digit forces, etc.) that show parallel changes in their magnitude with time and/or with changes in the magnitude of a salient performance variable. Other words are also used to address such patterns of parallel changes in a set of variables, including primitives, modes, modules, and factors (reviewed in Ivanenko et al. 2006; Hogan and Sternad 2012; Giszter 2015; Overduin et al. 2015; Latash 2020b). As we will see in the next section, this meaning of "synergy" is close to one of the features of synergies described by Bernstein (1947). Whether elements within such groups "work together" toward a goal or have a clear purpose remains unclear. So, we would rather address them with a different term; in this book, the term *mode* is used.

The third meaning of *synergy* implies a clear purpose: Synergies provide task-specific controlled stability of salient performance variables (reviewed in Latash 2008, 2021b). Since stability of movement is paramount in the unpredictable (or, at the very least, imperfectly predictable) environment, this purpose is highly functional. This meaning is also linked to Bernstein's understanding of synergies, which is the topic of the next section.

6.1. Bernstein's understanding of synergies and its development

Within his multilevel scheme of movement construction, the second-from-the-bottom level was termed by Bernstein "the level of synergies and patterns or the thalamo-pallidar level." We will address it, for brevity, simply as "the level of synergies" (see also Chapter 2). This level was assumed to be based on the lowest level, the level of tone, which was expected to be responsible for steady-state, slow actions related to states of individual muscles and muscle groups. The higher levels, responsible for the production of purposeful actions, were presumed to perform goal-directed functional movements and use the level of synergies as the foundation to ensure stability of salient performance variables.

Two features of the level of synergies were emphasized by Bernstein. First, this level was responsible for arranging large groups of elements into units, with each unit controlled by a single higher-level neural variable. Most commonly, Bernstein associated elements with individual muscles. Second, this level was responsible for ensuring dynamical stability of natural movements in the imperfectly predictable environment. Notably, the former feature of synergies was eagerly accepted and developed by the following generations of researchers, while the latter feature was all but ignored for half a century. Bernstein did not explicitly emphasize one of the two features as more important than the other. However, in his descriptions of skill acquisition and movement disorders, he rarely invoked the creation, development, or disintegration of muscle groups and very frequently emphasized the crucial importance of acquiring dynamically stable patterns and impairments caused by loss of dynamical stability. In his numerous descriptions of patients, one cannot find examples of cases with "too many elements to control" but can find many cases when dynamical movement stability had been lost.

The "grouping of elements" feature of synergies obviously depends on the level of analysis at which elements are defined. As we will see later, a biomechanically non-redundant task (e.g., producing a required force magnitude with a single finger) may be abundant at the level of control variables (referent coordinates), at the level of involved muscles, and at the level of recruited motor units (Chapters 7, 11, and 12). Since biomechanical variables, such as forces, coordinates, and their derivatives, cannot be prescribed by the central nervous system (see Chapters 1, 2, and 4), defining groups of biomechanical elemental variables is of limited usefulness for those interested in neural mechanisms of motor control. The same is true for muscle activation levels and patterns because these depend on movement biomechanics due to the action of reflex loops. If one is interested in processes of neural control, it makes sense to define elemental variables at a control level, even if this definition may not be shared by most colleagues. Analyzing synergies at the level of neural control is a necessary step toward making the concept of synergy useful within the field of motor control as an area of natural science.

Typical methods of identification and analysis of multi-element groups are based on so-called matrix factorization methods applied to matrices of correlation of covariation coefficients in spaces of elemental variables. These methods effectively reduce the original data in the multidimensional space of elemental variables to a lower-dimensional set of hypothetical higher-order variables. This process is based on a set of assumptions and relatively arbitrary criteria of accepting a particular number of higher-order variables. The two most frequently used methods, non-negative matrix factorization (NNMF) and principal component analysis (PCA), were described in Chapters 4 and 5. As the name suggests, NNMF is applicable to sets of values that are always non-negative, for example, levels of muscle activation or neuronal firing frequencies. The method produces a lower-dimensional set of eigenvectors (higher-order variables addressed further as *modes*), which are independent but not necessarily orthogonal. PCA allows both positive and negative values of elemental variables, which may be viewed as a drawback in analysis of non-negative variables but may turn into an advantage if deviations of those variables from some referent values are analyzed because such deviations can be both positive and negative. PCA produces a set of orthogonal modes, a feature very important for the next step—analysis of action stability.

Many studies applied the NNMF and PCA methods to analysis of indices of muscle activation (after proper processing of surface electromyography [EMG] signals) during whole-body tasks such as vertical standing and locomotion (Krishnamoorthy et al. 2003a; Ivanenko et al. 2004; Ting and Mcpherson 2005). Those studies provided evidence for a relatively small number of modes, much smaller than the original number of EMG signals, accounting for much variance in the muscle activation space. The number of modes and their composition were relatively consistent across healthy subjects as long as the tasks remained similar. When the tasks became more challenging, the number of modes increased, and their composition also changed, reflecting preference for agonist-antagonist coactivation patterns (Krishnamoorthy et al. 2004; Asaka et al. 2008; Danna-dos-Santos et al. 2008). This was observed, in particular, during vertical standing tasks associated with cyclical body sway and self-triggered perturbations and performed while standing on narrow supporting surfaces, with eyes open or closed, and with vibration applied to the Achilles tendons. Practicing such tasks over a few days restored the more typical mode compositions associated with reciprocal patterns of activation within major agonist-antagonist muscle pairs.

Typical movements are associated with relatively short-lasting changes in salient variables, which show non-stationary trajectories. In addition, biological systems are active and commonly react to quick perturbations (changes in the external forces) at relatively short time delays. This makes some of the methods commonly used to estimate dynamical stability of movements inapplicable. A very important step was made by Gregor Schöner (1995), who suggested that biological movements were characterized by task-specific (intention-specific) stability of salient performance variables and introduced a framework for quantitative analysis of action stability in multidimensional, abundant spaces of elemental variables.

Imagine that a person performs a series of movements trying "to do the same." Individual movements start from somewhat different initial conditions because it is impossible to standardize the initial conditions perfectly (see an example with only two elements in Fig. 6.1). In particular, the initial states of excitability of various neuronal pools within the subject's central nervous system will vary across repetitive trials. By definition of stability, trajectories in the space of elemental variables across individual trials will converge in stable directions and diverge in unstable directions. Higher stability is expected to correspond to faster convergence of the trajectories. Imagine now that all the trajectories are aligned by the time of their initiation (and, possibly, also by movement time), and a specific phase along the trajectories is selected. It is clear that inter-trial variance will be lower along stable directions than along unstable directions. Distributions similar to the ones illustrated in Fig. 6.1 at different phases of the task of reaching a certain value of the sum of two variables (E1 and E2) produced by two elements were reported in experiments (e.g., Solnik et al. 2015).

This logic results in a nontrivial conclusion: Variance indices computed *across trials* can be used as indices of stability *along each of the trajectories* assuming that stability is a robust feature of the set of trials. Based on the idea of task-specific stability, one

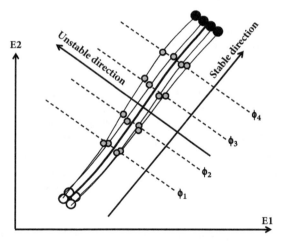

Figure 6.1 An illustration of trajectories in a sequence of trials at the same well-practiced task in stable directions and unstable directions assuming some variability in the initial states (shown with open circles, exaggerated). Note that measuring inter-trial variance at different phases ($\phi_1, \phi_2, ..., \phi_n$), shown with clusters of gray circles, is expected to produce smaller values in stable directions than in unstable directions. The desired trajectory is estimated as the mean of the actual trajectories (thick trace). E1 and E2 stand for elemental variables.

can expect that stability along directions in the space of elemental variables leading to changes in a potentially important performance variable will be relatively high (inter-trial variance low) and stability along directions that keep the important variable unchanged will be relatively low (inter-trial variance high). So, exploring the structure of inter-trial variance for potentially important performance variables (i.e., the relative magnitude of variance along different directions) can be used to infer their stability (see also Chapter 5).

The uncontrolled manifold (UCM) hypothesis formalized and summarized this line of reasoning (Scholz and Schöner 1999; reviewed in Latash et al. 2002d, 2007). It assumes that the central nervous system can organize stability of a task-specific salient variable in an abundant space of elemental variables, and as a result, higher amounts of inter-trial variance are expected in a subspace where this salient variable does not change (the UCM for that variable) as compared to variance in directions along which it changes (orthogonal to the UCM, ORT). Methods of analysis within the framework of the UCM hypothesis were described in Chapter 5. To remind, these methods most commonly include comparison of indices of inter-trial variance along the UCM and ORT, V_{UCM} and V_{ORT}, and also analysis of motor equivalence, that is, magnitudes of deviations along the UCM (motor equivalent [ME]) and along the ORT (non–motor equivalent [nME]).

6.2. Intra-muscle and multi-muscle synergies

In his definition and analysis of synergies, Bernstein considered muscles as the smallest controllable elements, resulting in the notion of multi-muscle synergies. In fact, the smallest controllable element of the neuromotor system is the motor unit (MU), which consists of a single alpha-motoneuron and all the muscle fibers innervated by its axon. Typical skeletal muscles contain hundreds or even thousands of MUs, which makes transformation from muscle activation to MU recruitment abundant. Even though the order of MU recruitment typically follows the classical size principle (Henneman et al. 1965), from the smallest to the largest, the possibility of varying the frequency of firing of recruited MUs creates room for an infinite number of recruitment patterns compatible with a desired magnitude of muscle activation. Potentially, this allows the central nervous system to organize intra-muscle, multi-MU synergies stabilizing muscle action.

The neural control of an MU can be described similarly to the neural control of a muscle within the equilibrium-point hypothesis (see Chapter 3). Figure 6.2A shows a dependence of the frequency of firing of an MU (f_{MU}) on muscle length defined by the stretch reflex. The MU becomes active ($f_{MU} > 0$) at a certain threshold length value, λ_{MU}. Changes in f_{MU} define the contribution of the MU to the muscle activation level and, in isometric conditions, to muscle force. Imagine now that a subset of MUs is controlled as a single group (MU-mode) and can be characterized by individual λ_{MU} values changing in parallel (Fig. 6.2B; see Madarshahian et al. 2021). The contributions of individual MUs within such an MU-mode to muscle force will sum up, resulting in an MU-mode force-length characteristic with the threshold (λ_{MODE}) equal to the smallest λ_{MU} value across the MUs within the MU-mode.

6. SYNERGIES

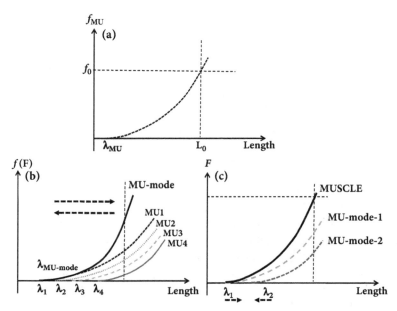

Figure 6.2 (a): A dependence of the frequency of firing of a motor unit (f_{MU}) on muscle length. Stretch reflex threshold is shown as λ_{MU}. (b): MUs form groups (MU-modes) characterized by parallel changes in the individual λ_{MU} values. The MU-mode characteristic is the sum of the individual MU characteristics originating from the smallest λ_{MU} value. (c): Thresholds for individual MU-modes can covary to stabilize a performance variable produced by the muscle (e.g., muscle force magnitude in isometric conditions).

If muscle action is viewed as combined action of a small set of MU-modes, there may be covariation of individual MU-mode involvement stabilizing task-specific muscle action. In other words, there may be synergies in the abundant λ_{MODE} space stabilizing λ_{MUSCLE} shifts that define salient task-specific performance variables, for example, muscle force in isometric conditions (Fig. 6.2C). Studies of patterns of MU recruitment, frequency (f_{MU}) modulation, and corresponding synergies have been based on MU identification and recording their action potentials with surface EMG methods (Madarshahian et al. 2021; reviewed in Latash et al. 2023). This analysis involved several nontrivial steps that could influence the outcome variables. In particular, the discrete sequences of MU action potentials had to be converted into smooth functions reflecting instantaneous frequency of firing, f_{MU}. Since some MUs generate action potentials at relatively low frequencies (about 4 to 5 Hz), the time resolution of this analysis is inherently poor: To estimate the frequency of firing within a time window, one has to have at least two action potentials in that window. There is no direct model that could link f_{MU} magnitudes to salient performance variables such as force. So, to link changes in the values of individual MU-modes (measured in units of frequency) to changes in force, multiple linear regression was used, resulting in a set of coefficients combined into a Jacobian (J) matrix. Similar methods were used in earlier studies with analysis of multi-muscle synergies (Krishnamoorthy et al. 2003b; Danna-dos-Satos et al. 2007). The null-space of J was used as a linear approximation of the UCM.

Despite all the mentioned complicating factors, the consistency of findings across subjects and studies suggests that none of the mentioned complicating factors is fatal (see also Madarshahian and Latash 2022a, 2022b; Madarshahian et al. 2022; Ricotta et al. 2023a, 2023b; De et al. 2024). Experiments with accurate force production have shown stable and robust MU groups (MU-modes) in the f_{MU} space defined with PCA. The MU-modes showed similar MU compositions during MU recruitment and derecruitment, that is, during the action phases with force increase and decrease. The same studies have provided evidence for MU-mode-based synergies stabilizing force magnitude quantified within the framework of the UCM hypothesis (i.e., leading to the inequality $V_{UCM} > V_{ORT}$).

There is substantial evidence suggesting that intra-muscle, multi-MU-mode synergies are primarily of a spinal origin. In particular, reflex-induced changes in finger force are stabilized by covaried involvement of the MU-modes defined over groups of MUs in both agonist and antagonist muscles (Fig. 6.3; Madarshahian et al. 2022). In contrast, no such synergies are seen in spaces of individual finger forces, which are likely based on transcortical loops involving the basal ganglia and cerebellum

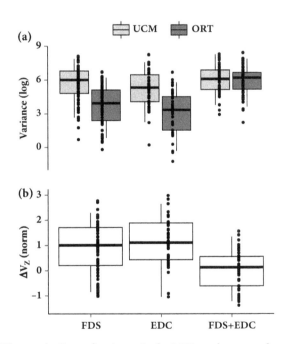

Figure 6.3 (a): The two indices of variance in the MU-mode space, along the uncontrolled manifold and orthogonal to it (UCM and ORT), computed with respect to finger force change produced by the stretch reflex. (b): The corresponding synergy indices. Analysis was performed using groups of MUs within the agonist (flexor digitorum superficialis [FDS]), within the antagonist (extensor digitorum communis [EDC]), and pooled over both muscles (FDS + EDC). $V_{UCM} \gg V_{ORT}$ (note the log scale!) for the FDS and EDC analyses, but not for the (FDS + EDC) analysis. Reproduced by permission from Madarshahian et al. 2022.

(described in more detail in the following sections). Additional findings (Ricotta et al. 2023a; De et al. 2024), discussed in more detail in Chapters 11 and 21, have provided evidence for a central role played in intra-muscle synergies by spinal circuitry.

Numerous spinal mechanisms can potentially contribute to intra-muscle synergies stabilizing muscle contribution to action. These include both reflex projections from sensory endings, which are typically negative feedback loops, and intraspinal projections, such as recurrent inhibition, which are also negative feedback loops. As described in more detail elsewhere, such loops tend to minimize the effects of any perturbation, intrinsic or extrinsic, on the overall level of muscle activation (cf. Uchiyama et al. 2003; Hultborn et al. 2004; Latash et al. 2005). More on the synergic control of single muscles, as well as agonist-antagonist muscle pairs, can be found in Chapter 10.

Similar computational methods have been used to explore multi-muscle synergies stabilizing salient variables during a variety of tasks including arm action and actions performed while standing in the field of gravity (Krishnamoorthy et al. 2003b, 2007; Danna-dos-Santos et al. 2007; Klous et al. 2011). In particular, PCA in muscle activation spaces was used to identify muscle modes (M-modes), and multiple linear regression was used to define the J matrix linking small changes in the M-modes to changes in salient performance variables such as coordinates of the center of pressure by standing persons or force applied by the hand to an external object. Analyses of both inter-trial variance and motor equivalence were used to quantify performance-stabilizing synergies (Falaki et al. 2017a). More on multi-muscle M-modes and synergies can be found in Chapter 14.

6.3. Synergies in kinematic and kinetic spaces

Following Bernstein's traditions, a large number of studies explored performance-stabilizing synergies in spaces of kinematic variables such as rotations in individual joints, translations of segments, and their derivatives (reviewed in Latash et al. 2007; Latash 2008, 2021b). Performing this analysis requires having a formal model linking small displacements of elemental variables to changes in typical performance variables such as coordinates of the limb endpoint. Such models involve trigonometric functions and can be linearized only for relatively small variations of the elemental variables and only for some of their values (e.g., far from the limits of joint rotation). In such cases, a formal mapping can be introduced with the help of the J matrix, and its null-space can be used as a linear approximation of the UCM for the selected performance variable. The J matrix could also be computed using multiple linear regression methods, as described in the previous section. Comparing the two methods led to qualitatively and quantitatively similar indices of the main outcome variables such as inter-trial variance and synergy indices (Freitas et al. 2010).

The very first studies of kinematic synergies demonstrated some of the important features of the analysis of inter-trial variance components within and orthogonal to the UCM, V_{UCM} and V_{ORT} (Scholz and Schöner 1999; Scholz et al. 2000). In particular, they confirmed the existence of performance-stabilizing synergies ($V_{UCM} > V_{ORT}$) for some of the potentially important performance variables but not for others. They also showed that the synergy index (ΔV) could be modulated across movement phases. In particular, these findings were reported in a study of very quick shooting from an infrared pistol at an infrared-sensitive target (Scholz et al. 2000). The salient performance variable stabilized by covaried adjustments of joint angles was the angle between the pistol barrel and the direction from the pistol back-sight to the target. This variable had to be close to 0° for an accurate shot at the time of pressing the trigger. The variable, however, was stabilized at other movement times, even at phases when the pistol pointed away from the target, suggesting that the synergy stabilizing this variable was established prior to or early in the shooting action and acted throughout the movement.

One of the important results was that, in conditions of an unexpected constraint applied to one of the moving joints (a rubber band crossing the elbow joint), more than half of the subjects performed an accurate shot during the very first trial. Given that the movement time was very short (about 300 ms), the corrections were viewed as reflecting synergic adjustments in other joints of the moving limb stabilizing the salient performance variable and keeping it relatively unaffected. First, these observations demonstrate the benefits of not selecting a single optimal solution by "freezing" apparently redundant degrees of freedom but instead using all those degrees of freedom to organize performance-stabilizing synergies. Second, this is an illustration of the phenomenon of motor equivalence (see also Chapters 5 and 18).

A follow-up study of pointing with a pointer at spatial targets in conditions of similar constraints applied to the elbow joint in random trials quantified ME and nME deviations in the joint configuration space and showed that the ME deviations were consistently larger than the nME ones during corrections of the unexpected movement deviations caused by the rubber band (Fig. 6.4). Since ME deviations are, by definition, unable to cause a change in the salient variable (pointer tip coordinates and/or orientation in those experiments), these results suggest that much of the movement in the joint angle space did not correct the salient variable during the correction phase. The large ME deviations look wasteful from the point of view of energy expenditure (and many other optimization principles; see Chapter 9). But they can be naturally expected given that the subspace where the pointer tip does not move is the UCM for the pointer tip coordinate, and it is characterized by relatively low stability. As a result, it accepts most of the kinematic effects of both external perturbations and quick corrective actions by the central nervous system, leading to the observed large-amplitude ME motion.

Studies of synergies in spaces of kinetic variables have been mostly limited to multi-digit force production, although a number of studies explored different tasks and spaces of kinetic variables (e.g., Yen et al. 2009). Elemental variables in multi-digit force and moment production tasks have been associated with digit forces or hypothetical variables, digit modes (Danion et al. 2003), which take into account the unintentional force production by all the fingers of a hand when one finger produces force intentionally (enslaving; Zatsiorsky et al. 2000; see Chapter 12). Multi-finger force production in pressing tasks is a particularly attractive task because the Jacobian matrix is trivial: $J = [1\ 1\ 1\ 1]$ for the four-finger force production studied in the space of finger forces.

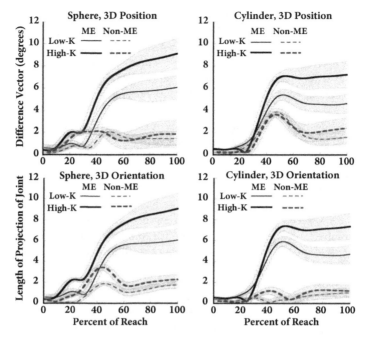

Figure 6.4 Two components of deviations in the joint configuration space, motor equivalent (ME) and non–motor equivalent (non-ME), computed with respect to the pointer orientation and pointer tip coordinates during corrections of fast pointing movements to a visual target, which could be spherical or cylindrical. The movements were perturbed unexpectedly by a rubber band of low or high stiffness crossing the elbow joint (Low-K and High-K, respectively). Note that ME (solid lines) > nME (dashed lines) across conditions and analyses. Reproduced by permission from Mattos et al. 2011.

Studies of multi-finger force production resulted in a number of nontrivial findings. In particular, the very first studies used visual feedback on total force (F_{TOT}) and required the subjects to produce accurate cyclical F_{TOT} changes between two visual targets while being paced by the metronome (Latash et al. 2001; Scholz et al. 2002). Four-finger F_{TOT}-stabilizing synergies were expected. The results, however, showed that such synergies ($V_{UCM} > V_{ORT}$) were present only within a relatively small range of forces, close to F_{TOT} peak values (Fig. 6.5). When the same data were processed with respect to another performance variable, total moment of force in pronation-supination, moment-stabilizing synergies were seen over the whole cycle, although the subjects were not instructed to produce specific moment magnitudes and received no visual feedback on the moment. Those results showed that task formulation did not necessarily predict performance-stabilizing synergies, and the central nervous system might reconsider the task in terms of its own priorities with respect to stability of performance variables.

Note that in two-finger tasks, stabilization of both total force and moment is impossible because the UCM for F_{TOT} coincides with ORT for moment and vice versa

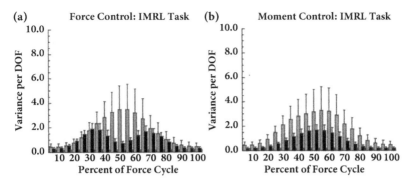

Figure 6.5 Two components of intercycle variance (V), along the uncontrolled manifold (UCM—gray bars) and orthogonal to the UCM (ORT—black bars), during the task of accurate total force production. (a): Computed for the total force. (b): Computed for the total moment of force in pronation-supination. Note $V_{UCM} > V_{ORT}$ over the whole force cycle for the moment of force, but not for total force. Modified by permission from Scholz et al. 2002.

(Fig. 6.6A). Studies of two-finger tasks always showed moment-stabilizing synergies despite the instruction and feedback that were formulated in terms of F_{TOT}. Three fingers can potentially stabilize both F_{TOT} and moment (as illustrated in Fig. 6.6B), but the experiments showed that adding a third finger made moment-stabilizing synergies stronger, while force-stabilizing synergies were still absent. And only the set of four fingers showed at least some stabilization of F_{TOT}, required by the explicit task.

Studies of force-stabilizing synergies during force production at different rates and over different magnitudes (Latash et al. 2002a; Friedman et al. 2009) have shown that V_{UCM} was strongly dependent on F_{TOT} magnitude but not on its rate, whereas V_{ORT} was strongly dependent on the rate of F_{TOT} production (as well as on F_{TOT} magnitude):

$$V_{UCM} = k_1 \bullet |F_{TOT}|$$

$$V_{ORT} = k_2 \bullet |dF_{TOT}/dt| + k_3 \bullet |F_{TOT}|, \quad (6.1)$$

where k_1, k_2, and k_3 are constants. Those results were interpreted within a model assuming that finger force magnitude and timing were defined by two coefficients set at a neural control level and characterized by mean values and dispersions (Goodman et al. 2005). Within the model, synergies were able to organize covariation of the magnitude-related coefficients to individual fingers stabilizing total force magnitude. There were, however, no signs of such covariation among the timing-related coefficients leading to the rate-dependent increase in V_{ORT}. This idea fits the general model by Schöner and colleagues (Martin et al. 2009, 2019) where the level of action timing is hierarchically higher than the level of synergies. An experimental study of force pulse production into a target that defined both peak force magnitude and timing provided additional support for this conclusion (Latash et al. 2004). This study documented peak F_{TOT} magnitude stabilization by covarying contributions of the individual finger

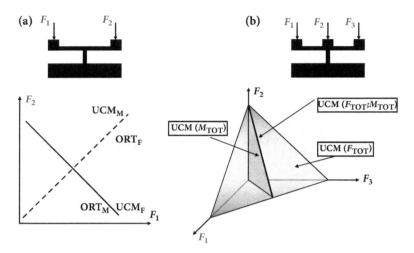

Figure 6.6 (a): The UCM (uncontrolled manifold) and ORT (orthogonal to the UCM) spaces for the total force (F_{TOT}) and total moment of force (M_{TOT}) produced by two fingers (see the insert). Note that stabilization of both F_{TOT} and M_{TOT} is impossible. (b): If the task is performed by three fingers, both F_{TOT} and M_{TOT} can be stabilized. The two UCMs are shown with triangles; the combined UCM is shown with the thick line.

force pulses, but no synergy stabilizing F_{TOT} peak timing by covarying adjustments of the timing of the individual finger force pulses. In other words, there seem to be magnitude-stabilizing synergies, but not time-stabilizing ones.

Studies of multi-finger actions confirmed the phenomenon of motor equivalence in experiments where the subjects were asked to produce accurate cyclical force changes (Mattos et al. 2015a). At some moment, one of the fingers was lifted smoothly (by 1 cm over 0.5 s) by the "inverse piano" (IP) device, leading to an increase in that finger's force. After a few seconds, that finger was lowered to its initial position. The subjects had visual feedback on the total force at all times and were able to keep total force deviations from the target to a minimum. The IP perturbations, however, led to major redistributions of total force across the four fingers, that is, to major ME motion with virtually no nME motion (see Fig. 5.11 in Chapter 5). These results were also confirmed in an experiment with accurate F_{TOT} and total moment (M_{TOT}) production (Mattos et al. 2015b) when the visual target on the screen jumped to a new {F_{TOT}; M_{TOT}} combination with visual feedback provided on only one of the two variables, followed by a jump to the initial {F_{TOT}; M_{TOT}} combination. Larger ME deviations in the space of commands to individual fingers were observed for the variable with visual feedback, while large nME deviations were seen for the variable without visual feedback (Fig. 6.7).

Another important result observed in studies of multi-finger pressing tasks was the competition among hierarchical levels in the corresponding force-stabilizing synergies. In particular, during two-finger steady force production by a hand, strong two-finger force-stabilizing synergies were observed. When two hands were

6.3. SYNERGIES IN KINEMATIC AND KINETIC SPACES 91

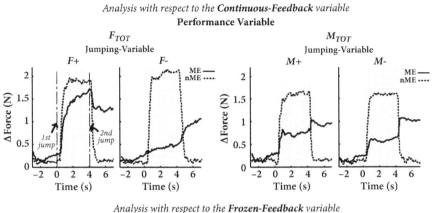

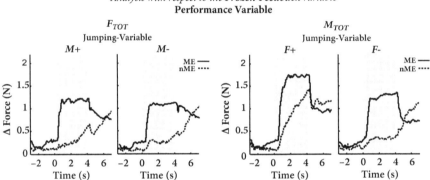

Figure 6.7 Two components of deviations in the finger mode space, motor equivalent (ME) and non–motor equivalent (nME), computed with respect to the total force (F_{TOT}) and total moment of force (M_{TOT}) during quick action to a new $\{F_{TOT}; M_{TOT}\}$ target ("first jump") and then back to the initial target ("second jump"). Top: The variable with visual feedback ("continuous feedback") showed ME > nME after returning back to the initial target. B: The variable without visual feedback ("frozen feedback") showed nME ≈ ME. Reproduced by permission from Mattos et al. 2015b.

involved in a similar four-finger task, two fingers per hand, two-hand synergies were seen, but there were no two-finger synergies stabilizing the contributions of the individual hands to the total force. The main result of this experiment is illustrated in Fig. 6.8. Imagine that there are two-hand force-stabilizing synergies (panel A1) characterized by typical hand force distributions elongated along the UCM. Note that individual hand force variance is defined by V_{UCM} and is proportionally large. At the lower level, with two fingers contributing to the hand force (panels A2 and A3), individual hand force variance is, by definition, V_{ORT}. Its large magnitude acts against force-stabilizing synergies at that level. Imagine now that individual hand forces are stabilized by the two-finger synergies at the lower level

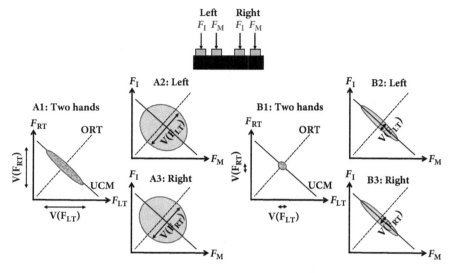

Figure 6.8 An illustration of the task of accurate force production with two hands, two fingers per hand, index and middle, I and M (top insert). The presence of a two-hand force-stabilizing synergy (A1) is reflected in large variance along the uncontrolled manifold (V_{UCM}), leading to large variance of each hand's force. At the level of fingers within individual hands (A2 and A3), hand force variance is, by definition, V_{ORT} acting against two-finger force-stabilizing synergies at that level. If individual hand forces are stabilized by the two-finger synergies (B2 and B3), V_{ORT} is small, resulting in small variance of the hand's forces. Such small hand force variances (B1) make it hard to organize a force-stabilizing synergy at the higher level.

(panels B2 and B3). Such synergies mean that V_{ORT} is relatively small, resulting in small variance of the individual hand forces. Such small hand force variances (panel B1) make it hard to organize a force-stabilizing synergy at the higher level because it may require the total force variance to be unrealistically small. We will consider implications of this trade-off for both intra-muscle synergies and prehension synergies in future chapters (Chapters 11 and 12).

The reviewed studies of kinematic and kinetic synergies share a major flaw: They use sets of elemental performance variables defined at the level of peripheral mechanics. As mentioned in earlier chapters (Chapters 1, 2, and 4), the central nervous system is in principle unable to prescribe peripheral mechanical variables, only parameters that lead to emergence of peripheral mechanics indirectly, depending on the external forces. Studies of force production in isometric conditions seem to avoid major changes in muscle length and, therefore, major changes in the contribution of the stretch reflex. This is, however, a questionable conclusion. Indeed, during force production in isometric conditions, there are changes in the activity level of gamma-motoneurons, which are expected to change the stretch reflex contribution to muscle activation. In addition, there is a redistribution between muscle length and tendon length. Another factor is changes in the activity levels of force-sensitive sensory endings, which may have strong effects on muscle activation levels (see reviews in Nichols 2002, 2018).

To explore possible covariation at the level of neural control variables, these variables have to be defined explicitly and methods of their quantitative estimation have to be developed. Earlier (Chapter 3), we suggested that the adequate variable for analysis of the neural control of a muscle is the threshold of the stretch reflex (λ). As discussed in Chapter 7, this general approach is applicable to multi-muscle systems, using the spatial referent coordinate (RC) for the effector. Although there are still no reliable "lambdameters" (see Chapter 24), several recent studies have designed methods to estimate the magnitudes of relevant neural commands under certain assumptions and explore synergies in those spaces (Latash and Gottlieb 1991; Ambike et al. 2016a). The results of those studies and their implications are described in the next section.

6.4. Synergies in spaces of control variables

Any natural task involves multiple muscles. Given that each muscle is controlled by at least one neural control variable (λ), any task is potentially abundant at the neural control level, even tasks that seem non-redundant at the level of mechanics. Consider, for example, accurate force production by a single effector along a coordinate X, for example, pressing with the tip of a finger. For simplicity, let us view all the contributing muscles as forming two groups, agonists (those contributing to the required force) and antagonists (those acting against the required force direction). These two muscle groups are controlled by two control variables, referent coordinates, RC_{AG} and RC_{ANT}, which define their force-coordinate characteristics as illustrated in Fig. 6.9. The algebraic sum of these characteristics defines the force-coordinate dependence for the fingertip, $F_X(X)$. It is illustrated in Fig. 6.9 as a straight line, which is a simplification assuming that the agonist and antagonist characteristics are perfectly symmetrical. While this is rarely the case, considering nonsymmetrical characteristics does not change the main message.

As described in Chapter 3, two control variables can be used to describe this system, the reciprocal command (R-command) and the coactivation command (C-command). At the level of mechanics, changing the R-command leads to shifts of the intercept of the effector $F_X(X)$ characteristic, that is, its referent coordinate, RC. Changing the C-command leads to changes in the slope of the $F_X(X)$ characteristic, that is, its apparent stiffness, k. If the coordinate of the fingertip is fixed—that is, it acts in isometric conditions (e.g., $X = 0$)—the fingertip force, in a linear approximation, is

$$F_{FT} = -k(C) \bullet RC(R) \qquad (6.2)$$

So, to produce a fixed magnitude of fingertip force, an infinite number of combinations of the R- and C-commands can be used, reflected in the infinite number of {RC; k} combinations. These solutions form the UCM for this task in the space of mechanical variables, RC and k, which are direct reflections of the two basic neural commands. The UCM is hyperbolic, which makes analysis of inter-trial variance

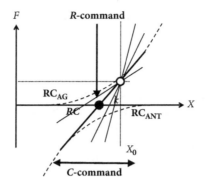

Figure 6.9 An illustration of the task of accurate force (F) production in isometric conditions (coordinate $X = X_0$) with an effector controlled by an agonist-antagonist muscle pair. The muscles are controlled by setting their referent coordinates (RC_{AG} and RC_{ANT}). The effector characteristic is shown with the thick slanted line. Its intercept (RC) reflects the reciprocal command (R-command), and the apparent stiffness (k) reflects the coactivation command (C-command). Note that multiple $\{RC; k\}$ combinations can be used to perform the task (their force-coordinate characteristics are shown with thin solid lines).

impossible. Therefore, different methods have been used to define and quantify force-stabilizing synergies in the $\{RC; k\}$ space.

First, multiple pairs $\{RC; k\}$ have to be collected in individual trials. This was done using the IP device (Martin et al. 2011) to lift the fingertip smoothly while recording its coordinate and force (Ambike et al. 2016a,b; Reschechtko and Latash 2017). The subjects were instructed and trained not to interfere voluntarily with the induced force changes. Then, linear regression was used on the F_X-X plane over the data collected during the lifting phase. The intercept and slope of the regression line were taken as estimates of RC and k in that trial. After multiple trials were collected, the most straightforward method was using hyperbolic regression across data points on the $\{RC; k\}$ plane collected in individual trials. Typically, such regressions resulted in very high coefficients of determination. However, when a similar method was used to explore the control of vertical posture (Nardon et al. 2022), the coefficients of determination were more modest (although still statistically significant), possibly due to the phenomenon of spontaneous postural deviations—postural sway (see Chapter 13).

An alternative method described in Chapter 5 used randomization of the magnitudes of elemental variables (RC and k) collected in different trials (cf. Müller and Sternad 2003). This procedure effectively removed possible inter-trial covariation between RC and k. Then, predicted force magnitudes were computed and their inter-trial variability was compared to the actual force variability observed in the experiment. This method also produced results pointing at strong synergies, that is, inter-trial co-variation of RC and k stabilizing force magnitude, reflected in much smaller inter-trial force variability in the actual data set compared to the surrogate one.

So far, the number of studies exploring synergies in spaces of control variables has been relatively modest. No studies explored such synergies in special populations (e.g., in the healthy elderly and neurological patients), which limits our current ability to draw conclusions regarding the practical utility of this method.

6.5. Possible neurophysiological mechanisms

Traditionally, the concept of synergies has been linked to such major brain structures as the cerebellum, basal ganglia, and cortex of large hemispheres. These views follow the traditions of classical neurology, in particular papers by Hughlings Jackson (1889) and Babinski (1899), and the description of the level of synergies by Bernstein (1947). Many earlier studies, however, used the word *synergy* without defining it or in a meaning different from the one accepted in this book. As of now, it seems clear that neurophysiological mechanisms of synergies are distributed among numerous structures and pathways within the central nervous system. Relative contributions of different neurophysiological mechanisms likely differ across tasks and levels of analysis.

Synergic signatures, such as large motor equivalent movements and stability of salient performance variables, have been described in animal preparations when only spinal circuits could contribute to the observed actions. For example, the wiping reflex in the spinal frog is associated with variable trajectories in the joint configuration space in conditions of loading the involved hindlimb and restrictions of joint motion (Berkinblit et al. 1986a; Latash 1993). The limb endpoint, however, was able to perform the wiping movement accurately under such conditions, reflecting its stability ensured by spinal circuitry. Intra-muscle synergies described earlier in this chapter and in Chapter 10 are also likely to be based on spinal circuits. On the other hand, a number of studies of multi-finger, multi-joint, and multi-muscle synergies in neurological patients have shown impairments of synergic control in patients with basal ganglia disorders (such as Parkinson disease), cerebellar disorders (e.g., olivopontocerebellar atrophy), and cortical disorders (e.g., stroke) (reviewed in Latash and Huang 2015; see Chapter 22). These studies suggest that subcortical supraspinal loops are important for the organization of such synergies.

The hierarchical scheme of control with RCs defined at different levels (Latash 2010a; Fig. 6.10) suggests that multiple synergies are possible, associated with the construction of a movement, stabilizing RC(t) functions at hierarchically higher levels by covaried adjustments of RC(t) at hierarchically lower levels. This scheme, however, is motivated largely by considerations of the body anatomy. For example, it is not obvious that joint-level control variables are stabilized during natural movements. In particular, studies of activity levels in spinocerebellar pathways during locomotion have shown correlations of those signals with global variables, such as leg length and leg orientation, but not with rotations in individual joints (Bosco and Poppele 2002).

A comprehensive scheme linking the basic steps in the production of a voluntary movement to brain structures has been developed by the group of Gregor Schöner (Martin et al. 2009). This scheme assumes the participation of various structures within the central nervous system in processes stabilizing salient performance variables including all the mentioned structures but with specific roles assigned to

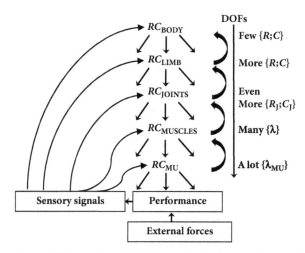

Figure 6.10 A schematic illustration of the hierarchical scheme of control with spatial referent coordinates (RCs) defined at various levels, from the whole body to individual muscles (where RC = λ), and motor units (MUs). Note the increase in the number of degrees of freedom (DOFs) for lower levels of the hierarchy compared to its higher levels. R and C stand for the reciprocal and coactivation commands.

individual structures. Within this model, such processes as task representation, selection of targets, and assignment of movement parameters related to its timing and magnitude involve supraspinal structures. Synergic processes are assumed to be based primarily on spinal circuitry with possible involvement of the cerebellum and cortex. This model includes unidirectional effects from the level of timing to hierarchically lower levels defining movement stability. The scheme was modified and expanded in a later paper (Martin et al. 2019), which also allowed back-coupling effects from proprioceptive signals to the level labeled as "movement preparation," which potentially allows adjustments of movement timing.

6.6. Missing pieces of the mosaic

The concept of synergy is still in the process of discussion and being used in different meanings across studies. Most studies still emphasize only one feature of synergies, that is, grouping of elements, while ignoring the other feature—stabilization of salient performance variables. Moreover, they accept with little doubt that their specific findings can be generalized across tasks and even populations. In a sense, these studies perform the first step in synergy analysis—the identification of elemental variables—and stop short of using the findings to explore stability of action.

Another major missing piece is information about neurophysiological circuits involved in both features of synergies. The current conclusions are based primarily on observations in various patient populations (reviewed in Latash and Huang 2015; Vaz et al. 2019). Since synergies are, by definition, task and intention specific, it is hard to explore them in animals. So far, there have been only a few studies of both features of synergies in animals (Klishko et al. 2014; Mangalam et al. 2018). As described in more detail later, spinal circuitry is likely to contribute to synergies at the level of motor units, whereas subcortical loops involving the cerebellum, basal ganglia, and thalamus are likely to be crucial for multi-effector (multi-muscle) synergies. The latter conclusion is very much in line with Bernstein's association of the level of synergies with such structures as the globus pallidus and thalamus. The role of the cortex remains unclear: Some studies described no detectable changes in synergy indices after mild cortical stroke (Reisman and Scholz 2003; Jo et al. 2016b), but impaired synergies have been reported in other studies (Gera et al. 2016a, 2016b).

Another unclear issue is the apparent trade-off between synergies at different hierarchical levels described earlier. Since transformation of control signals at the task level to recruitment of motor units typically involves multiple steps (see more details in Chapter 7), it is unclear at what levels synergies are maintained at the expense of not having them at other levels. This issue may be directly related to the issue of involved neurophysiological mechanisms.

We will discuss a few related issues in future chapters. These will include the concept of turning stability off when a previously stabilized variable has to be changed quickly, changes in synergies with practice and aging, and the highly nontrivial relations between the stability and optimality of movements.

7
Control With Spatial Referent Coordinates

All functional movements involve multiple muscles, and according to the equilibrium-point hypothesis (EP-hypothesis; Feldman 1966; see Chapter 3), the neural control of a muscle can be described with a neural variable corresponding to the threshold (λ) of its stretch reflex. Of course, muscles are themselves complex structures consisting of numerous motor units (MUs)—the smallest controllable output element of the central nervous system—which are likely to be organized into stable groups addressed as MU-modes (Chapters 6 and 11). Theoretically, the neural controller could act in an extremely high-dimensional space of λs to individual MUs, in a very high-dimensional space of λs to MU-modes, or in a high-dimensional space of λs to individual muscles. However, all these options look very much unlikely. More likely, the highest level of neural control is action specific and not very high-dimensional, although typically abundant (Chapter 5; cf. Bernstein 1947).

The idea that the brain contains representations of movement elements, from individual muscles to joints, to limbs, and to the whole body, dates back to classical works by Hughlings Jackson (1889). Which of those representations to use for purposes of control is at the discretion of the neural controller, and likely, this problem does not have a universal solution. All solutions are task specific.

The basic concept of indirect parametric control of peripheral variables, including muscle activations and movement mechanics, can be naturally developed for both intra-muscle and multi-muscle systems. By its very nature, λ is a spatial parameter that defines the dependence of muscle active force on muscle length. It may be viewed as a spatial referent coordinate for the muscle (i.e., a coordinate to which the muscle moves in the absence of external load). It seems intuitively natural to search for control parameters for intra-muscle and multi-muscle systems that would also have the meaning of spatial referent coordinates. A number of studies provided evidence for spring-like behavior of multi-muscle systems, from the endpoint of a limb to the whole body (e.g., Flash 1987; Domen et al. 1999), suggesting that the spatial origin of those experimentally observed force-coordinate characteristics can represent a control parameter specified by the central nervous system.

Testing this hypothesis and exploring possible neurophysiological mechanisms involved in the specification of referent coordinates for systems of different complexity has turned into a research field of its own (reviewed in Feldman 2015; Latash 2019). It led to the development of new methods to analyze synergies at the level of neural control variables and clinical applications. This approach, however, is not broadly accepted by the research community, likely for the same reasons as the resistance to accepting the EP-hypothesis (discussed in Chapter 3).

7.1. Referent coordinate as generalization of lambda

The spatial referent coordinate (RC), by definition, is a coordinate to which the system is attracted following the removal of external forces. RCs are rarely attained by natural systems because of the ever-present field of gravity and other common forces, including those originating within the body. For example, muscle length rarely stays at λ because of the non-zero forces of antagonist muscles. Of course, during movements, muscles commonly move through coordinates corresponding to the current values of λ but do not stop there because of a number of factors, such as inertia of the moving effectors. This fact was used to test one of the predictions of the theory of control of spatial RCs that muscle activation would reach a global minimum during movements involving large groups of muscles to a target and back, because, during some phases of such movement, muscle length and λ are both time functions changing in opposite directions and meeting at some coordinate (Feldman et al. 1998).

Generalizing the idea of control with thresholds of the stretch reflex to a joint controlled by a pair of opposing muscles is relatively straightforward. Consider Fig. 7.1, showing the moment-angle characteristics for an agonist muscle (producing positive moment values) and antagonist muscle (producing negative moment values). This system can be described with two λs converted into angular units, λ_{AG} and λ_{ANT}. Mechanical behavior of the joint is defined by the algebraic sum of the two muscle characteristics. Assuming nearly symmetrical characteristics (not a very realistic assumption!), the joint characteristic is linear within the spatial range where both muscles show non-zero activation levels and, correspondingly, non-zero active moment values. The joint characteristic can be described with its intercept with the angle axis—its RC (i.e., the joint angle where the system would rest in the absence of external load) and slope (k), which can be viewed as apparent stiffness of the joint.

Two basic commands have been introduced to describe the control of movements at the joint level, the reciprocal command (R-command) and the coactivation

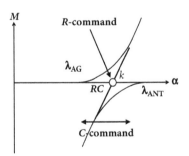

Figure 7.1 The control of a joint with one kinematic degree of freedom. The agonist and antagonist muscles are controlled by setting their stretch reflex thresholds, λ_{AG} and λ_{ANT}. The joint moment-angle (M-α) characteristic (thick line) can be described with the intercept (referent coordinate, RC) and slope (k)—apparent stiffness of the joint. These reflect two basic commands, reciprocal (R-command) and coactivation (C-command).

command (C-command). A change in the R-command produces unidirectional shifts of both λ_{AG} and λ_{ANT}, resulting in a spatial shift of the RC (Fig. 7.2A). A change in the C-command produces counterdirectional shifts of λ_{AG} and λ_{ANT}, leading to a change in the range where both muscles show non-zero activation simultaneously; changing the C-command causes changes in the slope of the moment-angle joint characteristic, k (Fig. 7.2B). Note that changing R- and C-commands does not prescribe joint moment of force or angle. Actual changes in joint angle and moment will depend on the external load characteristic. The same change in the two basic commands can lead to movement in isotonic conditions, moment change in isometric conditions, and changes in both angle and moment in more realistic conditions when the external load is a function of joint angle.

The two pairs of commands $\{\lambda_{AG}; \lambda_{ANT}\}$ and $\{R; C\}$ are equivalent to describe the control of the cartoon joint illustrated in Fig. 7.1. This is not so if one considers more realistic joints crossed by multiple muscles for each rotational degree of freedom. For example, the elbow joint with one kinematic degree of freedom is crossed by three flexors and three extensors. In such situations, the number of commands at the $\{R; C\}$ level remains two, while the number of commands at the muscle level $\{\lambda_1, \lambda_2, ..., \lambda_N\}$ is larger, and the mapping from joint-specific commands to muscle-specific ones becomes abundant (Latash 2012a; Chapter 5).

If one considers the neural control of the endpoint of a multi-joint effector, each of the spatial coordinates of the endpoint may be associated with an $\{R; C\}$ pair based potentially on a large number of muscles that can be classified as agonists (if they produce active force in the desired direction) or antagonists (those producing force in the opposite direction). The description of the control along each coordinate is similar to the one illustrated in Fig. 7.1, and the mapping from the endpoint-related $\{R; C\}$ level to muscle-level control variables may be highly abundant (few-to-many). The same logic applies to the control of whole-body movements.

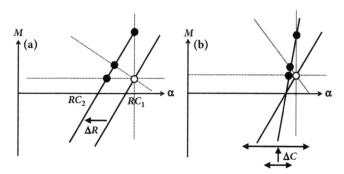

Figure 7.2 Effects of changes in two basic commands to a joint with one kinematic degree of freedom, the reciprocal command (R-command) and the coactivation command (C-command). Changes in the two commands lead to different mechanical consequences depending on the load characteristic (three load characteristics are shown with the dashed lines; the initial load is assumed non-zero). (a) Effects of a shift in the R-command (ΔR). (b) Effects of a shift in the C-command (ΔC).

7.2. Hierarchical control with referent coordinates

The last figure in the previous chapter (Fig. 6.10) illustrates a possible hierarchy of RC commands at various anatomical levels, from the task level that can involve the whole body to limb, joint, muscle, and MU levels. The dimensionality of the control space increases in the top-to-bottom direction. Back-coupling loops, both within the central nervous system and from peripheral sensory endings, have been postulated as the means of stabilizing behaviors specified at higher levels by covaried adjustments of RC commands at the lower levels (Latash et al. 2005; Martin et al. 2009, 2019).

One of the models involving back-coupling (Latash et al. 2005) was motivated by the following observation from the inanimate nature. Imagine that you pour water at a constant rate into an old rusty bucket with holes in the bottom. There are also pieces of garbage on the bottom, which partly block some of the holes. At some level of water, the pressure-related outflow will be exactly equal to the inflow, and the system will be in a dynamic equilibrium. Imagine now that one of the holes gets blocked by a random piece of garbage. Its contribution to the water outflow will become zero, the level of water will start to increase, and, as a result, the outflow from other holes will increase. This will compensate for the lost contribution of the blocked hole to total water outflow and a new equilibrium state will be achieved. This example shows the feature of error compensation (cf. Jaric and Latash 1998; Latash et al. 1998) or dynamical stability (cf. Scholz and Schöner 1999) without any controller. It functions based on the laws of conservation of matter and hydrostatics.

Such laws are in general only metaphorical if one is interested in the functioning of biological systems. In the model, their action was replaced with explicit short-latency feedback loops similar to the system of recurrent inhibition mediated by Renshaw cells (Fig. 7.3). In the model, the gains to individual alpha-motoneurons of the pool could be modified by the central controller (including a change in their sign!). Using different gain matrices allowed simulating multi-finger force-stabilizing and moment-stabilizing synergies.

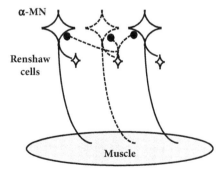

Figure 7.3 The system of recurrent inhibition stabilizes the output of the alpha-motoneuronal pool. If one neuron (dashed lines) stops contributing to muscle activation, other neurons of the pool will receive less inhibition and compensate partly for the drop in the output of the pool.

Experimental exploration of the hierarchy shown in Fig. 6.10 has not been easy, primarily because time changes in RCs are not easily measurable in real time. This topic is described in more detail in Chapter 24. The situation is less problematic if only steady states of the effector are considered. Indeed, as can be seen from Fig. 7.2, at the level of mechanics, the R-command modifies RC, and the C-command modifies the slope of the force-coordinate dependence, k. Hence, if a smooth external force change is applied to the effector, measuring the ensuing coordinate (X) and force (F) changes is expected to produce a close-to-linear dependence between these two variables. If this is true, the intercept of the regression can be viewed as a proxy of the R-command, and its slope as a mechanical reflection (k) of the C-command (described in more detail in Ambike et al. 2016a; Nardon et al. 2022).

This method is based on two important assumptions. First, it is expected that the subject in this experiment does not change the R- and C-commands during the process of measurement, that is, that the values of force and coordinate belong to a single dependence $F(X)$. This is a strong assumption, which has been under debate for many years. A typical strategy is to instruct the subjects not to interfere with possible changes in the effector position and force produced by the external perturbation, that is, not to resist effector movement and not to relax. After some training, subjects learn how "not to react" and show consistent behavior, more consistent than when they are instructed to react to the perturbation (Latash 1994). Second, if the perturbation is too fast, unintentional phasic reflex and reflex-like responses can be seen in the stretched muscles distorting the $F(X)$ dependence. So, parameters of the perturbation have to be selected wisely, not too brief to avoid transient reflex-like reactions, and not too long to minimize chances of corrections introduced by the subject. More on these issues can be found in Chapter 24.

Figure 6.10 implies possible synergies at multiple levels of the illustrated hierarchy. Note, however, that this figure has been motivated to a large degree by the anatomy of the human body. In addition, as described in Chapter 6, synergies at different levels within a hierarchy are in competition (see also Gorniak et al. 2007, 2009). This makes it hardly possible to organize synergies at all the levels in Fig. 6.10. Are there levels where synergies are more likely to be seen and those that are sacrificed? This question remains open. Numerous studies have documented synergies stabilizing task-related salient performance variables analyzed in various spaces (reviewed in Latash et al. 2007, 2021b). These synergies have been associated with functioning of subcortical loops involving the basal ganglia and cerebellum (reviewed in Latash and Huang 2015). On the other hand, a series of more recent studies provide evidence for intramuscle synergies (Chapters 6 and 10) that are likely based on spinal circuitry (Madarshahian et al. 2021, 2022; reviewed in Latash et al. 2023). At this time, it is unclear whether synergies stabilizing intermediate variables implied by the hierarchy in Fig. 6.10 exist.

7.3. Synergies in spaces of referent coordinates: Analysis of mechanics

Synergies in the space of the neural commands can be defined as covaried adjustments in the mechanical reflections of the R-command (referent coordinate, RC)

7.3. ANALYSIS OF MECHANICS   103

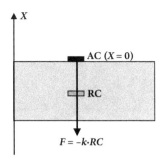

Figure 7.4 The task of producing a certain force magnitude (F) by an effector at a coordinate ($X = 0$) can be performed using an infinite number of force-coordinate characteristics corresponding to different combinations of the referent coordinate (RC) and apparent stiffness (k) values.

and C-command (apparent stiffness, k) stabilizing the salient performance variable. Such synergies have been studied primarily during isometric force production tasks, which allow applying smooth positional perturbations to the effector and measuring changes in both the force and coordinate of the effector. During isometric force production against a fixed surface, RC of the effector is under the contact surface (Fig. 7.4; see Pilon et al. 2007). The difference between the actual effector coordinate (set at zero in Fig. 7.4) and RC results in active force production in the direction of RC. Force magnitude, in a linear approximation, can be computed as $F = -k \cdot RC$. Note that the mechanical effects of the C-command, which is measured in spatial units, are reflected in k, which is measured in units of apparent stiffness, N/m.

Imagine that a person is asked to produce a certain force level with the help of visual feedback and "not to react" to possible changes in the state of the effector (e.g., a finger or a hand). Potentially, this task can be performed using an infinite number of combinations of the R- and C-commands, resulting in different $F(X)$ characteristics of the effector. Three of those are illustrated in Fig. 7.5. A smooth external change in the coordinate of the effector is produced by a programmable linear motor (as in the "inverse piano" device; Martin et al. 2011, Ambike et al. 2016a) lifting the effector away from the RC. This results in a smooth increase in force in parallel with the induced change in the coordinate following the $F(X)$ characteristic. Using linear regression allows identifying the intercept (RC) and slope (k) of the characteristic in that particular trial. Repeating the task multiple times produces a set of $\{RC; k\}$ data points.

These points are expected to be always close to the solution space (uncontrolled manifold [UCM]) for the task, $RC = F_0/k$, where F_0 is the prescribed force level. Their deviations from the UCM are expected to be small, reflecting unavoidable variability of performance, while deviations along the UCM are not constrained by the task: They can be of about the same magnitude or smaller or larger than the deviations orthogonal to the UCM. As illustrated in Fig. 7.6, acceptable performance (within the same error margin) may be associated with a fixed value of RC and varied k, a fixed value of k and varied RC, RC vs. k covariation along the UCM, or RC vs. k covariation orthogonal to the UCM. Experiments have shown large inter-trial variation of both

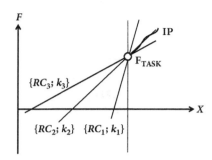

Figure 7.5 Applying a smooth positional perturbation to an effector involved in an accurate force production task (F_{TASK}) is expected to follow the force-coordinate, $F(X)$, characteristic, under the assumption of "non-interference" (inverse piano [IP]). Linear regression of this characteristic (black curve) allows reconstructing the referent coordinate and apparent stiffness, $\{RC, k\}$, combination.

RC and k, mostly confined to the UCM, with very high values of the coefficient of determination computed for the hyperbolic regressions (Ambike et al. 2016a, 2016b; Reschechtko and Latash 2017, 2018).

Note that using the traditional method of analysis associated with the UCM hypothesis (Chapter 5) (i.e., analysis of the inter-trial variance) is inapplicable to the data illustrated in Fig. 7.6, because the hyperbolic UCM cannot be linearized. So, two different metrics have been used. One of them is the aforementioned coefficient of determination for hyperbolic regression analysis. The other method uses the randomization method producing covariation-free surrogate data sets with the same means and standard deviations as in the original data set (see Kudo et al. 2000; Müller and Sternad 2003). This is done by selecting elemental variables, RC and k, from different trials and computing the expected force values. If a synergy in the $\{RC; k\}$ space helps to reduce the intertrial force variability, the surrogate data set is expected to show

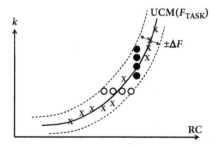

Figure 7.6 Accurate force production within an acceptable error margin is compatible with variable patterns of variation of the referent coordinate and apparent stiffness, $\{RC, k\}$. Only RC can vary (open circles), only k can vary (filled circles), or both can vary with covariation (crosses). The hyperbolic line reflects the uncontrolled manifold (UCM) in the $\{RC, k\}$ space.

significantly higher force variability, for example, SD(F_{SUR}), compared to the actual data set. Both methods have been used and led to similar results; that is, both provided evidence for strong synergies at the control level stabilizing force magnitude.

A similar method has been applied in studies of the neural control of vertical posture with respect to the position of the center of mass (COM) of the standing person (Nardon et al. 2022). That study produced very high R^2 values for the $RC(k)$ hyperbolic regressions in some subjects and more modest values for other subjects. This could partly be related to the phenomenon of postural sway, that is, rather large spontaneous deviations of the body during quiet standing (Chapter 13). Additional analysis has shown that lower R^2 values were associated with higher mean k values across trials. In other words, subjects who used larger C-commands (higher muscle coactivation) showed worse synergies in the space of neural commands stabilizing COM coordinates. These observations support a more general view that increased coactivation is not always good for stability of salient variables (see Yamagata et al. 2019a, 2021; Chapter 19).

Similar results have been obtained in studies of motor equivalence (see Chapter 18), providing evidence for larger deviations along the UCM compared to deviations orthogonal to the UCM between two steady states separated by a voluntary action. In those studies, RC and k were quantified before and after a quick voluntary whole-body sway to a target and back to the initial coordinate. The data points across trials before and after the voluntary sway belonged to about the same hyperbolic regression, but pairs of data points within a single trial deviated from each other primarily along the UCM (ME > nME; cf. Mattos et al. 2011, 2013, 2015a, 2015b; see Chapters 5 and 18).

7.4. Synergies in spaces of referent coordinates: Analysis of muscle activations

A different approach to quantitative analysis of the referent coordinates to muscles and muscle pairs has been developed based on studies of MU recruitment during force production tasks (see Chapters 6 and 11). This method views individual muscles as complex structures consisting of MU groups (MU-modes) with parallel scaling of the frequencies of firing of the individual MUs (f_{MU}) comprising the MU-mode. The control of individual MU-modes is based on specifying their RCs, and the control of a muscle as specifying its λ. The force-length characteristic of a muscle is the algebraic sum of the MU-mode characteristics. This scheme is abundant at the level of control of a single muscle and is characterized by intramuscle synergies in the space of RCs ($λ_{MU-mode}$) to individual MU-modes stabilizing muscle force in isometric conditions. Such synergies were seen in the MU-mode spaces defined for both agonist (flexor) and antagonist (extensor) muscle separately in spite of the fact that the antagonist produced force against the required direction (Madarshahian and Latash 2022a; Madarshahian et al. 2022). This counterintuitive fact reflects the nontrivial role of the C-command associated with parallel changes in the activation of both agonist and antagonist. As illustrated in Fig. 7.7, an increase in the C-command can lead to a net increase in the resultant force in spite of an increase in the force produced by the antagonist muscle due to the nonlinearity of the force-length muscle characteristics.

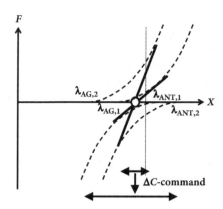

Figure 7.7 An increase in the coactivation command (*C*-command) during force production in isometric conditions leads to an increase in both agonist and antagonist force with an increase in the net force magnitude. ΔC is a change in the C-command; the stretch reflex threshold for the agonist and antagonist muscles is shown as λ_{AG} and λ_{ANT}.

When MUs from both agonist and antagonist muscles were pooled together, two MU-modes were consistently seen (Madarshahian and Latash 2022a). One of them had the same sign of the loading factors for the MUs from both muscles and, therefore, was associated with the *C*-command. The other one had opposite signs of the loading factors for MUs from the flexor and from the extensor and was associated with the *R*-command (Fig. 7.8). Indeed, an increase in the former MU-mode (MU_C-mode) corresponded to an increase in the firing frequencies of MUs from both agonist and antagonist muscles (i.e., coactivation). An increase in the latter mode (MU_R-mode) corresponded to higher firing frequencies of MUs in one of the two muscles and lower frequencies of MUs in the other one (i.e., reciprocal activation).

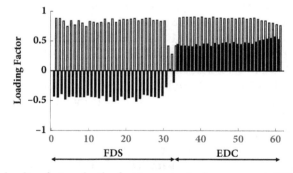

Figure 7.8 The loading factors for the first two principal components (PCs) in the space of firing frequencies of motor units pooled over the agonist (flexor digitorum superficialis [FDS]) and antagonist (extensor digitorum communis [EDC]) muscles. One of the PCs shows loading factors of the same sign reflecting the coactivation command (light bars). The other PC shows loading factors of opposite signs reflecting the reciprocal command (dark bars). Reproduced by permission from Madarshahian and Latash (2022b).

7.4. ANALYSIS OF MUSCLE ACTIVATIONS 107

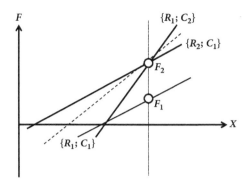

Figure 7.9 Finger force in isometric conditions can be changed (from F_1 to F_2) by a change in the reciprocal command (R-command) from R_1 to R_2, in the coactivation command (C-command) from C_1 to C_2, or both (thin dashed line).

If a person is instructed to change the net force from a certain value F_1 to another value F_2, this task can be accomplished by a change in the R-command, a change in the C-command, or a coordinated change in both (Fig. 7.9). Experiments have shown, however, that the MU_C-mode was modulated within the force cycle consistently across subjects. However, the MU_R-mode showed inconsistent modulation, which, on average, showed no change within the force cycle. This may be viewed as a rather unexpected result, since the R-command has been considered as hierarchically higher than the C-command (Levin and Dimov 1997; Feldman 2015). For example, during voluntary joint movement, the R-command defines the new desired position, and the coactivation level (C-command) is transferred to that new position.

Analysis of MU-mode covariation defined in the combined space of MUs from both muscles showed no force-stabilizing synergies. This may be related to the aforementioned trade-off between synergies at different hierarchical levels (see Chapter 6). However, the result looks counterintuitive: The task of accurate resultant force production is more directly related to the neural control at the effector level, not at the control at the individual muscle level. Nevertheless, force-stabilizing synergies are seen within spaces of MU-modes defined for each muscle separately, but not for those defined for the agonist-antagonist pair. The reason may be that intra-muscle synergies are reflecting spinal reflex circuitry, which is acting across tasks and leading to stability of action by individual muscles. While there are agonist-antagonist reflex effects, in particular the well-known system of reciprocal inhibition, those may be dominated by the autogenic reflex circuits. If single-muscle contributions to force are stabilized, the hierarchically higher agonist-antagonist pairs may be unable to show such synergies. A more recent study (Benamati et al. 2024) has shown force-stabilizing synergies at both intra-muscle and agonist-antagonist levels, although, even in that study, the synergy index for the analysis based on MUs from both muscles was smaller than the synergy index for the analysis based on each set of MUs from each muscle taken separately.

There is, of course, a potentially important open question: Why does the analysis in the space of mechanical variables show strong force-stabilizing synergies and the analysis in the space of MU-related variables (MU-modes) fails to show such synergies? It is possible that the mechanical analysis is addressing the issue at the effector level and is a more direct reflection of covariation between the R- and C-commands that are later projected on a relatively large number of muscles. In contrast, analysis of only a subset of two muscles contributing to the effector force is performed at a lower level, that of MUs within individual muscles, even if the MUs are pooled across a muscle pair. So, the former analysis is immune to the potential trade-off effects inherent to hierarchical systems, while the latter analysis may suffer from such effects.

7.5. Synergies stabilizing referent coordinates

Consider a task with several effectors contributing to a task-specific variable—a very common situation during natural movements. The control of such a task is illustrated schematically in Fig. 7.10 using, as an example, constant force production by the four fingers of a hand. At the highest control level (Task level), the neural control can be described as setting two variables, R_{HAND} and C_{HAND}. These variables are used to generate pairs of control variables to each of the fingers: $\{R_I, C_I\}$, $\{R_M, C_M\}$, $\{R_R, C_R\}$, and $\{R_L, C_L\}$, where the subscripts refer to the index, middle, ring, and little finger, respectively. In isometric conditions, each of the four pairs results in a certain force magnitude produced by each finger. These forces sum up to produce the hand force.

Traditional analysis of synergies is organized from a higher-dimensional level at the level of elemental variables to performance variables (reviewed in Latash et al. 2002d, 2007). For example, one can ask questions with respect to four-finger synergies in the space of finger forces stabilizing total force, with respect to two-dimensional $\{R_{HAND}; C_{HAND}\}$ covariation stabilizing total force, with respect to individual finger $\{R; C\}$ pairs covarying to stabilize that finger's force, or even with respect to covariation within the eight-dimensional space of $\{R; C\}$ pairs to individual fingers stabilizing total force. But

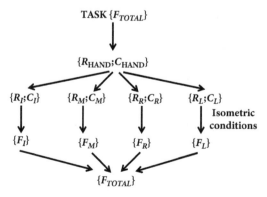

Figure 7.10 The hierarchical control of the hand with the reciprocal and coactivation commands $\{R; C\}$ to the individual fingers (I, index; M, middle; R, ring; L, little) and to the hand, $\{R_{HAND}; C_{HAND}\}$.

one can also ask a question in the "opposite direction": Do the individual {$R; C$} pairs to individual fingers covary to stabilize the hand level {R_{HAND}; C_{HAND}}? The former types of analysis have been referred to as "descending synergies," and the latter as "ascending synergies" (Reschechtko and Latash 2018).

All the examples presented in this chapter so far belong to the group of descending synergies. There has been a single study exploring potential ascending synergies during four-finger accurate force production. The mapping between the eight-dimensional level of commands to fingers to the two-dimensional level of commands to the hand is trivial for the apparent stiffness variable k reflecting the action of the C-command. Indeed, for four linear springs acting in parallel, total stiffness $k_{HAND} = \Sigma k_i$, where i stands for the individual fingers, and the Jacobian is [1 1 1 1]. This is not so simple for the RC variables, and there is no simple linear model mapping changes in the RC_i to changes in RC_{HAND}. This mapping had to be discovered using multiple linear regression methods similar to those used in mapping of integrated muscle activation indices on mechanical variables (see Chapters 6 and 14).

The results of this analysis were somewhat unexpected. There were moderate, not very strong, synergies stabilizing RC_{HAND} by covaried changes in the finger-specific RC_i. There were, however, no such synergies in the space of finger-specific k_i stabilizing k_{HAND}. The latter finding could result from the organization of muscles involved in finger force production, specifically from the action of multi-finger extrinsic hand muscles, which produced positively covarying magnitudes of k_i to individual fingers. Given that k_i summed up to produce k_{HAND}, such positive covariation acted against possible synergies stabilizing k_{HAND}.

7.6. Missing pieces of the mosaic

Much of the information presented in this chapter is obviously incomplete, leading to quite a few guesses rather than firm conclusions. It remains unclear whether the hierarchical control of action can be associated with synergies at multiple levels and, if so, which levels get priority given the described trade-off between synergies in a hierarchy. It seems safe to conclude that intra-muscle synergies are common across muscles and based primarily on the spinal circuitry including reflexes from peripheral sensory endings. Strong stabilization of action by individual muscles leaves little room for synergies at the next level related to the control of muscle pairs because of the overall low variance, which naturally limits V_{UCM}. However, most salient mechanical variables get contributions from multiple muscles, which may lead to enough variance in such multi-muscle systems to allow for the signature inequality $V_{UCM} > V_{ORT}$. It is also possible that, depending on the task, gains in the spinal circuits may be modulated by descending pathways, leading to more or less variance at the muscle level. For example, a study of toy basketball throwing showed large indices of multi-joint synergies stabilizing ball-in-hand speed and direction over most of the movement (Hasanbarani and Latash 2020). However, just prior to the ball release, both synergy indices dropped, resulting in higher variance of the ball speed and angle of release, which were necessary to organize covariation of these two variables beneficial for throw accuracy.

On the other hand, synergies stabilizing task-specific salient performance variables have been confirmed across tasks and effector sets. They seem to be crucially dependent on the most salient source of sensory information, typically vision (see Chapter 16), and on intact functioning of subcortical loops involving the cerebellum and basal ganglia (see Chapter 22). Are there synergies ensuring stability of intermediate variables produced by effectors contributing to the task and involving multiple muscles? If so, what neural circuits and sensory modalities are likely to contribute to such synergies? These questions remain open.

Another unclear issue is the obvious contrast between strong synergies among control variables estimated at the level of mechanics, $\{RC; k\}$, and lack of such synergies estimated at the level of MU firing frequency. The explanation offered earlier is clearly ad hoc. To look for a better answer one has to apply both methods of synergy estimation in the same study. This was done in a recent study (Benamati et al. 2024), which confirmed the existence of strong force-stabilizing synergies in the $\{RC; k\}$ space and in MU-mode spaces defined for the agonist and antagonist muscle separately. Uniting MUs across the two muscles led to a significant drop in the synergy index (short of its complete disappearance!). Finger force production involves multiple muscles, and the analysis of only two of them might be misleading. Indeed, as shown earlier (Scholz and Schöner 2014), analysis within a subset of elements contributing to a potentially important performance variable can lead to spurious results. For finger force production, the mapping $\{R; C\} \approx \{RC; k\} => \{\lambda_i\}$ involves more than two λ values ($i > 2$). The low synergy indices might result from ignoring the contributions of other muscles. Definitely, this is an area begging for more research.

8
Anticipatory Control of Action

It is well known that animals frequently behave with respect to future, expected goals. For example, a predator tries to intercept the prey, while the prey tries to avoid the predator by predicting its future movements. Anticipatory actions are common in everyday human life and in athletic movements. This ability of biological systems was emphasized by Bernstein (1947) and his student, Josef Feigenberg (1998), who introduced the concept of *desired future* as the driving force of voluntary actions. Feigenberg developed this concept using the idea of probabilistic prognosis and performing series of ingenious experiments (reviewed in Feigenberg 1969), which were close in spirit to the actively developed field of action planning based on Bayesian statistics in recent years (reviewed in Körding and Wolpert 2006; Heald et al. 2023).

The concept of anticipatory control is also close to the idea of feedforward control within the control theory, when control variables are set by the controller before their effects on behavior become known. Note that the equilibrium-point hypothesis and its generalization to the control of multimuscle systems with spatial referent coordinates (see Chapters 3 and 7) are based on feedforward specification of neural variables to muscles and effectors (λ and reference coordinates [RCs]) and, in this sense, formally represent examples of feedforward control. Further, feedback loops from sensory endings participate in the execution of the neural commands and their transformation into muscle activation and mechanical variables. However, at the time neural control variables are generated by the central nervous system, their effects on performance remain unknown because performance depends equally on the unpredictable or, at the very least, imperfectly predictable changes in external forces.

In this chapter, we discuss anticipatory components of action that are not explicitly specified by the task goals but rather optimize conditions for planned goal-related actions and/or minimize their destabilizing effects on the state of the actor. These anticipatory adjustments may be nonobligatory, and under some conditions, similar actions can be performed in the absence of such adjustments. Taken together, these observations raise questions related to purposes of those adjustments, their relations to the explicit goals of planned actions, and maybe other factors such as, for example, predictability of goals and external conditions of movement execution.

Anticipatory adjustments to action have shown high sensitivity to the state of the actor. In particular, they show modulation with the natural process of aging, neurological disorders, and treatment (reviewed in Nadin 2015; Delafontaine et al. 2019; Vaz et al. 2019). They are also sensitive to perceived difficulty of the task and time available to initiate the action. There has been a degree of confusion about the terminology used to describe different types of anticipatory adjustments, which is not an uncommon situation in the field; see, for example, the different definitions of *synergy* (Chapter 6). Clarifying the situation, which is absolutely necessary for productive development of the field in the subjective view of the author, is a major purpose of this chapter.

Although in most studies anticipatory adjustments are quantified using muscle activation and/or mechanical variables, we will always keep in mind that the central nervous system is in principle unable to prescribe those variables. As is the case for voluntary movements and their components, adequate language for the description and analysis of anticipatory adjustments is based on time profiles of neural control variables, which we continue to associate with spatial referent coordinates for the effectors. These variables, however, are not easily measured in real time (see Chapter 24). Anticipatory adjustments in muscle activation levels seen in predictable conditions may be seen as relatively direct reflections of changes in the control-level variables because they are associated with minimal changes in kinetic and kinematic variables, which modify the contribution of reflexes from proprioceptors to muscle activation patterns.

8.1. Anticipatory postural adjustments

When a standing person performs a fast arm movement (e.g., a quick bilateral arm extension), the first signs of changes in the background muscle activation level are seen not in the prime movers for the planned action but in muscles of the legs and trunk (Fig. 8.1). These early muscle activation changes are accompanied by shifts in mechanical variables such as the coordinate of application of the resultant vertical ground reaction force, the center of pressure (COP). Such anticipatory postural adjustments (APAs) have been described in many studies, starting with the pioneering publication by the group of Victor Gurfinkel (Belen'kii et al. 1967). APAs can be seen as early as 150 ms prior to the first signs of muscle activation in the prime mover with respect to the intended action. Given typical conduction and electromechanical delays (Corcos et al. 1992), any changes in the baseline muscle activation up

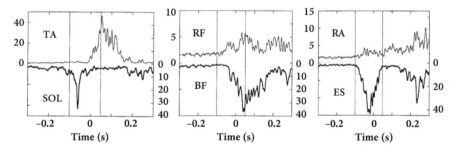

Figure 8.1 An illustration of typical muscle activation patterns in the lower body associated with a fast, bilateral arm movement forward by a standing person. The arm movement initiation time is at $t = 0$. The thin vertical lines show a typical time range associated with anticipatory postural adjustments. TA, tibialis anterior; SOL, soleus; RF, rectus femoris; BF, biceps femoris; RA, rectus abdominis; ES, erector spinae. Modified by permission from Slijper and Latash 2000.

to 50 ms after the initiation of the first electromyographic (EMG) burst in the prime mover have been considered as APAs.

As the name suggests, APAs are seen only in conditions when the timing of a postural perturbation is predictable (e.g., the perturbation is triggered by the actor or the actor watches an approaching object expected to cause a postural perturbation). Since APAs are initiated prior to the action initiation and prior to the associated perturbation, they are based on an estimate of the perturbation magnitude and, as such, are always suboptimal. The residual actual perturbations are handled by feedback-based posture-stabilizing mechanisms, which are sometimes collectively addressed as compensatory postural adjustments (CPAs). Negative covariation between the magnitudes of APAs and CPAs when the explicit perturbation characteristics remain unchanged has been shown in several studies (Krishnan and Aruin 2011; Kaewmanee et al. 2020).

APAs have been described in a variety of tasks and associated actions, most commonly during fast arm actions or load interactions, such as catching and dropping the load, performed by standing persons (reviewed in Massion 1992). Other examples include tasks where the postural component is limited to a single limb or a single joint (Cavallari et al. 2016) and when tasks associated with postural perturbations are performed by sitting persons (reviewed by Chikh et al. 2016). Catching a load or releasing a load from extended arms has been used to standardize the changes in the external forces acting on the standing actor, in particular to compare APAs across populations with different abilities to perform fast voluntary movements (e.g., Bazalgette et al. 1986; Latash et al. 1995a).

The traditional view on APAs has been that these phenomena reflect preparation of the central nervous system to the predicted perturbation, and their function is to generate forces and moments of force directed against those expected from the perturbation (e.g., Cordo and Nashner 1982; Bouisset and Zattara 1987; Massion 1992). This is a natural expectation with respect to feedforward adjustments. However, a few studies have shown that APA patterns do not always follow this simple rule. In particular, if the mechanical effects of APAs could themselves turn into posture-destabilizing factors, the direction of forces during APAs may be the same as the direction of the expected external forces (Hirschfeld and Forssberg 1991; Krishnamoorthy and Latash 2005). For example, if a person stands in precarious conditions (e.g., close to the edge of a drop), generating APAs moving the body toward the drop may be more dangerous than acting away from the drop, even if the external forces are also expected to move the body away from the danger.

APA magnitude has shown modulation with factors other than the magnitude and direction of the external forces associated with the action. These include, in particular, stability of the initial posture: APAs are delayed and decreased in conditions of both higher and lower stability (Nardone and Schieppati 1988; Nouillot et al. 1992; Aruin et al. 1998; see Fig. 8.2). When stability is high, posture may not be threatened, and APAs are reduced. If stability is low, strong APAs themselves may turn into a destabilizing factor.

APAs are delayed and decreased during movements performed under a typical simple reaction time instruction, to move "as soon and as fast as possible" following an imperative signal (Lee et al. 1987; De Wolf et al. 1998), which is not completely

8. ANTICIPATORY CONTROL OF ACTION

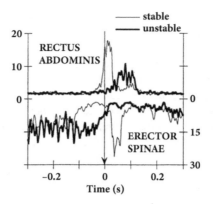

Figure 8.2 When the same action is performed by a standing person during standing in natural and unstable conditions (on a board with narrow support area), anticipatory postural adjustments are reduced and may be delayed. Action initiation time is $t = 0$. Modified by permission from Aruin et al. 1998.

unexpected given that typical simple reaction times to prime mover activation are relatively short, on the order of 150 ms, and leave little room for incorporating postural adjustments of about the same duration. Somewhat more surprisingly, APAs also scale with the magnitude of the action triggering a standard perturbation (Aruin and Latash 1995). This may reflect everyday experience suggesting that a more vigorous action is associated with a stronger postural perturbation. Consider, for example, lifting quickly light and heavy loads or performing slow and fast arm movements. As a result, a strong perturbation triggered by a minor action (e.g., when a standing person shoots from the rifle) is associated with inadequately weak APAs and large effects of the recoil on the vertical posture.

A number of studies have provided evidence for lower APAs in a variety of special populations ranging from the healthy elderly to neurological patients and persons with atypical development (Inglin and Woollacott 1988; Delafontaine et al. 2019; Vaz et al. 2019; Ferrario et al. 2023). Interpretations of reduced APAs varied from assuming that the central nervous system in certain groups of people was impaired in the ability to use feedforward postural control to considering reduced APAs as reflections of an adaptive strategy. The latter interpretation is based on a number of observations of reduced APAs in young healthy persons in conditions of uncertainty with respect to characteristics of an expected perturbation, its timing, and possible detrimental effects of APAs on postural stability. Populations with impaired postural stability may view everyday situations as challenging and reduce APAs to minimize their possible destabilizing effects on the body.

Traditionally, APAs have been studied in terms of performance variables such as forces, coordinates of force application, displacement of body parts, and muscle activation levels. As discussed earlier, these variables cannot in principle be prescribed by the central nervous system because they depend on the unpredictable actual states of the body, that is, actual values of mechanical variables and excitability of relevant neuronal pools. APAs are less sensitive to possible changes in the mentioned variables because they are observed before any visible action or mechanical effects of a perturbation, that is, when the body remains in a relatively steady state. There are, however, relevant variables that are likely to vary even during quiet standing. One of them is the effects of spontaneous postural sway, which makes body states time-varying (reviewed in Collins and De Luca 1993; Riley et al. 1997). Another one is changes in activation of gamma-motoneurons and other neurons that modulate characteristics of reflex loops and can lead to changes in their contributions to muscle activation even in the absence of measurable changes in mechanical variables.

A more adequate way to characterize APAs is to use the language of neural control variables (see Chapter 7) such as RCs for the muscles, joints, limbs, and whole body and associated reciprocal and coactivation commands (*R*- and *C*-commands). Since these commands are not directly observable in typical human experiments, they have to be inferred from performance variables, which has been done in a number of studies, of course under a number of assumptions related to transformations from RCs to variables quantified in typical behavioral experiments. In particular, those studies suggested scaling primarily of the *C*-command to produce different APAs in anticipation of a self-triggered load release across conditions that differed in stability of the initial posture (Slijper and Latash 2000).

8.2. Early postural adjustments

A different group of postural adjustments has also been addressed as APAs, although their characteristics and features differ rather dramatically from the "classical APAs" described in the previous section. These adjustments are seen in preparation of actions that are not themselves associated with explicit postural perturbations. These adjustments are initiated much earlier as compared to typical APAs, 400 to 1000 ms prior to the action initiation. Typical examples include postural preparation to step initiation and to large-amplitude, fast voluntary body sway (Elble et al. 1994; Lepers and Brenière 1995; Wang et al. 2006).

When a person stands quietly, the projection of the center of mass (and of COP) falls in front of the ankle joints, more or less aligned with the body midline. Making a step requires lifting one of the feet and rotating the trunk forward with respect to the ankle joint of the trailing (supporting) leg. This is possible only if the COP (and body weight) is transferred to the trailing foot and, at the same time, the COP is moved backward to generate a properly directed moment of force (Fig. 8.3).

Such adjustments are seen typically prior to making a step (reviewed in Bouisset and Do 2008). In the mediolateral direction, the COP is transferred first to the leading foot, thus creating a moment of force moving the center of mass toward the trailing foot, and then the COP moves to the trailing foot. This sequence of events unloads the leading foot and allows lifting it off the ground. At the same time, in the anterior-posterior direction, the COP moves backward. Of course, these mechanical events are associated with appropriate muscle activation patterns. If a person is asked to shift the COP to a different initial coordinate (e.g., toward the trailing foot and backward or toward the leading foot and forward), the pattern of postural preparation changes

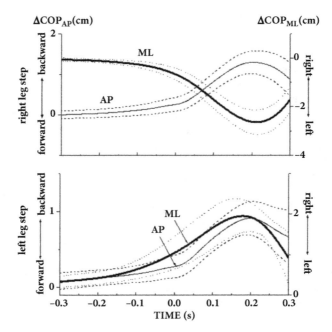

Figure 8.3 Typical changes in the center of pressure (COP) in the anterior-posterior (AP) and mediolateral (ML) directions. Reproduced by permission from Wang et al. 2005.

in a predictable way to comply with the requirements of basic mechanics (Hansen et al. 2016).

COP shifts with similar timing are observed when a person occupies a body posture leaning forward or backward and then tries to perform a quick body sway, forward to backward or backward to forward. Under such conditions, the COP shifts in the opposite direction as compared to the planned body movement, thus helping to generate a proper moment of force, and then it moves quickly in the sway direction.

These sequences of postural adjustments to making a step or to performing fast body sway are mechanically necessitated and cannot be avoided. In contrast, typical APAs help to attenuate the expected effects of an anticipated perturbation, but they are not obligatory. One can perform a similar action triggering a standard postural perturbation under the simple reaction time instruction, and APAs will be shifted toward the action initiation time with changes in their characteristics. Postural preparations to stepping or swaying show relatively minor changes under the simple reaction time instruction because the action cannot be initiated without them. Given the differences in the appearance (e.g., timing), purpose, and other features between the two groups of postural adjustments to action, it makes sense to address them using different terms. In some studies, they have been addressed as early postural adjustments (EPAs). Overall, APAs are a luxury, and EPAs are a necessity.

Both EPAs and APAs can be observed within a single action. This was shown in a study where the subjects watched a pendulum swinging toward them and hitting their

8.3. GRIP ADJUSTMENTS TO PLANNED ACTIONS   117

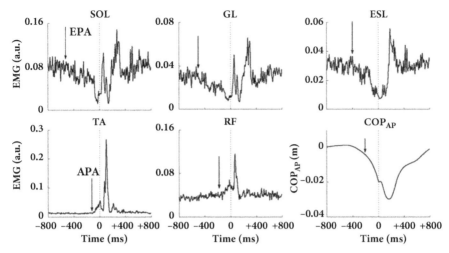

Figure 8.4 Two types of postural adjustments seen as changes in the lower-body muscle EMGs. Time *t* = 0 corresponds to the moment of load impact. Note the early slow drop in the activation level of dorsal muscles (early postural adjustments [EPAs]), followed by its abrupt drop accompanied by an EMG burst in the ventral muscles (anticipatory postural adjustments [APAs]). SOL, soleus; BF, biceps femoris; ES, erector spinae; TA, tibialis anterior; RF, rectus femoris; COP$_{AP}$, center of pressure coordinate in the anterior-posterior direction. Reproduced by permission from Krishnan et al. 2011.

shoulders (Fig. 8.4; Krishnan et al. 2011). Although the subjects were instructed not to move the body, all of them showed an EPA starting about 500 ms prior to the impact and an APA starting about 100 to 150 ms prior to the impact. The EPAs induced slow motion of the body forward, toward the approaching pendulum, thus making sure that the impact moved the body backward, closer to its initial, natural posture. The APAs attenuated the effects of the impact on posture by generating force in the opposite direction.

8.3. Grip adjustments to planned actions

Anticipatory adjustments are seen also in tasks not involving keeping body posture in the field of gravity. A commonly studied example is grip adjustments to predictable changes in external forces acting on a hand-held object. Imagine an object held vertically in a prismatic grip, with the thumb opposing the four fingers (Fig. 8.5). The digits generate force normal to the contact surface (F_{GRIP}) and tangential to the surface (F_{TAN}). The role of F_{TAN} is to counteract the gravity force, while the purpose of F_{GRIP} is to provide sufficient normal force to avoid slippage, given the friction conditions: $|F_{GRIP}|\cdot k > |F_{TAN}|$. Typically, there is a substantial safety margin reflected in higher F_{GRIP} than strictly necessary, which ensures that any spontaneous changes in F_{TAN} are not leading to slippage.

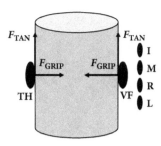

Figure 8.5 During the prismatic grasp, the thumb (TH) opposes the four fingers (I, index; M, middle; R, ring; L, little). The combined mechanical action of the four fingers is sometimes represented with an imagined digit—virtual finger (VF). The grip force (F_{GRIP}) has to be over the slippage threshold, that is, $|F_{GRIP}| \cdot k > |F_{TAN}|$, where F_{TAN} is tangential to the surface of contact force, and k is the friction coefficient.

Under a variety of conditions, the external force acting on the hand-held object can change. Such changes may be produced by external forces acting on the object as well as by motion-related forces. For example, if the object is lifted quickly, F_{TAN} increases with acceleration (a): $F_{TAN} = m(g + a)/2$ (assuming equal sharing of the load between the thumb and opposing fingers) and requires an increase in F_{GRIP} to preserve the safety margin or even increase it given the uncertainty of the magnitude of a. Such an increase in F_{GRIP} and associated changes in muscle activation levels are observed as feedforward adjustments in anticipation of predicted changes of forces acting on the object. They are seen across tasks including walking with a hand-held object (Flanagan and Wing 1993, 1995).

Similar feedforward adjustments are seen in other components of the force-moment vector produced by the hand on the gripped object in anticipation of other components of the external forces and moments. For example, lifting a handle connected to a set of asymmetrical loads (Fig. 8.6) leads to acceleration-dependent changes in the external force and moment. Feedforward adjustments in the moment of force applied to the object are seen after a few repetitions of this task as the subject learns the mechanical consequences of the actions (Fu and Santello 2015). These adjustments modify not only F_{GRIP} but also the total moment of force acting on the object and prevent its tilting during the action.

Another example of F_{GRIP} adjustments shows that the central nervous system tries to keep a desired level of safety margin without exceeding it by too much. If a person holds an object and then applies vertical force to the bottom of the object, thus helping the other hand to hold the object in the air against the gravity force, F_{GRIP} shows a drop before or simultaneously with the helping force application (Fig. 8.7; Scholz and Latash 1998). As the supporting force is removed, F_{GRIP} returns close to its initial magnitude. Similar changes in F_{GRIP} are seen when the helping force is provided by another person, but these F_{GRIP} adjustments come at a time delay typical of F_{GRIP} adjustments to unexpected perturbations. Interestingly, persons with Down syndrome show an atypical pattern of F_{GRIP} adjustments when they apply helping force

8.3. GRIP ADJUSTMENTS TO PLANNED ACTIONS 119

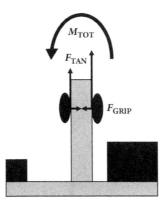

Figure 8.6 During vertical motion of an asymmetrical object, both vertical load force and moment of force change as functions of object acceleration. With minimal practice, the grip force (F_{GRIP}) and total moment of force (M_{TOT}) applied by the digits show feedforward modifications to compensate for the expected changes in the external load and moment.

with the other hand: They show an increase in F_{GRIP}, suggesting that they view their other hand not as a helper but as a source of potential perturbing forces.

In terms of the neural control with RCs, force application to the surface of the gripped object is produced by specifying RCs for the opposing digits inside the gripped object: The digits virtually penetrate the object (cf. Pilon et al. 2007). The difference between the actual coordinate of a digit (AC in Fig. 8.8) and its RC produces force with the magnitude defined by the C-command, which translates into the apparent stiffness (k) of the digit tips. Taking the object abruptly from the hand leads to motion of the opposing digits toward each other, that is, toward the respective RCs.

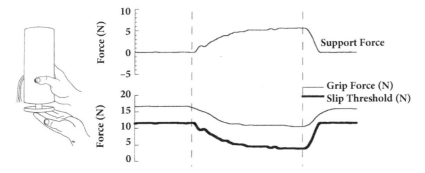

Figure 8.7 Feedforward adjustments are seen in the grip force in anticipation of the application of vertical force (support force) by the other hand to the bottom of the hand-held object. Reproduced by permission from Scholz and Latash 1998.

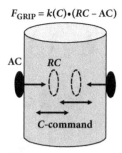

Figure 8.8 The control of the grip force in terms of the digit referent coordinates (RCs) and the C-command that helps to convert the spatial discrepancy between the RC and digit actual coordinate (AC) into force units with a coefficient k.

Other force-moment components are also associated with RCs different from actual coordinates in linear (for force) or rotational (for moment) units, which are illustrated in Fig. 8.8.

Grip adjustments are associated with changes in RC components based on predicted changes in external forces. These may include changes in the R-command and/or C-command. So far, no reliable experimental data are available on changes in the two basic commands associated with grip force adjustments. The proposed hierarchy of the two basic commands, with the R-command being hierarchically higher (Levin and Dimov 1997; Feldman 2015), suggests that F_{GRIP} adjustments are primarily produced by changing the R-command while keeping the C-command relatively unchanged. On the other hand, a recent study of changes in the recruitment patterns of motor units in opposing muscles has suggested that force modulation may be associated with more reproducible changes in the C-command, thus modulating the apparent stiffness of the digits and producing force changes with little consistency in the modulation of the R-command (Madarshahian and Latash 2022a; see also Chapter 10).

8.4. Anticipatory synergy adjustments

Another class of anticipatory adjustments to action involves no obvious change in the magnitude of salient performance variables but a change in their stability. These phenomena, termed anticipatory synergy adjustments (ASAs), are seen in conditions when a person performs a steady-state action and gets ready to initiate a fast change in performance. One of the typical tasks demonstrating ASAs involves asking a subject to press with the four fingers of a hand on force sensors and keep a steady level of total force (F_{TOT} in Fig. 8.9) followed by a very quick force pulse into a target initiated at a self-selected time or a perfectly predictable time (Olafsdottir et al. 2005a; Shim et al. 2005b; Park et al. 2012). Under such conditions, a strong synergy stabilizing F_{TOT} in the individual finger force space (or finger mode space; see Chapters 6 and 11) is seen during the steady-state phase. It is reflected in the signature inequality,

8.4. ANTICIPATORY SYNERGY ADJUSTMENTS

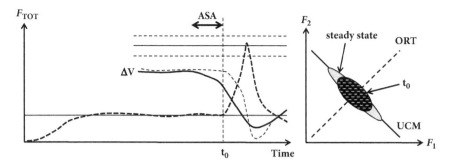

Figure 8.9 Left: Anticipatory synergy adjustments (ASAs) are seen as a drop in the synergy index (ΔV) in preparation of the production of a self-paced quick change in the total force (F_{TOT}) produced by two fingers of a hand (F_1 and F_2) into a target (shown with thin horizontal lines). The time of force pulse initiation is t_0. Note the lack of ASAs when the same action was performed under the simple reaction time instruction (thin dashed line). Right: A schematic illustration of the data distributions across trials for a two-finger task during the steady state (gray ellipse) and at the time of force pulse initiation (black ellipse). Note a drop in the variance along the uncontrolled manifold (UCM) and its increase along the line orthogonal to it (ORT).

$V_{UCM} > V_{ORT}$, and positive values of the synergy index, ΔV, estimated across repetitive trials. In this analysis, individual trials are aligned by the time of F_{TOT} initiation. A few hundred milliseconds, typically 300 ± 100 milliseconds, prior to this time (t_0 in Fig. 8.9), a smooth drop in ΔV starts, leading to a substantial drop in the synergy index by t_0. This phenomenon has been interpreted as purposeful attenuation of stability of the salient variable (F_{TOT}) to facilitate its planned quick change. Note that this task can be performed without ASAs, as demonstrated in experiments when the subjects were instructed to perform the same action as quickly as possible after an imperative auditory signal (Olafsdottir et al. 2005a). Under such a simple reaction time instruction, ASAs disappear, and ΔV starts to drop simultaneously with action initiation.

ASAs have been demonstrated in other tasks and in other spaces of elemental variables. In particular, they are seen as a drop in the synergy index, ΔV, reflecting stabilization of the COP coordinate in the space of muscle modes in a standing person in preparation of an anticipated event, a self-initiated quick action by the person or an imminent, predictable external perturbation (Klous et al. 2011; Krishnan et al. 2011). ASAs are also seen about 200 to 300 ms prior to the initiation of a step as a drop in the index of COP-stabilizing synergy estimated in the space of muscle modes within the leading leg, but not within the trailing (supporting) leg (Wang et al. 2005; Falaki et al. 2023).

Another group of adjustments of action stability is seen when a person expects an event requiring a quick change in a salient performance variable even if this event does not occur. This was shown in experiments when trials requiring a quick action (or a quick change in the ongoing action) were intermixed randomly with trials when no signal to perform the action took place (Freitas et al. 2007) as well as during a steady-state phase in trials where a signal to act happened at an unpredictable time in the future (Tillman and Ambike 2018, 2020). In one of the very first experiments of this type (Fig. 8.10; Freitas et al. 2007), the subjects were asked to point at a target on the screen as quickly as possible. In some trials, the target could jump after the movement initiation, requiring a very quick movement correction. Trials were presented in two blocks, when the target never jumped and when the target could jump in about half of the trials. Synergies stabilizing the fingertip trajectory were quantified in the space of individual joint rotations across trials when the target did not jump selected from each block. The index of synergy (ΔV) was shown to be lower across trials selected from the block when the target could jump, that is, when the subjects expected a possibility of quick movement correction. These phenomena may be viewed as "steady-state ASAs," in contrast to the earlier-described "transient ASAs," which are timed to the planned action.

The phenomenon of ASAs, both steady state and phasic, demonstrates that humans can modify stability of an ongoing action in a task-specific way and adjust action stability based on their expectations and planned actions. It is easy to imagine situations when the effects of ASAs, particularly the steady-state ones, are seen with the naked eye. Consider, for example, a tennis player getting ready for a powerful serve or a goalkeeper preparing for a penalty kick. In both situations, a visible increase in postural sway is observed, which may be interpreted as partial attenuation of postural stability leading to facilitation of future quick whole-body movements required by the situation.

So far, ASAs have not been studied in spaces of neural control variables (i.e., RCs). The reason is that measuring RCs requires the application of a controlled smooth perturbation and, therefore, the currently used methods have very poor time resolution (see Chapter 7). Potentially, it is possible to estimate indices of performance-stabilizing

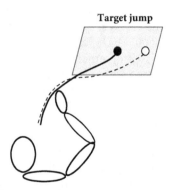

Figure 8.10 A schematic illustration of the task when subjects were instructed to point at a target in the middle of the screen. In one of the series of trials, the subjects were aware that the target would never jump. In the other series, the target could jump in some of the trials unexpectedly after the movement initiation. The index of multi-joint synergy stabilizing the endpoint trajectory dropped in the latter series, even in trials when no target jump occurred.

synergies in the relevant RC space during steady-state APAs, but such experiments have not been performed yet.

Characteristics of ASAs have been interpreted as reflections of an ability to facilitate fast actions (i.e., agility). This conclusion was based, in particular, on observations of delayed and reduced-in-magnitude ASAs in populations who show an impaired ability to initiate quick actions, from the healthy elderly to patients with subcortical disorders and cortical stroke survivors (Olafsdottir et al. 2008; Park et al. 2012, 2013; Jo et al. 2016b, 2017; for more detail see Chapter 22).

8.5. Distinguishing APAs from ASAs

APAs and ASAs show certain similarities in their changes with the task and state of the actor. In particular, as reviewed earlier, both are delayed in tasks involving the typical simple reaction time instruction as compared to similar actions performed in a self-paced manner. Both are reduced and delayed in the healthy elderly and in patients with early-stage Parkinson disease. These similarities were viewed as possible reflections of common neurological mechanisms involved in APAs and ASAs.

On the other hand, the timing of these two types of feedforward adjustments is obviously different: In experiments where both APAs and ASAs were quantified in standing persons during such actions as fast arm movements and load release from extended arms, the former started visibly later than the latter (Klous et al. 2011; Krishnan et al. 2011). Their hypothesized functions were also quite different. In particular, APAs are expected to be sensitive to changes in the direction of postural perturbation. Indeed, assuming that their main function is to generate forces counteracting the expected effects of the perturbation on the body, a change in the perturbation direction requires a change in the APAs. APAs that produce forces acting in a wrong direction could potentially exacerbate the effects of the perturbation and lead to loss of postural stability. In contrast, ASAs are more universal: They destabilize salient variables to facilitate future action independently of action direction.

The predicted differences in the direction specificity of APAs and ASAs were tested in an experiment when the subjects were required to stand and hold a bar with two loads attached to the bar through electromagnetic locks and acting in opposite directions, along the field of gravity and against the field of gravity (via a system of pulleys; Fig. 8.11A) (Piscitelli et al. 2017). The subjects stood quietly and then pressed a button on the bar, which released one of the loads. In one of the two main conditions, the subjects knew in advance which of the two loads would be released. In the other condition, however, pressing the button led to release of one of the loads selected randomly by the experimenter without the subject's knowledge. So, the timing of the load release was always controlled by the subject, while the direction of the triggered perturbation could be known (condition 1) or unknown (condition 2).

APAs and ASAs were seen under both conditions. ASAs showed no difference in their characteristics, such as the initiation time, across the two conditions (Fig. 8.11B). In contrast, under condition 2, APAs became significantly shorter

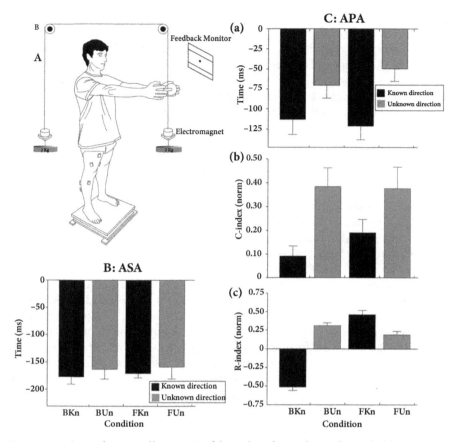

Figure 8.11 A: A schematic illustration of the task with standing subjects holding a bar attached to two loads acting up (via pulleys) and down. The subject pressed the button, releasing one of the loads. In some trials, the subject knew in advance which load would be released. In other trials, either load could be released unpredictably. B: Anticipatory synergy adjustments (ASAs) showed no differences between the two conditions. C: Indices of anticipatory postural adjustments (APAs) changed significantly between the two conditions. APAs were characterized by a lower index reflecting the reciprocal command (R-index) and a higher index reflecting the coactivation command (C-index). Modified by permission from Piscitelli et al. 2017.

(delayed) and changed the pattern of muscle activation (Fig. 8.11C): When the perturbation direction was unknown, subjects predominantly used changes in the coactivation of agonist-antagonist muscle pairs, while when the direction was known in advance, these muscles showed more reciprocal patterns of EMG changes. These results suggest that, in spite of some of the mentioned commonalities in changes of both APAs and ASAs under some conditions and in some populations, these two types of adjustments are likely to be of different neurophysiological origins.

8.6. Missing pieces of the mosaic

Only a handful of studies have so far attempted to explore anticipatory adjustments to action at the level of neural control variables, such as the R- and C-commands. The main problem is the lack of tools able to track time changes in these commands at a sufficient time resolution. As mentioned earlier, changes in muscle activation during such adjustments seem to be relatively immune to changes in the contributions from reflex pathways. But this is only seeming immunity because of the phenomenon of alpha-gamma coactivation leading to changes in the activity of gamma-motoneurons and, as a result, to changes in responses of spindle sensory endings to unchanged values of muscle length and velocity. There may also be changes in the state of spinal interneurons mediating the effects of afferent signals on muscle activation patterns.

One of the questions discussed in earlier studies is: Are APAs tightly linked to action by the person? Can APAs be observed in the absence of any voluntary action? On the one hand, the dependence of APA magnitude on voluntary action magnitude triggering a standard perturbation suggests that the two are tightly coupled. On the other hand, APAs are observed during catching a falling object or watching an approaching object expected to produce a postural perturbation when no action is required to trigger the perturbation. However, prior to impact on a body segment (up to the whole body), small voluntary movement by this segment is typically seen, which is not consciously perceived by the subject and cannot be reliably suppressed. Examples include small hand motion toward the falling object and small body sway toward the approaching object seen a few hundred milliseconds prior to the impact. It is not clear whether an experimental study can be designed with a predictable perturbation, which would rule out any voluntary action by the subject.

A relatively new issue has emerged related to possible anticipatory adjustments (ASAs) in intra-muscle synergies based on motor unit groups, MU-modes (see Chapter 11). This is a potentially important issue. On the one hand, intramuscle synergies are likely of a spinal origin (Madarshahian et al. 2022; reviewed in Latash et al. 2023; see Chapter 11). On the other hand, ASAs reflect anticipation by the subject, which is expected to be based on supraspinal processes. Can anticipatory processes lead to modulation of signals along descending pathways, leading to controlled attenuation of synergies based on spinal circuitry?

So far, a major problem has been the relatively poor time resolution of all studies exploring intra-muscle synergies. The root of the problem is that some of the motor units generate action potentials at relatively low frequencies, with inter-spike intervals on the order of 200 to 300 ms, which makes it hard to estimate the instantaneous firing frequency over time intervals smaller than those where at least two action potentials are observed. This time is on the order of ASA duration. One method to deal with the problem is to estimate the firing frequency of individual MUs over multiple trials. However, estimating ASAs requires analysis across at least 20 data points (see de Freitas et al. 2018). So, if each data point requires analysis over multiple trials, the total number of trials may be prohibitively large, on the order of 100 to 200 trials. It is possible that bootstrapping techniques or similar approaches could help in solving this problem.

9
Stability, Agility, and Optimality

All human movements involve multiple elements, from limbs to joints to muscles to motor units. On the other hand, typical tasks are associated with relatively few constraints as compared to the number of involved elements. The last statement may be viewed as trivial because the number of the smallest controllable output units of the central nervous system, motor units, is at least an order of magnitude larger than the number of task-specific important performance variables defined by the task constraints. As described in Chapter 2, following traditions set by Nikolai Bernstein (Bernstein 1947; translation in Latash 2020a), such problems have been addressed as problems of motor redundancy. There are two major groups of such problems. The first one, the problem of *state redundancy*, is related to the aforementioned excess of elements. The second one, the problem of *trajectory redundancy*, reflects the fact that a single element can move from an initial state to a desired final state over an infinite number of trajectories. During natural movements, both problems coexist and are commonly addressed together.

Depending on one's attitude to the apparent excess of elemental variables (those produced by elements at the selected level of analysis), their availability is viewed as a source of computational problems for the central nervous system or as a highly useful feature of the design of the body affording flexibility and adaptability of behavior. The former attitude—the apparent redundancy is a *bug*—leads to using such concepts as elimination of redundant degrees of freedom and optimization defining observed patterns of behavior. The latter attitude—the apparent redundancy is a *feature*—emphasizes structured variability of behavior at the level of elemental variables as a reflection of controlled stability of movements. However, the range of combinations of elemental variables during movements is typically limited to a relatively small area within the solutions space, suggesting that some soft optimization constraints may still be used.

Another feature of movements, which rarely gets into the spotlight, is agility. Intuitively, it is clear that movements that are agile with respect to time changes in a set of performance variables are unlikely, at the same time, to be very stable with respect to the same performance variables. It is less clear how agility is related to movement optimality: Can one change an ongoing action very quickly and precisely while conforming to a single criterion of optimality? Addressing these questions requires defining the three concepts explicitly and, if possible, operationally, that is, in such a way that movements can be classified into stable vs. unstable, optimal vs. suboptimal, and agile vs. non-agile and, if we are lucky, to introduce methods to quantify these concepts (reviewed in Latash 2023).

9.1. Definitions and metrics

Stability is the ability of a system of interest to return to a desired state or trajectory following a small transient change in the external forces (perturbation) and/or in its intrinsic state, or a small change in its state at action initiation. This definition reflects our understanding of stability as *dynamical stability*, with static stability being a particular example when no change in the state of the system is desired.

During any natural movement, all the mentioned factors—external forces, intrinsic states, and initial conditions—vary unpredictably or, at least, imperfectly predictably. As a result, even the best-trained subjects participating in the best-controlled experiments show inter-trial variability of states and trajectories (see Bernstein 1930 and Chapter 2). This allows using inter-trial variance in a specific direction in the space of elemental variables as an index of stability in that direction: Trajectories in unstable (less stable) directions are expected to diverge more (converge less) compared to trajectories in more stable directions. The uncontrolled manifold (UCM) hypothesis assumes that stability of a salient performance variable is reflected in higher inter-trial variance computed in the space of elemental variables at a certain phase of the movement along directions spanning the solution space (UCM) for that performance variable as compared to variance orthogonal to the UCM (ORT in Fig. 9.1; reviewed in Latash et al. 2002d, 2007), $V_{UCM} > V_{ORT}$. Various metrics reflecting the difference between V_{UCM} and V_{ORT} have been used as indices of stability including the ratio V_{UCM}/V_{ORT} and the normalized difference ($V_{UCM} - V_{ORT}$). Of course, any variance-based comparison has to be done taking into consideration the dimensionality of the UCM and ORT spaces and is valid only for linearized systems.

Another method of quantifying stability has been based on the phenomenon of motor equivalence (Scholz et al. 2007; Mattos et al. 2011, 2013, 2015a, 2015b): Any quick action by a multielement system leads to a task-specific change in the salient performance variable reflected in motion along the ORT direction in the space of elemental variables, addressed as non–motor equivalent [nME] displacement. Displacement along the UCM (motor equivalent [ME] displacement) is, by definition, unable to change the salient performance variable. In particular, a change in a

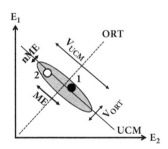

Figure 9.1 An illustration of two metrics of stability for the task of producing a value of the sum of two elemental variables, E_1 and E_2. The first is based on comparing inter-trial (intercycle) variance within the uncontrolled manifold (UCM) and orthogonal to it (ORT), V_{UCM} and V_{ORT}, each quantified per dimension in the corresponding subspace. The second compares the deviations along the UCM (motor equivalent [ME]) and along the ORT (non–motor equivalent [nME]) per square root of the number of dimensions. A typical data point distribution is shown with a gray ellipse, and two data points with circles (1 and 2).

performance variable to a new value and back to the initial state is expected to result in a small nME displacement between the initial and final states. In contrast, ME displacement is unconstrained: It can be smaller than, equal to, or larger than the nME displacement. In such tasks, the properly quantified ratio ME/nME has been used as an index of stability (Mattos et al. 2015b). Note that this ratio can be measured in individual trials, which gives it an advantage as compared to analysis of V_{UCM} and V_{ORT}, which require analysis over relatively large sets of trials (cf. De Freitas et al. 2018). The two methods of analysis are expected to lead to correlated results if pairs of measurements in the analysis of motor equivalence are sampled from the same distribution (cf. Leone et al. 1961; Falaki et al. 2017a). This assumption, however, is not trivial and can be violated as shown in experiments with measurements performed before and after a quick force pulse production by a set of fingers (Cuadra et al. 2018).

Imagine that you walk along the street with a mug of coffee in your hand. The mug orientation should be close to vertical to avoid spilling the coffee. Each step leads to imperfectly predictable contact forces with the ground, which propagate along the body and perturb all the joints of the body and arm with the mug. Low stability along the UCM for the mug vertical orientation in the joint configuration space channels the kinematic effects of the perturbations into the UCM, thus keeping the mug orientation relatively unchanged. The same qualitative analysis applies to any external perturbation (e.g., if another person bumps unexpectedly into your arm with the mug). Overall, low stability along the UCM reflected in high V_{UCM} is functionally important, which is supported indirectly by observations of reduced V_{UCM} with little change in V_{ORT} in neurological patients who show impaired stability of posture and movement (e.g., Falaki et al. 2017a; Jo et al. 2017).

Optimality is defined operationally as minimization of a *cost function* (CF) of elemental variables across their possible combinations and/or along their possible trajectories. Optimization is implied here as a computational tool used by researchers, not as a reflection of computations within the brain. A typical cloud of data points, such as the one illustrated in Fig. 9.2 (the gray ellipse), is limited in its size along the UCM, and the location of the center of the data cloud may be viewed as defined by an optimization criterion (i.e., a minimum of a CF). We assume here that such a minimum exists, that is, that the CF is not perfectly flat within a range of values of the elemental variables. In other words, even if there is a ravine along the UCM (cf. Gelfand and Tsetlin 1962, 1971), the bottom of this ravine has a minimum corresponding to the minimum of the function ϕ_{UCM} illustrated in Fig. 9.2. In other words, a degree of stability is implied along the UCM although smaller compared to that along ORT, as illustrated schematically by the two potential fields, ϕ_{UCM} and ϕ_{ORT}, in Fig. 9.2. The illustration in Fig. 9.2 is supported by data showing that, when a subject is asked to perform a task starting from a point on the UCM displaced as compared to the naturally preferred point, turning visual feedback off leads to a drift of the performance in the space of elemental variables toward that preferred point (Parsa et al. 2017).

Most commonly, the cost is guessed based on intuitive considerations (e.g., minimum energy expenditure, minimum joint wear, or minimum fatigue) and/or personal theoretical views of the researcher. Alternatively, the cost can be defined using methods of inverse optimization (Bottasso et al. 2006; Terekhov et al. 2010). Relatively large inter-trial variance along the UCM (as in Fig. 9.1) suggests that strictly optimal

9.1. DEFINITIONS AND METRICS   129

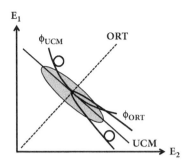

Figure 9.2 An illustration of potential fields within the UCM (ϕ_{UCM}) and within ORT (ϕ_{ORT}) for the same system as in Fig. 9.1: $E_1 + E_2 = C$. A typical data point distribution is shown with a gray ellipse. Hypothetical states of the system different from the center of the distribution are shown with open circles.

movements are rare if one assumes a CF with a single minimum (which is true for most CFs in the field of motor control).

A large number of CFs are able to account for behavioral patterns during simple voluntary movements reasonably well (reviewed in Seif-Naraghi and Winters 1990; Prilutsky and Zatsiorsky 2002; Latash and Zatsiorsky 2016). On the other hand, a single CF may be able to account for movement patterns across various behaviors (e.g., minimum fatigue; Harris and Wolpert 1998; Prilutsky 2000; Anderson and Pandy 2001; reviewed in Prilutsky and Zatsiorsky 2002). It is easy, however, to imagine a movement violating any a priori CF, for example, any movement performed purposefully in an unusual way (e.g., for esthetic purposes). Taken together, these observations suggest that CFs are *not prescriptive but softly constraining*: They can be violated during natural movements as long as movements are within an acceptable range—*good enough* (cf. Loeb 2012; Akulin et al. 2019).

There is also an important distinction between evolutionary optimal behaviors and optimal behaviors of individuals in specific situations. It is obvious that some movements performed by humans may be suboptimal from the evolutionary standpoint, for example, dangerous actions such as those by circus performers, some athletes, and mountain climbers, which may lead to lower chances of long life and having healthy offspring. Evolutionary costs of movements are a special topic that would take us far away from the main topic: how the brain controls voluntary movements.

To summarize: (1) *No single CF is universally applicable as a prescriptive criterion.* (2) *Perfectly optimal actions are rare; "good enough" ones are common.* (3) *Behavior may be viewed as defined by a confluence of optimization criteria acting with weights that are task and person specific.*

Agility is an ability to change a salient performance variable quickly and accurately in a desired direction. Sometimes an additional requirement is implied of doing this with minimal *effort*, which is not a clearly defined concept (Full et al. 2002; Sefati et al. 2013; Dakin et al. 2020). If one equates effort with an index of changes in all elemental variables, deviations along the UCM are clearly violating the minimal-effort criterion. In other words, if an optimality criterion defines a preferred sharing pattern of a salient performance variable across elemental variables, preservation of the original sharing pattern during movements contributes to its quick change with minimal effort.

Recent studies (reviewed in Latash and Huang 2015) associated agility with the phenomenon of anticipatory synergy adjustments (ASAs; see Chapter 8). ASAs represent a drop in the index of stability for a salient variable, for example, the normalized difference ($V_{UCM} - V_{ORT}$) in preparation for a planned quick change of that variable (Olafsdottir et al. 2005a). Indeed, trying to change quickly a performance variable, which is highly stable, is expected to lead to large deviations along the unstable directions (i.e., those spanning the UCM). For example, if you take a spring from a typical pen, place it between the tips of the thumb and index finger, and try to compress it quickly along its main axis, the spring will buckle and jump away (cf. Valero-Cuevas et al. 2003). This happens because the direction of action is never perfectly aligned with the desired direction and has measurable orthogonal components that lead to relatively large deviations if the system is unstable in those directions (as the spring is in directions orthogonal to its main axis). The brain has the ability to reduce the stability of a steady-state action in preparation for a quick change of a salient variable, which helps to mitigate this problem and reduce negative interference of strong performance-stabilizing synergies with quick changes in the salient variables.

Accurately timed destabilization of salient variables is reflected in a drop of the synergy index, which can be seen in young, healthy persons about 300 to 400 ms prior to the action initiation (see Chapter 8). ASAs are reduced in both duration and magnitude across populations who present mild problems with motor coordination such as the healthy elderly (Olafsdottir et al. 2008) and patients after a mild cortical stroke (Jo et al. 2016b), with early-stage Parkinson disease (Park et al. 2012), and with cerebellar disorders (Park et al. 2013b). Associating agility with ASAs links it to *preparation* for a quick movement, rather than to the movement itself. A few studies have suggested that larger ASAs could indeed be associated with smaller destabilization of the salient performance variable during the following movement and quicker stabilization of a new steady state after the movement (Kim et al. 2006; Zhang et al. 2006). The importance of ASAs is also exemplified by observations of athletes, such as tennis players getting ready for a powerful serve: A visible increase in postural sway is observed, a reflection of controlled destabilization of the initial posture. Although ASAs are likely to be important contributors to agility, our current definition is broader and involves factors beyond ASAs.

9.2. Optimization in human movements

Traditional approaches to problems of apparent motor redundancy have used the concept of optimization, including ideas of optimal feedback control (reviewed in Prilutsky and Zatsiorsky 2002; Diedrichsen et al. 2010). All such approaches share a number of features that may be viewed as problematic. The first is subjective selection of an assumed CF (objective function) that the central nervous system tries to minimize. CFs are chosen based on the theoretical views of researchers, possibility of reasonably accurate measurement and/or computation, intuitive considerations, and other factors.

The second problem is the necessity to predict changes of all the elemental variables contributing to changes in the salient performance variable and, therefore, involved in the computation of the CF value over future movement time. This obviously requires a nearly perfect prediction of future changes in all those variables prior to a movement—a highly questionable assumption, particularly given the well-documented large variability in the elemental mechanical and electrophysiological variables over repetitive attempts at the same task by highly trained individuals (Bernstein 1930; reviewed in Latash 2008).

Third, typically, CFs are assumed to be universal, applicable across sets of effectors, actions, and sometimes even species. This assumption also looks counterintuitive: It is hard to imagine that an athlete performing a high jump, another athlete running the marathon race, and a chess player moving a piece on the board would all optimize the same CF, although energy and its components, in particular free energy, have been suggested as a universal currency (Yufik and Friston 2016). It seems, however, more natural to assume that relevant body currencies are selected based on task-specific and actor-specific criteria.

Arguably, one of the most influential optimization criteria is that of "minimum jerk" (Flash and Hogan 1985), based on minimization of a CF representing an integral over movement time of the third derivative of coordinate (or, equivalently, derivative of acceleration) squared: $C_J = \frac{1}{MT}\int_0^{MT}\left(\frac{da}{dt}\right)^2 dt$, where C_J stands for the jerk-related cost, a for acceleration, and MT for movement time. The minimum-jerk criterion is kinematic in nature, and as such, it is limited to a range of tasks. In particular, it cannot describe force production tasks not associated with kinematic changes. However, reproducible deviations from the optimal solutions defined by the minimum-jerk criterion have been reported in studies of persons with motor impairments, suggesting its potential practical usefulness (Osu et al. 2011; Hu et al. 2019).

A variety of optimization approaches have been suggested based on minimization of intuitively reasonable CFs such as those related to metabolic energy expenditure, fatigue, discomfort, and effort, as well as more complex functions representing combinations of those mentioned earlier. All of the mentioned functions were reasonably accurate in describing movement patterns over specific classes of tasks (otherwise, those papers would not have been published!). None of them, however, were universal enough to reflect a global principle applicable across tasks and effector sets.

Recently, ideas of optimal feedback control have been actively developed for problems of motor coordination (Todorov and Jordan 2002; Todorov 2004; Diedrichsen et al. 2010). These ideas are also based on minimizing a CF, which is, however, recomputed in the course of a movement based on feedback information on the actual movement trajectory. Typical CFs have represented the sum of two components. One of them reflects the trajectory of a salient performance variable toward the task-related goal. The other component reflects changes in control variables, that is, "effort." As a result, unpredictable changes of external forces or within the body cause effects in the space of elemental variables, which are corrected only if they affect the salient performance variable. This leads to variable-across-repetitive-trials trajectories with more variance in directions not affecting the salient variable, that is, along its UCM (see Chapter 5). The ability of optimal control schemes to account for the synergic signature of inter-trial variability, $V_{UCM} > V_{ORT}$, is a very attractive feature. However, the arbitrary selection of the CF, in particular of its "effort" component (commonly associated with changes in muscle activation levels or forces), and the necessity to perform computations over movement time into the future during the course of the movement remain serious drawbacks.

A major problem inherent to all the optimization approaches is in its first step, that is, selection of a space of elemental variables where optimal solutions are searched for. Indeed, an optimal solution in a kinematic space cannot incorporate such important phenomena as muscle coactivation, which may have little direct effect on movement kinematics. An optimal solution in a space of muscle activations implies that the central nervous system can prescribe those variables, which is obviously wrong given the imperfectly predictable effects of reflex feedback loops (see Chapter 4). If we are interested in processes of the neural control of movements, it is natural to select elemental variables that can be manipulated by the central nervous system across tasks and external conditions, that is, referent coordinates (RCs) for the effectors at the selected level of analysis (see Chapter 7).

9.3. Inverse optimization

Recently, attempts to infer CFs from experimental data have been addressed as *inverse optimization* (Bottasso et al. 2006; Terekhov et al. 2010). Such attempts, however, have been so far relatively impractical, requiring large data arrays, involving many unknown parameters, and being applicable only to specific groups of motor tasks. Nevertheless, these studies represent an important step toward resolving one of the problems inherent to optimization approaches: the subjective, arbitrary selection of CFs. One of the approaches, termed analytical inverse optimization (ANIO), searched for an analytical solution to the problem within certain constraints imposed on CF and tasks. In particular, these functions were assumed to be differentiable and additive; that is, total cost $CF(x)$ was assumed to represent the sum of costs, $cf_i(x_i)$, incurred by changing each of the elemental variables (x_i), and task constraints were assumed to be linear. An example is the production of a combination of total force and total moment-of-force magnitudes by four fingers pressing in parallel (Fig. 9.3).

The necessary conditions for solving the direct optimization problem are known as the Lagrange principle. For the inverse optimization problem with additive CF and linear constraints, the following formulation of the Lagrange principle is valid: If g_i variables are continuously differentiable, then their derivatives for observed solutions are equal to the derivatives of the solution space. Estimating CF from observations does not lead to a unique solution. For example, multiplying a CF by a positive number or adding a number does not change the optimal solution. The *theorem of uniqueness* provides sufficient conditions for uniqueness, up to linear terms, of solutions to the inverse optimization problem (Terekhov et al. 2010). In this case, the

9.4. OPTIMALITY-STABILITY TRADE-OFF 133

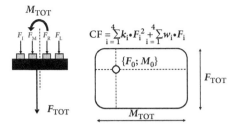

Figure 9.3 An illustration of the task of accurate total force (F_{TOT}) and total moment of force (M_{TOT}) production by the four fingers of a hand. A typical cost function, shown as CF, for such tasks reconstructed with analytical inverse optimization is a second-order linear polynomial function. k and w are constants; i stands for fingers: index, middle, ring, and little. This function is reconstructed based on multiple trials with the production of different $\{F_{TOT}; M_{TOT}\}$ combinations. One of them is shown as $\{F_0; M_0\}$.

equation provided by the Lagrange principle can be used to determine the objective function.

After a CF has been defined, one can use it to compute predicted optimal solutions for the same set of task constraints as those used at the ANIO step. Further, the goodness of fit can be estimated comparing the spaces of actual data and of computed optimal solutions. For example, in force-moment production tasks by four fingers, both spaces across variations in task constraints are close to hyperplanes in the four-dimensional finger force space. Then, measuring the dihedral angle ($α_D$) between the two planes has been used as a goodness-of-fit metric. Experiments have shown rather low values of $α_D$ (commonly under 1°) in young, healthy persons and much higher values in the healthy elderly (about 5°) and patients with neurological disorders such as Parkinson disease (about 17°) and cerebellar disorders (25°) (e.g., Park et al. 2011, 2013a; Niu et al. 2012; Latash 2012c). These results suggest that ANIO may be practically useful in providing metrics of how well persons within various populations follow a single self-imposed optimization criterion.

9.4. Optimality-stability trade-off

The introduced definitions of optimality and stability put them into competition with each other. Indeed, consider, for simplicity, the task of accurate production of a certain total force magnitude with two fingers (Fig. 9.4). There are two groups of findings related to this behavior. First, repeating this task multiple times typically produces clouds of data points, shown in Fig. 9.4, elongated along the solution space (the UCM for the task; cf. Chapter 5). Second, changing the target force magnitude leads to about proportional scaling of the finger forces, preserving their sharing pattern, as shown in Fig. 9.4 (cf. Li et al. 1998). The former phenomenon suggests a two-finger synergy stabilizing total force. The second phenomenon may be interpreted as following a particular optimization criterion defining the location of the center of the cloud of data points. Within each cloud of data points, deviations from the center

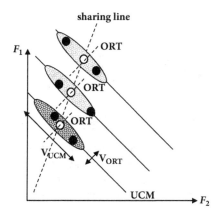

Figure 9.4 The task of accurate total force (F_{TOT}) production with two fingers pressing in parallel with forces F_1 and F_2. Note that changing the force magnitude preserves the sharing pattern between the fingers. Optimal solutions are shown with open circles, non-optimal ones with filled circles. UCM, uncontrolled manifold; ORT, orthogonal to the UCM space.

represent violations of the optimality criterion, even if relatively small ones. So, large spread of the inter-trial data along the UCM (large V_{UCM}) is a sign of imperfect optimization and, simultaneously, a sign of high stability.

Consider now the task of changing quickly the magnitude of total force produced by two fingers. An optimal solution would preserve the sharing pattern; that is, the trajectory in the individual finger force space would follow the constant sharing line. If the optimal sharing is not 50:50, this strategy would lead to components both orthogonal to the UCM and along the UCM (Fig. 9.5A). Since stability along the UCM is low, even a small component along that space is expected to produce large deviations (large ME motion; see Chapters 5 and 18), that is, a change in the sharing— a violation of the optimality criterion. This problem is expected to be solved if the preferred sharing is exactly 50:50 (Fig. 9.5B). Then, the component of action along the UCM is close to zero, and the task can be performed with minimal change in the sharing. These illustrations suggest that a large spread of inter-trial data along the UCM—a signature of stability!—is problematic if one wants to keep the sharing pattern at a single, optimal value. Deviations from an optimal sharing mean that a quick change in the performance variable is expected to produce large destabilizing effects along the UCM, which amplify those deviations.

Sharing of total force across fingers may be close to 50:50 for pairs of fingers that are nearly equally strong (e.g., the index and middle fingers). This is not the case when the fingers differ significantly in their force-generating capabilities (e.g., the index and little fingers). In four-finger tasks, the sharing patterns are consistently different from 25:25:25:25 (e.g., Li et al. 1998; Zatsiorsky et al. 1998). In such tasks with steady force production, total force is strongly stabilized by inter-trial adjustments of individual finger forces; that is, V_{UCM} is consistently larger than V_{ORT}. This implies the possibility of large destabilizing effects along the UCM during voluntary force changes.

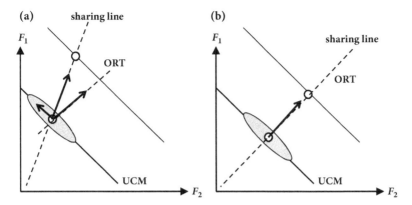

Figure 9.5 (a): During quick total force production by two fingers producing forces F_1 and F_2, if the force sharing is not 50:50, a substantial vector component is directed along the uncontrolled manifold (UCM). (b): This problem is avoided if the sharing is 50:50 and the force vector is directed along ORT (orthogonal to the UCM space).

Nevertheless, the sharing pattern seems to be preserved relatively well over a large range of force magnitudes. This observation allows two interpretations. First, visual and somatosensory feedback is used to keep the sharing pattern unchanged during voluntary force modulation. This interpretation is valid for relatively slow force changes but is questionable for quick force pulses. Unfortunately, so far, no studies have explored changes in the sharing pattern over a wide range of force rates.

An alternative interpretation is based on the theory of control with spatial RCs and assumes that, at the control level, the sharing of RCs across the four fingers is indeed close to equal, and the differences in the force magnitudes are due to the differences in the apparent stiffness of the individual fingers (k; see Chapter 12). This situation is illustrated in Fig. 9.6. Note that the differences in individual finger k values transform the 50:50 sharing of RCs into an unequal sharing of forces. This interpretation remains speculative since no data are available to support or refute it.

9.5. Agility-stability trade-off

Another type of trade-off can be seen in such phenomena as specialization of human hands and fingers. Handedness has traditionally been viewed as a consequence of brain lateralization, and the two hands viewed as "clumsy" (non-dominant) and "agile" (dominant). This view has been challenged by the dynamic dominance hypothesis (Sainburg 2002, 2005; see Chapter 10), which suggests that the hands are specialized for particular features of everyday tasks. In other words, the non-dominant hand is not a "bad hand" but a hand that may be better than the dominant hand for particular tasks. Imagine, for example, that you are putting a nail into a board. Assuming that you are right-handed, you will likely hold the nail with the left hand and move the hammer with the right hand. Trying to switch the hands leads to feeling

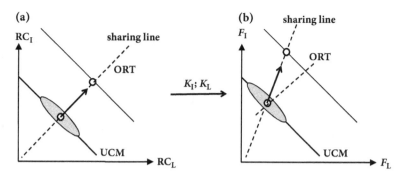

Figure 9.6 If the sharing of referent coordinate (RC) across the fingers (I, index; L, little) is 50:50 (a) but the k values (defined by the C-command) differ, $k_I > k_L$, the force sharing is expected to differ from 50:50 (b). UCM, uncontrolled manifold; ORT, orthogonal to the UCM space.

insecure and clumsy in both task components: The left hand is not as good at moving the hammer quickly and accurately, and the right hand is not as good at keeping the nail motionless. We can reformulate these intuitive observations as follows: The left hand is better at steady-state tasks requiring high stability of a salient performance variable, and the right hand is better at tasks that require fast and accurate changes in a performance variable.

This conclusion has been corroborated in a series of experimental studies with accurate steady force and force pulse production by the dominant and non-dominant hands. During steady-state tasks, the non-dominant hand has consistently shown higher indices of total force-stabilizing synergies compared to the dominant hand (Park et al. 2012; de Freitas et al. 2019). The dominant hand, however, showed larger indices of ASAs prior to the quick force pulse and smaller destabilizing effects on total force during the pulse (Zhang et al. 2006; de Freitas et al. 2019). These results are readily compatible with the dynamic dominance hypothesis and suggest that hand specialization is a particular example of a trade-off between ensuring high stability and being highly agile.

Somewhat similar conclusions have been reached in studies of individual fingers during prehensile and force production tasks. When a person holds a handle in the air, individual finger forces form two groups. Within one of the groups, individual force variables covary to stabilize the grip force, and within the other group, they covary to stabilize the total moment of force applied to the handle (Shim et al. 2003; Zatsiorsky et al. 2004). In other words, there are two synergies, force-stabilizing and moment-stabilizing ones. Interestingly, the two "central fingers" (middle and ring) vary their normal forces primarily with the grip force, while the two lateral fingers (index and little) vary their normal forces with the resultant moment of force. Note that stabilization of the moment of force is crucial for a variety of functional tasks such as taking a sip from a glass, using a spoon to eat soup, using implements to write, and using a variety of other hand-held tools. The importance of stabilization of the moment of force has also been confirmed in studies of multi-finger

9.6. MISSING PIECES OF THE MOSAIC 137

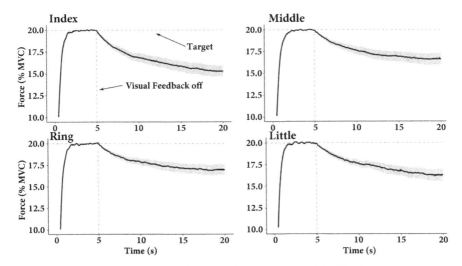

Figure 9.7 Unintentional force drifts caused by turning visual feedback on the force magnitude off are larger in the index finger compared to the middle and ring fingers. The initial force level was always 20% of the maximal voluntary contraction (MVC) force of the finger. Visual feedback was turned off 5 s after the trial initiation. The lines show the averages across the subjects with standard error shades. Reproduced by permission from Abolins and Latash 2022a.

synergies during pressing tasks (Latash et al. 2001; Scholz et al. 2002; see more in Chapter 12).

Recent studies of unintentional force drifts (see Chapter 16) have shown that the drifts were larger during tasks performed by the most agile index finger as compared to similar tasks performed by the middle and ring fingers (Abolins and Latash 2022a; see Fig. 9.7). Moreover, when the subjects were provided with visual feedback on their performance after a series of trials, they were able to avoid force drifts in future trials by the middle and ring fingers. The index finger, however, showed residual drifts even after a sequence of six trial blocks with visual feedback provided after each block. These observations resemble the difference between the dominant and non-dominant hands and suggest that effector specialization for stability or agility may be a general phenomenon, unrelated to brain lateralization because, obviously, all the fingers of a hand are controlled primarily by the same, contralateral hemisphere.

9.6. Missing pieces of the mosaic

One of the obvious missing pieces is the lack of an exact definition for agility. Optimality is defined relatively exactly, as long as a CF has been selected. Stability is also a well-defined concept. The ability to move fast in a desired direction in the space of elemental variables, however, is not as well defined. If moving fast is the only requirement, then the only constraints on this ability are properties of the load and muscle strength. On the other hand, if one wants to combine such an action with proper adjustments of stability of the salient variable and minimal violations of an optimality principle, the three concepts seem to be intimately linked together.

Indeed, if a variable is highly stable in the initial state, this corresponds to relatively large inter-trial variance within the UCM for the salient variable. Large V_{UCM} increases chances that a quick action in a particular trial starts from a non-optimal sharing and produces a perturbation along the UCM, thus violating the sharing even more. If the action is not completely unexpected (the actor knows that it may be required and/or performs it in a self-paced manner), it is possible that a quick action from a non-optimal sharing is preceded by ASAs (see Chapter 8), that is, a drop of stability, which helps avoid major displacements along the UCM. This is a conundrum requiring experimental exploration.

So far, the effects of dominance on indices of stability have only been explored in experiments involving the two hands of right-handed persons. Left-handers and ambidextrous persons have not been studied to a similar extent, which is an untapped source of potentially very useful information on the relations among stability, agility, and optimality. In addition, comparison to footedness could be very useful; a recent study (Rannama et al. 2023) remains the only one exploring these concepts in the lower extremities. Footedness has been reported as a better predictor of brain lateralization compared to handedness (Elias et al. 1998; Tran et al. 2014). On the other hand, the reviewed studies of finger force production suggest that specialization within the three-dimensional space of stability-agility-optimality may be unrelated to hemispheric lateralization.

Of course, as emphasized in a number of earlier chapters, all studies in spaces of mechanical variables have to be taken with a huge grain of salt if one is interested in processes of the neural control of movement. Unfortunately, studies in spaces of control variables (or their direct reflections in mechanical variables such as RC and k) have been sparse. There are signs of ranges of preferred solutions in spaces of control variables (see Ambike et al. 2016a; Reschechtko and Latash 2017; Nardon et al. 2022), which may point at some kind of an optimality principle. There are also signs that selecting a specific range may reflect on the stability of the salient performance variable. So far, these variables have not been used to quantify aspects of agility such as ASAs.

10
Brain Circuitry

Studies of the relations between brain activity and movements form two large groups. Within the first group, researchers focus on individual neurons and neuronal populations in a particular brain structure assumed to play a major role in the production of movements, for example, the primary motor cortex, the cerebellum, the thalamus, and so forth, and explore the relations between activation patterns in that area and movement characteristics. Within the second group, researchers are interested in the role of interactions among various brain structures, for example, in the loops involving the basal ganglia and the cerebellum, in the performance of movements. Within both groups, studies include observing neuronal patterns during prescribed motor tasks, quantifying the effects of stimulation of neuronal populations, and observing changes in movement patterns under injuries to or dysfunction of specific brain structures. As emphasized later, all three methods provide indirect information, which requires a theoretical framework to be deciphered.

Most direct information has been obtained from invasive studies in animals with direct recording from or stimulation of brain neurons. In addition, well-controlled injuries to brain structures can be used in animals. However, generalizing findings from animal studies to humans is questionable. In addition, it is harder to set specific motor tasks for animals. Using invasive methods in human studies is limited for obvious ethical reasons. Indirect sources of information include electrophysiological methods, such as electroencephalography; brain imaging methods, such as magnetic resonance imaging (MRI) and positron emission tomography (PET); and stimulation methods, such as transcranial magnetic stimulation (TMS) and transcranial direct current stimulation (tDCS). All these methods suffer from shortcomings, for example, the poor spatial resolution of EEG, even the high-density version of EEG, and the poor temporal resolution of MRI and PET. Studies of the effects of injuries to specific brain structures also have limitations. First, such injuries in humans are rarely local and they lead to secondary plastic changes across a variety of apparently unaffected brain structures. Second, interpretation of movement deficits following even a well-controlled injury (e.g., in an animal study) is also not straightforward. For example, similar effects on a class of movements can be expected with an injury to a structure generating relevant neural control variables, to pathways carrying this information to intermediate structures within the central nervous system, to intermediate nuclei involved in further processing of neural control variables, and to structures involved in delivering information from other structures, including sensory information, necessary for this class of movements.

Within this brief review, we will contrast two theoretical views. According to the first, more popular, one, the brain structures can prescribe peripheral variables recorded and analyzed in typical studies—kinetic, kinematic, and electromyographic (EMG)—with the help of "internal models" or some invisible strings from brain structures to effectors. Some studies formulate such assumptions explicitly, but more commonly, they are implied in interpretations of observations frequently representing correlations between neuronal activation in the brain and peripheral variables such as muscle activations and movement mechanics. The alternative view is that the brain uses indirect, parametric control of movements (see Chapter 7), and observed peripheral patterns are results of a dynamic interaction between the neural control signals and external force field mediated, in particular, by spinal reflexes. Within this theoretical view, patterns of peripheral variables cannot be seen as direct reflections of patterns of brain activity.

10.1. What variables are encoded by brain signals?

Exploring the relations between neuronal activity in brain structures and induced movements in humans dates back to Penfield's classical studies performed on awake patients during open-brain surgery. In particular, Penfield (reviewed in Penfield and Rasmussen 1950) stimulated different sites within the motor cortex with brief electrical stimuli, observed the induced jerky contractions in various parts of the body, and put these observations together into a distorted human image drawn over the motor cortex, addressed as the motor *homunculus*. Further, similar distorted human images were reported for other brain structures involved in the motor and sensory functions (Rijntjes et al. 1999; Yamada et al. 2007). Later, however, the concept of a coherent image of the body was challenged by better-controlled studies with better spatial resolution, which reported much more mosaic representations (Schieber 2001; Schieber and Santello 2004; reviewed in Lemon and Morecraft 2023) with numerous phenomena of convergence and divergence when stimulation of different groups of neurons produced action by the same effector (convergence) while stimulation of apparently the same group of neurons induced action by various effectors (divergence).

The method of local electrical stimulation was later used in studies of monkeys by the group of Evarts (1968) and Asanuma (1973). The stimulation produced muscle contractions in the forearm, which was fixed by an external device and could only generate forces against the stop. Changing the strength of the stimulation produced changes in the force response, and these results were interpreted as cortical neurons encoding muscle forces. Clearly, this conclusion could not be taken as general, in particular, because of the well-known dependence of muscle force on its length and velocity (i.e., on the actual external force during the induced muscle contractions). Indeed, further studies, in non-isometric conditions, showed that the stimulation of cortical neurons could generate different mechanical effects, leading to conclusions that cortical neurons encode a range of mechanical and electrophysiological variables including force direction in isometric conditions, movement extent, levels of muscle activation, and more complex variables reflecting interjoint coordination (Sergio and Kalaska 1997, 1998; Kakei et al. 1999; Cabel et al. 2001; Holdefer and Miller 2002).

As discussed earlier (Chapters 2 and 4), descending signals cannot encode peripheral mechanical variables because these depend on the external force field, which is never perfectly predictable. This was emphasized by Bernstein nearly 100 years ago (Bernstein 1935, translation in Bongaardt 2001; Bernstein 1947, translation in Latash

10.1. WHAT VARIABLES ARE ENCODED BY BRAIN SIGNALS?

2020a). An alternative view developed within the theory of control with spatial referent coordinates (RCs) suggests that descending signals to an effector encode its RC changes with time (Feldman 2015; Chapter 7). For a single muscle, the RC is equivalent to the threshold of the stretch reflex, λ (Feldman 1966, 1986; see Chapter 3). For a multimuscle effector, the RC is multidimensional and can be described with two basic commands, the reciprocal and coactivation commands (R- and C-commands).

This theory allows drawing predictions with respect to muscle responses to a standard TMS stimulus applied over the primary motor cortex, which differ from traditional predictions that muscle response is defined by the background muscle activation level. Figure 10.1 illustrates the basic idea. The muscle activation level is defined by descending signals and reflex feedback, in particular from length-sensitive spindle endings. If the muscle activation level is matched at two different positions, the descending signal is smaller for a position where the muscle is longer because of the larger reflex contribution. This means that the activity of neuronal populations in the primary motor cortex is smaller and, respectively, their excitability to a standard external stimulus is also expected to be smaller. These predictions were confirmed experimentally in a study of the TMS-induced responses in wrist muscles quantified during matched muscle activation conditions at different wrist angles (Raptis et al. 2010; Ilmane et al. 2013).

A number of later studies have suggested that activity along other descending pathways, in particular the vestibulospinal pathway, also leads to changes in RCs for the agonist-antagonist muscle groups during quiet standing (Zhang et al. 2018). A similar conclusion has been generalized for spinal circuitry, in particular for the output of the central pattern generator for locomotion (Feldman et al. 2021). These conclusions, however, are valid for output structures of the central nervous system targeting segmental neuronal apparatus involved in the production of movements (i.e., the sources of major descending pathways). The question remained open with respect to neuronal populations across various brain structures.

As a transition to the next section, it makes sense to quote a statement from one of Bernstein's early publications (1935): "In the higher motor centers of the brain (very

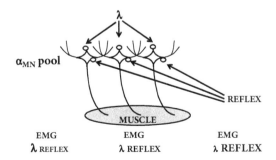

Figure 10.1 Convergence of descending (λ) and reflex pathways on an alpha-motoneuronal pool (α_{MN}, top). Matching overall output of the pool (same EMG level) is associated with negative covariation of the two inputs as reflected in the different font sizes (bottom).

probably in the cortex of the large hemispheres) one can find a localized reflection of a projection of the external space in a form in which the subject perceives the external space with motor means" (p. 80 in Bongaardt 2001). This statement suggests that activity patterns in local neuronal populations in the brain are likely to reflect not peripheral mechanics but perception of the external space (and the task) by the actor (compare to the main ideas within the field of ecological psychology, Gibson 1979). This insight has been confirmed by a series of classical studies linking activation in neuronal populations to action characteristics (reviewed in Georgopoulos et al. 1992; Georgopoulos and Carpenter 2015; Feldman 2015).

10.2. What is encoded by neuronal populations?

A highly important step in studies of the brain and its role in motor function was made by Apostolos Georgopoulos and his research team (Georgopoulos et al. 1982, 1983, 1986, 1989). They switched from analysis of individual cortical neurons or very small neuronal groups, as in the aforementioned studies by Evarts and Asanuma, to analysis of neuronal activity over large populations. The method involved recording large sets of cortical neurons (and later neurons in other brain structures) with chronically implanted arrays of electrodes in monkeys trained to perform specific motor tasks. In the most classical paradigm, the monkeys performed center-out movements with a forelimb, that is, movements from a fixed initial position of the hand to targets arranged along a circle (Fig. 10.2A). The modulation of activity of individual neurons just prior to movement initiation followed the so-called cosine pattern: It was maximal for a certain movement direction ("preferred direction" for that neuron in Fig. 10.2B) and it smoothly reduced for directions deviating from the preferred one. When the monkey moved in the opposite direction (π in Fig. 10.2B), the baseline level of activation could even show a drop.

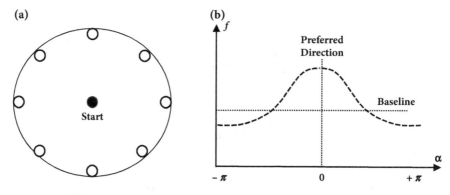

Figure 10.2 (a): An illustration of the center-out task. (b): A typical cosine modulation of the activity of a cortical neuron (its firing frequency, f) with deviations of movement direction (α) from a certain, "preferred" direction ($\alpha = 0$).

10.2. WHAT IS ENCODED BY NEURONAL POPULATIONS?   143

When such cosine-like tuning curves were measured for each of the recorded neurons, at the next step each neuron's activity was replaced with a vector pointing in its preferred direction with the length equal to the number of action potentials generated by the neuron just prior to the action initiation. Across the population, these vectors formed hedgehog-looking patterns, and the resultant vector pointed in the direction of movement (Fig. 10.3). A few conclusions can be drawn from these findings. First, individual neurons do not encode anything task specific. As a group, the neurons encode movement direction by the forelimb. Note that movement direction is a complex variable, reflecting changes in activation of many muscles. So, using the earlier Bernstein's quote, it reflected "a projection of the external space."

Another important step was made in experiments when the monkeys were trained to point not at the presented target but at a specific angle from it (Georgopoulos et al. 1989). In other words, they had to mentally rotate the movement direction after the target presentation prior to movement initiation. Those studies showed neuronal population vectors first pointing in the direction of the presented target and then rotating toward the movement directions. These observations fit the later portion of the Bernstein quote: "subject perceives the external space with motor means."

Later, this method was used to explore activation of neurons in other brain areas including other cortical areas, as well as other brain structures involved in motor function, such as the basal ganglia and cerebellum (Fortier et al. 1989). The overall result has been confirmed across brain structures, and properties of neuronal population vectors have been shown to depend not only on movement direction but also on a number of other factors such as the initial arm posture (Scott and Kalaska 1995) and the external force field (Arce et al. 2010).

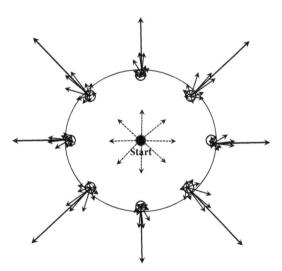

Figure 10.3 When the cosine distribution (see Fig. 10.2B) for each cortical neuron is replaced with a vector pointing in the preferred direction of that particular neuron (thin vector lines), the sum of these vectors (thick arrows) across the population of neurons points in the direction of movement (dashed arrows).

The correlations between characteristics of activation in cortical neuronal populations and movement parameters led to a series of studies relevant to recovery of the lost motor function in certain groups of neurological patients (Nicolelis and Lebedev 2009; Birbaumer et al. 2012; Wolpaw et al. 2018). The idea is relatively straightforward: If a population of neurons can encode movement direction but implementation of the movement is impossible due to a neurological condition, activation patterns in the population can be used to drive a robotic or prosthetic device performing the movement. Proof-of-concept studies were performed on monkeys with arrays of electrodes implanted into the motor cortex. Patterns of activation of the neurons during forearm movements in different directions were recorded and later used to drive a robot performing the movement for the monkey (Carmena et al. 2003; Wahnoun et al. 2006; Ma et al. 2017).

Success in those studies led to studies in paralyzed patients (e.g., those suffering from amyotrophic lateral sclerosis) with the help of implanted arrays of electrodes connected to a computer (Wolpaw et al. 2018). In the early trials, such devices, addressed as brain-computer interfaces, were marginally successful, but recent improvements in the involved technologies promise their broader application for motor rehabilitation in the future.

The concepts of neuronal populations and their directional tuning have been reconsidered within the theory of control with RCs (Feldman 2019). This article suggested that neuronal populations encode planned changes in the RC for the involved effectors, including the direction of RC shifts for movement production by changes in the R-command. In addition, neuronal populations can encode changes in the C-command, which by itself may not lead to a major overt action in the environment but is important for other movement characteristics, such as its stability (see Chapter 19). In reproducible experimental conditions, these RC(t) shifts are expected to lead to consistent changes in a variety of peripheral variables such as effector force vector, displacement, muscle activation, and so forth. If the external force field changes, however, the same neuronal population vectors can correlate with changes in a different performance variable. This hypothesis is a logical extension of the theory of control with RCs, which is compatible with a body of published data but requires more experimental exploration.

10.3. Relations between brain structures and functions

Over the last two centuries, two polar views have dominated the discussion on the relations between brain anatomy and body functions. According to one of the views, there are "places" in the brain responsible for specific functions. This view has been supported by observations of reproducible functional deficits after an injury to a particular brain area and, more recently, by effects of controlled brain stimulation. In the 19th century, an extreme version of this view led to phrenology, a "science" of the relations between bumps on the skull and features of one's character and abilities. The alternative view is that the relations between brain structures and functions are plastic, time varying, and shaped by experience. This view has been supported by phenomena of functional recovery following a localized brain injury and well-documented phenomena of neural plasticity, experience-dependent changes in neural projections from peripheral sensory endings to brain structures and from the brain to the spinal cord. At an extreme, this view led to the idea that the brain at birth represented *tabula rasa*, that is, unstructured neural media able to form any connections to support any function. Both views were criticized by Bernstein in his first book written in the mid-1930s and published about 70 years later (Bernstein 2003).

Bernstein viewed individual brain structures and circuits as plastic *operators*, that is, elements with a specific role that can be modified by experience and shared by various functions (Bernstein 1935). Later, Houk (2005) introduced the notion of distributed processing modules (DPMs), which is close in spirit to Bernstein's notion of operators. DPMs for the production of movements were supposed to involve transcortical loops via such structures as the basal ganglia, cerebellum, and thalamus. Relatively recently, another related concept was introduced, that of a *heksor*, a neural network supporting a functional behavior in spite of possible plastic changes in involved structures (Wolpaw and Kamesar 2022). The concept of heksor is closely related to the stability of salient performance variables and may be viewed as a neural substrate of performance-stabilizing synergies (see Chapter 6).

It is obvious that unstable behaviors are not functional. For example, every time we perform a task, such as picking up an object with a hand, we guess important characteristics of the object, such as its weight, friction of the surface of contact, and so forth, based on the visual information and earlier experience with similar objects. The actual values of those characteristics commonly differ from those guesses, leading to unpredictable perturbations of the ongoing action. Nevertheless, such tasks are accomplished successfully because of the dynamical stability of salient performance variables. Indices of stability (synergy indices such as ΔV; see Chapter 6) change with relatively minor, subclinical changes in the person's ability to perform movement, for example, with healthy aging and under fatigue (Shinohara et al. 2003b; Olafsdottir et al. 2007b; Singh et al. 2010; Singh and Latash 2011). Such changes have been reported early in patients with a number of neurological disorders involving the basal ganglia and cerebellum (reviewed in Latash and Huang 2015; for more details see Chapter 22) in support of the role of those subcortical structures in motor synergies hypothesized many years ago by Bernstein (1947).

Figure 10.4 illustrates two important aspects of the control of stability of a multi-finger action involving the production of a certain level of total force followed by a quick, self-paced force pulse into the target. The z-transformed synergy index (ΔV_z; see Chapter 6; Jo et al. 2016b) was averaged over the subjects within three groups: patients with Parkinson disease (PD), cortical stroke survivors, and control subjects. Only the group with PD showed a consistently reduced synergy index during the steady-state part of the task, suggesting the importance of the circuits via the basal ganglia for multi-finger force-stabilizing synergies. No such effects were seen after mild cortical stroke. In contrast, both patient groups showed significantly reduced anticipatory synergy adjustments (ASAs) in preparation for the force pulse. These observations suggest that anticipatory adjustments of stability of performance may involve different circuitry compared to that involved in ensuring steady-state stability.

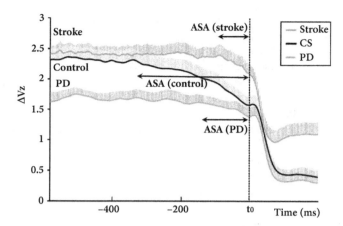

Figure 10.4 The index of force-stabilizing synergy during four-finger accurate total force production followed by a quick, self-paced force pulse into the target. The data for three groups are shown: patients with Parkinson disease (PD), cortical stroke survivors, and control subjects (CS). ASA, anticipatory synergy adjustment. Reproduced by permission from Latash 2019.

An important step was made by the group of Gregor Schöner, who suggested a general scheme for the production of movements with tentative mapping on neurophysiological structures (Fig. 10.5; Martin et al. 2009, 2019). This scheme starts from selecting a target and an end-effector to perform the movement (e.g., a hand to grasp the target or a finger to point at the target). This process has been described within the dynamic field theory (Erlhagen and Schöner 2002), which incorporates both intrinsic (e.g., memory-based) and extrinsic (e.g., sensory signals of different modalities) time-varying signals. Further, timing of the planned action is defined, including its initiation time and duration. At the next step, the planned action is converted into time courses of RCs for the task-specific end-effector. Further, the task-level RC patterns are converted into RC patterns for the involved elements down to muscles and motor units, for example, as described earlier (see Fig. 6.10 in Chapter 6). Interactions between the external force field and RC patterns produce movement mechanics, which is sensed by peripheral sensory endings.

The stability of salient task-specific variables is ensured by back-coupling loops both within the central nervous system and from sensory endings. This can happen at different levels, involving or not involving changes in the task-specific RC(t) patterns (i.e., updating the neural process at different levels). In other words, synergic changes in the neural control process can happen at different levels within the central nervous system, from the spinal cord (see the discussion of intramuscle synergies in Chapter 11) to subcortical loops and, possibly, to intracortical loops. According to the scheme in Fig. 10.5, it seems feasible that synergies at the level of R- and C-commands generated at the highest, task-related, level (see Ambike et al. 2016a; Nardon et al. 2022) emerge at the cortical level. Note that the other basic feature of synergies, the organization of multiple elements into a relatively small number of groups (modes), has been reported in studies of cortical neuronal populations (Shenoy et al. 2013;

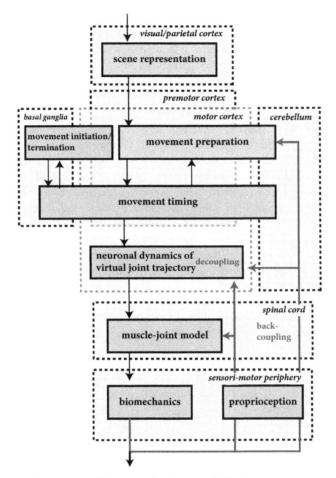

Figure 10.5 An illustration of the hierarchical control of voluntary movement and involved brain structures. Reproduced by permission from Martin et al. 2019.

Gallego et al. 2020; Zimnik and Churchland 2021). In contrast, synergies at the level of sharing the task across elements, for example, across digits in multidigit prehensile tasks, depend crucially on subcortical loops (reviewed in Latash and Huang 2015). Of course, all the processes illustrated in Fig. 10.5 should be viewed not as computational but as reflecting dynamical interactions among the involved structures happening in the course of movement preparation and execution.

10.4. The role of spinal circuitry

Recent series of studies on cats have suggested that spinal reflex loops are organized in a way that contributes to stabilization of important mechanical variables. In particular, the patterns of force-sensitive reflex projections across the muscles of a hindlimb suggest important contributions to such common actions as locomotion (Nichols 2002, 2018). On the other hand, studies of both nonhuman animals and humans have suggested that characteristics of reflex projections are plastic and can change with training (Wolpaw 2007; Thompson and Wolpaw 2015, 2021). How is the functional role of reflexes maintained given that their properties change with practice? The ability of the spinal cord to support multiple classes of actions while avoiding their negative interference has been addressed as negotiated equilibrium, and neural circuits involved in the process have been viewed as examples of heksors, neuronal circuitry providing stability for salient elements (Wolpaw and Kamesar 2022).

The spinal cord is known to contain neuronal machinery able to generate patterned activity of alpha-motoneuronal pools (e.g., typical activation patterns supporting locomotion). This conclusion has been supported by observations of induced locomotion in spinal quadruped preparations and of the so-called fictive locomotion, that is, patterned activity of alpha-motoneurons in the absence of muscle contractions suppressed by curare (see Chapter 14). In addition, humans with clinically complete spinal cord injury can show cyclical leg movements in response to stimulation of the lumbar enlargement of the spinal cord (reviewed in Shapkova 2004; Gerasimenko et al. 2008). The spinal neuronal structures able to produce patterned activity have been addressed as central pattern generators. The exact neuronal circuits of central pattern generators in mammals remain unknown; they have been described only for animals with relatively simple central nervous systems, such as lamprey (Orlovsky et al. 1999).

The scheme in Fig. 10.5 contains back-coupling loops that do not involve supraspinal structures, both from peripheral sensory endings (reflex loops) and within the spinal cord, for example, the systems of recurrent inhibition (cf. Hultborn et al. 2004; Latash et al. 2005). The importance of such loops for movement stability has been confirmed in studies of intra-muscle synergies stabilizing the muscle output by covarying contributions of motor unit groups (MU-modes; see Chapter 11). Spinal circuitry revealed in intramuscle synergies may be seen as a reliable mechanism involved in all functional movements. The spinal nature of intra-muscle synergies has been supported by several groups of findings. First, these synergies stabilize unintentional, reflex-induced changes in muscle force (Madarshahian et al. 2022). This is in contrast to the lack of such synergies stabilizing the hand force produced by several fingers by covaried contributions of the individual fingers. Second, these synergies in hand muscles show no effects of hand dominance (Madarshahian and Latash 2022b; Madarshahian et al. 2022), which have been traditionally associated with specialization of the two brain hemispheres—the main topic of the next section. Third, unintentional drifts in performance are associated with disappearance of multifinger synergies without detrimental effects on intra-muscle synergies (De et al. 2024).

10.5. Effects of dominance

Traditionally, the two hands have been viewed as a "good one" (dominant) and a "not so good one" (non-dominant). Moreover, right-handedness was viewed as normal and left-handedness as deviant in some societies, leading to re--educating left-handers to use their right hands for writing and eating with a spoon. Both these views are now obsolete. In particular, both hands (arms) are currently viewed as "good" and specialized for particular aspects of everyday behaviors. For example, when a person wants to slice a piece of bread, the dominant hand moves the knife while the non-dominant hand holds the loaf of bread. Similarly, if one wants to hammer a nail into a board, the dominant hand moves the hammer while the non-dominant hand holds the nail. If one switches the hands, both task components feel awkward. Not only does the non-dominant hand "feel" clumsy moving the knife or the hammer, but also the dominant hand is not so good at keeping the loaf or the nail motionless.

These intuitive considerations and a series of experiments on the kinematics of arm movement led to the emergence of the dynamic dominance hypothesis (Sainburg 2002, 2005), which emphasizes specialization of the two hemispheres and the corresponding contralateral arms. In its original formulation, the dynamic dominance hypothesis accepts the idea that signals from the cortical hemispheres encode peripheral mechanical variables, components of joint moments of force. The dominant hemisphere has been assumed to take responsibility for feedforward generation of neural signals translating into forces (Bagesteiro and Sainburg 2002; Yadav and Sainburg 2014). The non-dominant hemisphere has been assumed to be responsible for stabilization of steady state, formulated as the control of effector impedance (Bagesteiro and Sainburg 2003; Yadav and Sainburg 2014), which is an ambiguous term in biomechanics.

This hypothesis received support in numerous studies (reviewed in Sainburg 2005, 2014). For example, during quick hand movements performed across a variety of conditions, including unusual force fields, the dominant hand shows more straight trajectories, while the non-dominant hand is better at reaching the target accurately in spite of the more curved trajectories (Mutha et al. 2012; Sainburg et al. 2016). In addition, predictable changes in features of hand trajectories have been reported in studies of reaching movements performed by patients after cortical stroke affecting the dominant or non-dominant hemisphere (Schaefer et al. 2007, 2012; Maenza et al. 2020). Moreover, these changes are seen in both more affected (contralesional) and less affected (ipsilesional) extremities.

Hemispheric specialization reflected in movements of the dominant and non-dominant hands is an established fact. However, given another well-established fact—that the brain cannot prescribe patterns of peripheral mechanical variables (see Bernstein 1947 and Chapter 4)—the dynamic dominance hypothesis requires reformulation in terms compatible with the physiology of the body and current views on the control of movements. As of now, there are no data suggesting differences between the hands in patterns of the R- and C-commands (e.g., Ambike et al. 2016a). On the other hand, indices of enslaving—unintentional force production by fingers (Zatsiorsky et al. 2000; see Chapter 12)—are higher in the non-dominant hand, suggesting worse finger individuation in that hand.

Information relevant to the dynamic dominance hypothesis has come from studies of multi-finger synergies in the two hands. In particular, during accurate steady force production tasks, the non-dominant hand shows an advantage reflected in higher indices of force-stabilizing synergies (Park et al. 2012; de Freitas et al. 2019). In contrast, when the task requires quick force pulse production, the dominant hand shows larger synergy adjustments prior to the force change, avoiding major force destabilization during the force pulse (Zhang et al. 2006; de Freitas et al. 2019). These observations allow reformulating the dynamic dominance hypothesis: The dominant hand has an advantage in avoiding movement destabilization during fast actions, while the non-dominant hand has higher synergy indices with respect to salient performance variables in steady-state tasks. Note that hand differences are not seen in the indices of intra-muscle synergies (Madarshahian and Latash 2022b; Madarshahian et al. 2022), an expected result given that hand dominance is a supraspinal phenomenon while intra-muscle synergies are likely based on spinal circuitry (see Chapter 11).

Unintentional force drifts in the absence of visual feedback on the force magnitude (Vaillancourt and Russell 2002; Ambike et al. 2015a; see later in Chapter 16) have been discussed as consequences of partial loss of stability of force magnitude caused by turning the visual feedback off. Such drifts are smaller in the nondominant hand, as illustrated in Fig. 10.6 showing the time series of total force and total moment of force during four-finger pressing tasks over the period without visual feedback in the

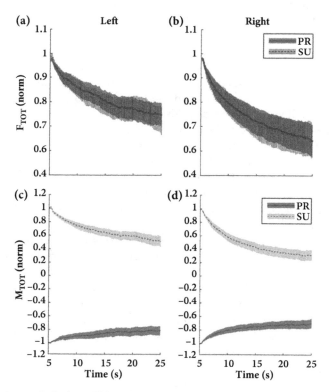

Figure 10.6 Force drifts (a and b) and moment drifts (c and d) in the tasks of accurate force-moment production tasks by the dominant (right) and non-dominant (left) hands. Note the larger drifts in both force and moment in the right hand for both moment directions, pronation (PR) and supination (SU). Means across subjects with standard deviation shades are shown. Reproduced by permission from Parsa et al. 2016.

force-moment accurate production tasks. Note that the force drifts were similar for the initial pronation and supination moments, and they were consistently smaller in the left hand of the right-handed subjects. Similar conclusions were drawn based on other studies, including those of patients with PD (Park et al. 2012; Jo et al. 2016a). Figure 10.6 (panels C and D) also shows that the total moment-of-force drifts in the absence of visual feedback toward lower magnitudes and that these drifts are also larger in the dominant hand. Taken together, these results confirm the advantage of the non-dominant hand in ensuring stability of salient variables in steady-state tasks.

10.6. Missing pieces of the mosaic

Current knowledge of the brain mechanisms of motor control is very modest. It is based primarily on correlation studies, which are inherently limited, unless they are designed within a feasible theory on the neural control of movements based on physics and physiology, not on available data processing techniques. So far, only a handful of studies have used such a theory (in our opinion, there is currently only one such theory—the control of movements with spatial referent coordinates), which is a major factor slowing down progress in this field. To use the puzzle metaphor, most studies try to solve the puzzle without any idea of the ultimate image, based only on properties of individual small pieces. The problem is not so much in missing pieces but in the lost or, more precisely, ignored by many researchers, cover of the box showing the ultimate image.

Nevertheless, some of the available pieces of the puzzle are very informative. In addition to the already mentioned studies, we should mention the studies of two neural streams through the cortical areas related to visual perception (Goodale et al. 1991; Goodale and Milner 1992; Kravitz et al. 2011). The two streams, addressed as ventral and rostral, are involved in two aspects of visual information processing, perception to report and perception to act, which are differentially affected in certain cases of cortical stroke. Somewhat similar conclusions have been drawn with respect to somatosensory perception, in particular based on qualitatively different effects of certain factors on force perception as assessed in verbal reports vs. force-matching tasks (Cuadra et al. 2020, 2021b). However, physiological streams potentially involved in these two aspects of kinesthetic perception remain unknown.

Although the importance of subcortical loops through the basal ganglia and cerebellum in performance-stabilizing multi-muscle synergies may be viewed as well established, the role of the cortex remains obscure. Observations in patients after cortical stroke are ambiguous: Some of those studies reported impaired synergic control, while other studies failed to find stroke-related differences (Reisman and Scholz 2003; Gera et al. 2016a, 2016b; Jo et al. 2016b; see Chapter 22). Brain mechanisms of synergies at the level of sharing motor tasks at the highest level of the hierarchy between the R- and C-commands remain unknown.

Evidence for signals from the brain to spinal structures encoding time patterns of referent coordinates for the involved muscles comes from both animal and human studies (Feldman and Orlovsky 1972; Raptis et al. 2010; Zhang et al. 2018). How these signals are shared among the main descending pathways, however, remains unknown. Is there specialization of descending pathways for encoding the R- and C-command time changes? Are these pathways specialized for movement tasks, postural tasks, force production tasks, and so forth? What neurophysiological structures and circuits are involved in sharing the task-specific control patterns at the highest control level among $\{R; C\}$ pairs of commands to apparent effectors, such as joints, limbs, digits, muscle groups, and so forth? All of these major questions remain missing pieces of the puzzle.

PART III
EFFECTORS AND BEHAVIORS

11
Synergic Control of a Muscle

The concept of *synergy* was introduced by Bernstein to describe the neural control of large muscle groups assigned to the second-from-the-bottom level of control, within the multilevel hierarchy, with two functions: uniting elements (associated with muscles) into a few groups and ensuring dynamic stability of movements with respect to salient performance variables (Chapters 2 and 6). The control of single muscles was assigned to the lowest control level, the level of tone. This scheme assumes that the concept of *muscle* is well defined, which is not obvious. There are muscles consisting of so-called compartments, for example, the extrinsic muscles of the hand, with multiple distal tendons attached to different digits. Compartments are assumed to represent anatomically and physiologically distinct groups of muscle fibers dedicated to the control of individual digits (Jeneson et al. 1990; Serlin and Schieber 1993; Mariappan et al. 2010). On the other hand, there are groups of limb muscles that are sometimes viewed together as single entities, such as quadriceps femoris, triceps surae, triceps brachii, and biceps brachii.

The smallest controllable element of the central nervous system is the *motor unit* (MU), a single alpha-motoneuron with all the muscle fibers it innervates. In healthy muscles, individual muscle fibers show no multiple innervation; that is, a single fiber is innervated only by a single branch of the axon of a single alpha-motoneuron, which makes the concept of MU defined rather unambiguously. The behavior of a muscle reflects superposition of the effects of changes in the output of numerous individual MUs. Formally, this is a problem of motor redundancy, and the notion of synergy is directly applicable. The well-known size principle (Henneman principle; Henneman et al. 1965) imposes constraints on possible patterns of MU recruitment but does not prescribe a unique solution because each MU can be recruited with varying frequencies of action potential generation. Recent studies reviewed in this chapter have suggested that both features of synergies indeed can be identified at the intra-muscle level.

Expanding the range of applicability of the concept of synergy not only to more complex actions and non-motor functions, such as perception and cognition, but also "downstream" to more simple systems is an important step in understanding the neural control of natural, functional actions. The hierarchical structure of control introduced earlier (see Fig. 6.10 in Chapter 6) assumes stabilization of salient variables at multiple levels. The results of studies of intra-muscle synergies suggest, however, that there may be only two control levels where synergies stabilizing performance can be observed: the task level and the intra-muscle level. The former reflects actors' understanding of the task and identification of important performance variables and then using available sensory modalities to establish desired levels of stability of those variables. The latter reflects the spinal machinery, including reflexes from proprioceptive endings, which may be viewed as relatively task independent, although there is a possibility of task-specific descending control of the gains in the involved circuitry. Other levels in the mentioned scheme may be apparent, motivated by anatomical factors, not actively participating in the process of neural control.

Analysis of intra-muscle synergies may reflect action by the relatively well-known spinal circuitry of feedback loops, both within the central nervous system and from peripheral sensory endings. Such studies may represent a much-needed window into mapping synergic control onto specific neurophysiological mechanisms with implications for both the basic neurophysiology of movement and movement disorders associated with dysfunction of specific structures within the central nervous system.

11.1. Steps and challenges in analysis of motor unit–based synergies

The behavior of a single MU is, in some basic aspects, similar to the behavior of a muscle described in Chapter 3. The alpha-motoneuron receives synaptic input from descending pathways (we imply here inputs from various structures including those mediated by spinal interneurons) and from reflex projections from peripheral sensory endings, in particular those sensitive to muscle length (Fig. 11.1A). If the descending input is fixed, the MU is recruited at a particular muscle length and shows an increase in its frequency of action potential generation (firing frequency, f_{MU}) with length increase, as illustrated in Fig. 11.1B). This curve is similar to the force-length characteristic of the muscle (e.g., see Fig. 3.5 in Chapter 3), and we are going to address the threshold length when MU recruitment starts as λ_{MU}. The contribution of an MU to muscle force depends on its frequency of firing and size; note that, typically, the f_{MU} of smaller MUs is higher.

Two mechanisms contribute to changes in muscle activation level and force, recruitment of new MUs and frequency modulation of the already recruited ones. The relative contributions of these two mechanisms differ across muscles. In some muscles

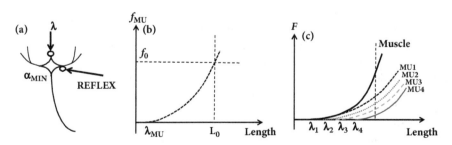

Figure 11.1 (a): Convergence of central and reflex inputs on alpha-motoneurons (α-MN). (b): The dependence of the firing frequency of a motor unit (f_{MU}) on muscle length given a value of the stretch reflex threshold (λ_{MU}). For example, at length L_0, the firing frequency is f_0. (c): The muscle force-length characteristic reflects the sum of the contributions of individual motor units. Frequency-length characteristics are shown for four MUs (MU1, MU2, MU3, and MU4) with different stretch reflex threshold values ($\lambda_1, \lambda_2, \lambda_3,$ and λ_4).

(e.g., hand muscles), recruitment is complete by the time the person produces about 50% of the maximal voluntary contraction (MVC) level in isometric conditions and further force increase is due to frequency modulation (Kukulka and Clamann 1981; Iyer et al. 1994). In other muscles (e.g., large trunk muscles), recruitment continues for nearly up to 100% of MVC. Figure 11.1C illustrates the contributions of both mechanisms to the muscle force-length characteristic, which represents the algebraic sum of the individual MU characteristics.

To analyze synergies at the level of multiple MUs composing a muscle, one has to ask two questions. First, do individual MUs form stable groups with parallel changes in f_{MU} within each group? Second, does involvement of those groups covary, for example, across repetitive trials at the same task, reflecting stabilization of a task-specific salient variable? Addressing these problems experimentally, for example, similar to the way multimuscle synergies have been explored within the framework of the uncontrolled manifold (UCM) hypothesis (see Chapters 5 and 6), faces a number of challenges. First, each MU produces an essentially discrete output, a sequence of individual action potentials (spikes) with varying inter-spike intervals. This makes defining the apparent elemental variable in this analysis, f_{MU}, difficult and ambiguous. Second, there are essential nonlinearities in the contributions of individual MUs to muscle action related to their threshold properties, as illustrated in Fig. 11.1. Third, some MUs show relatively low f_{MU} values, on the order of a few hertz, which makes time resolution of any analysis based on those values rather poor. Fourth, mapping f_{MU} values on potentially important performance variables such as, for example, muscle force is problematic due to a host of obvious factors including varying sizes and contraction properties of individual MUs.

The first step of any analysis of synergies is identifying and measuring elemental variables. Recording the f_{MU} of individual MUs can be done either with indwelling intramuscular electrodes or with arrays of surface electrodes (reviewed in Merletti et al. 2008; Farina et al. 2014; Enoka 2019). Both methods have their advantages and disadvantages. Intramuscular recording can be uncomfortable, it leads to recording MUs within a relatively small space in the vicinity of the tip of the inserted electrode, and it may lose some of the signals with changes in muscle fiber geometry during its contraction. Surface recording avoids these problems but adds an element of ambiguity in identifying individual MU action potentials, which can only be ensured with some probability depending on the type of electrode array and signal processing software.

The two mechanisms of changes in MU contribution to muscle action, recruitment and frequency modulation, have been combined at the level of data processing with the help of a Hann filter, which also helps to produce smooth differentiable functions of f_{MU} from the discrete sequences of MU action potentials. This filter replaces a discrete event with a sine function of a particular width centered about that event. To avoid discontinuities, the width of the function has to be sufficient to overlap with the next function centered about the next action potential. This effectively reduces the time resolution to the maximal inter-spike interval for the lowest f_{MU}. The last problem, mapping changes in f_{MU} values onto changes in a performance variable, has been solved in a way similar to the one used earlier in the analysis of multi-muscle synergies, that is, with the help of multiple linear regression (cf. Krishnamoorthy et al. 2003b; Madarshahian et al. 2021).

Principal component analysis in the space of filtered, smooth f_{MU} functions revealed at least two MU groups with parallel scaling of f_{MU} within each group addressed as MU-modes (Madarshahian et al. 2021). MU-modes showed robust composition during force increase (recruitment) and decrease (derecruitment) phases of action as revealed by close correlation of the loading factors for the MU-modes defined using the f_{MU} functions recorded during these phases of the action separately. In addition, the UCM analysis performed in the space of f_{MU} within a single MU-mode showed strong positive covariation of the individual f_{MU} values across repetitive trials, that is, $V_{ORT} \gg V_{UCM}$, also suggesting the robust, inflexible composition of the MU-modes.

In contrast, when the UCM-based analysis was performed across cycles of force production in isometric conditions, it revealed strong synergies in the abundant space of MU-modes stabilizing force magnitude, that is, $V_{UCM} \gg V_{ORT}$. Hence, both features of synergies have been confirmed in the analysis of individual f_{MU} functions within a muscle. These results are illustrated in Fig. 11.2 showing the force-length characteristics of individual MUs contributing to a MU-mode (panel A) and the contributions of MU-mode characteristics to the muscle force-length characteristic (panel B). Note that the λ_{MU} values for individual MUs vary in the same direction, thus leading to effective shift in λ_{Mode}. In contrast, the λ_{Mode} values for the two modes typically vary in opposite directions, thus stabilizing the muscle characteristic and its force in isometric conditions (shown with the vertical dashed lines in both panels).

The first studies of MU-based synergies were performed using the tasks of cyclical accurate force production during pressing with one finger or several fingers acting in parallel and MU identification in the extrinsic flexor muscle, the flexor digitorum superficialis (FDS), a compartmentalized muscle (Madarshahian et al. 2021; Madarshahian and Latash 2022a). This led to a suspicion that MU-modes reflected the contributions from individual muscle compartments. One of the studies tried to test this hypothesis indirectly and got a negative answer. Later studies (Ricotta et al.

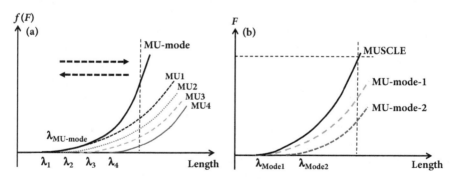

Figure 11.2 (a): Frequency-length (f_{MU}-L) characteristics of individual motor units (MU) sum up to produce the MU-mode characteristic. The dashed arrows show parallel changes in the individual λ values for the four MUs (MU1, MU2, MU3, and MU4) comprising the MU-mode. (b): Two MU-mode characteristics sum up to produce the muscle force-length characteristic. The corresponding λs (λ_{Mode-1} and λ_{Mode-2}) typically show counterdirectional shifts.

2023a, 2023b) generalized the main findings, the existence of robust MU-modes and stabilization of muscle force by MU-mode covariation, using MUs recorded in the tibialis anterior, a noncompartmentalized muscle, thus showing that both features of synergies do not depend crucially on muscle compartmentalization.

11.2. Agonist-antagonist interactions at the motor unit level

All human movements involve muscles or muscle groups acting in the desired direction (agonists) and those resisting motion in the desired direction (antagonists). As described earlier (Chapter 3), the neural control of movement along a single spatial coordinate can be described with time shifts of λs for the agonist and antagonist muscles, $\lambda_{AG}(t)$ and $\lambda_{ANT}(t)$, or, alternatively, with two commands representing time shifts of the spatial coordinate where the resultant force is zero (reciprocal command or R-command) and of the spatial range where both muscles are active simultaneously (coactivation command or C-command). The presence of two basic commands at the neural level makes control of movement along a single direction by a single effector abundant. For example, force change from an initial to a final value (Fig. 11.3) can be produced by a change in the R-command only, the C-command only, or both.

Studies of intramuscle synergies within single muscles during accurate cyclical force production during finger pressing in isometric conditions documented MU-modes and MU-mode-based synergies stabilizing force magnitude in both agonist and antagonist muscles analyzed separately (Madarshahian and Latash 2022b; Madarshahian et al. 2022). Synergies within the antagonist muscle (extensor digitorum communis [EDC]) look counterintuitive because this muscle produces force directed against the desired resultant force. However, if both agonist and antagonist muscles increase force simultaneously (e.g., due to a change in the C-command), changes in the resultant force are expected to correlate with changes in both agonist

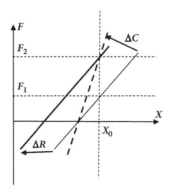

Figure 11.3 A change in force magnitude (from F_1 to F_2) produced by an effector in isometric conditions at a fixed coordinate X_0 can be produced by a change in the reciprocal command (ΔR), coactivation command (ΔC), or both.

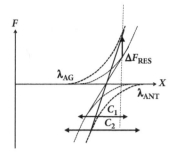

Figure 11.4 An increase in the resultant force magnitude (ΔF_{RES}) produced by an increase in the coactivation command (C-command from C_1 to C_2) in isometric conditions is accompanied by an increase in the force of both agonist and antagonist muscles. The agonist and antagonist muscle characteristics are shown with thin lines and thick dashed lines for the smaller and larger C-command, respectively.

and antagonist force as illustrated in Fig. 11.4. Indeed, as shown in one of the following illustrations (Fig. 11.6), during cyclical force production, subjects show much more consistent modulation of the C-command within the force cycle as compared to the modulation of the R-command, which was subject specific and inconsistent across tasks.

When MUs recorded from the agonist and antagonist muscles were analyzed together, principal component analysis revealed two consistent MU-modes with contrasting patterns of the loading factors. One of those two MU-modes had all MUs, from both the FDS and EDC, loaded with the same sign (Fig. 11.5). This implies that a change in the magnitude of that MU-mode led to parallel changes, an increase or a decrease, in the firing frequency of MUs in both the agonist and antagonist, that is,

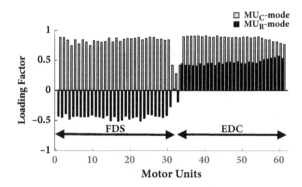

Figure 11.5 A typical set of the loading factors for the agonist (flexor digitorum superficialis [FDS]) and antagonist (extensor digitorum communis [EDC]) muscles for the first two MU-modes. Note that one of the MU-modes (MU_C) is characterized by the loading factors of the same sign (light bars), while the other MU-mode (MU_R, dark bars) has loading factors of the same sign within each muscle but opposite between the muscles. Reproduced by permission from Madarshahian and Latash 2022b.

11.2. AGONIST-ANTAGONIST INTERACTIONS AT THE MOTOR UNIT LEVEL 161

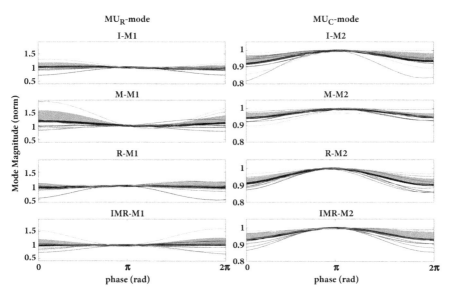

Figure 11.6 The magnitude of one of the MU-modes (reciprocal mode, MU_R) shows inconsistent modulation within the force cycle across subjects (thin lines), while the magnitude of the other MU-more (coactivation mode, MU_C) shows consistent modulation within the force cycle. This is consistent across tasks performed by individual fingers (I, index; M, middle; R, ring) and by the three fingers together (IMR). Reproduced by permission from Madarshahian and Latash (2022b).

changes in the amount of muscle coactivation. Hence, this MU-mode may be seen as a peripheral reflection of the C-command (MU_C-mode in Fig. 11.5). The other MU-mode had MUs from the flexor loaded with opposite signs to those for the MUs from the extensor. A change in the magnitude of that MU-mode led to a decrease in f_{MU} in one of the opposing muscles and an increase in f_{MU} within the other muscle. This is a peripheral reflection of the R-command (MU_R-mode in Fig. 11.5).

Analysis of the time profiles of the two MU-modes within the force cycle showed consistent patterns of the MU_C-mode magnitude expected if the force modulation was performed by changing the C-command. In contrast, the patterns of the MU_R-mode magnitude were variable across subjects and not consistent across tasks performed by different fingers acting alone and several fingers pressing together (Fig. 11.6). These findings led to the conclusion that force modulation in the task was performed primarily by modulating the C-command, a somewhat unexpected conclusion given that earlier studies suggested the existence of a hierarchy between the two commands, with the R-command being hierarchically higher (Levin and Dimov 1997). Indeed, if you coactivate muscles acting at a joint voluntarily and then perform a quick movement to a new spatial coordinate, the amount of coactivation will be about the same in the initial and final positions: The coactivation zone will be carried to a new spatial coordinate defined by the R-command.

The opposite conclusion reached in the force production experiments suggests that, possibly, the control of tasks performed in isometric conditions is organized differently to movement tasks. In particular, as illustrated earlier in Fig. 11.3, one can perform force modulation by changing the C-command only. In contrast, it is impossible to perform movement in isotonic conditions without a resisting load by changing the C-command only. As illustrated in Fig. 11.3, changing the C-command can only change the slope of the force-coordinate characteristic without changing its intercept; that is, no movement would take place if the external load is zero.

Analysis of force-stabilizing synergies in the spaces of MU-modes defined across the flexor and extensor muscles led to somewhat surprising results. Indeed, the task variable (resultant force) represented the combined action of the opposing muscles. Hence, synergies defined in the space of MU-modes across the muscles were expected to be more directly relevant to the salient performance variable. However, no force-stabilizing synergies were found in the space of MU_R- and MU_C-modes in the presence of strong synergies in the spaces of MU-modes defined for the flexor muscle and extensor muscle separately (Madarshahian and Latash 2022b; Madarshahian et al. 2022; illustrated in Fig. 11.7). There are two non–mutually exclusive interpretations of this result. First, the inconsistent modulation of the MU_R-mode with force suggests that the subjects used consistently only one of the two basic commands, namely the C-command, to modulate force across cycles. Such a strategy turns the originally abundant task into a non-abundant one, which has no room for performance-stabilizing synergies. Second, one can view the control of an

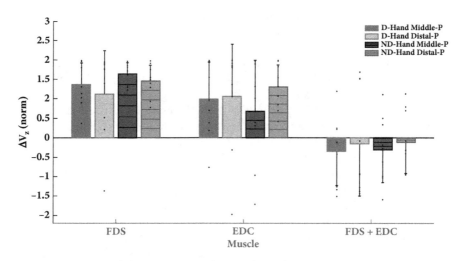

Figure 11.7 Force-stabilizing synergies ($\Delta V_Z > 0$) are observed within the spaces of motor unit modes (MU-modes) defined for the agonist muscle (flexor digitorum superficialis [FDS]) and for the antagonist muscle (extensor digitorum communis [EDC]) separately. No such synergies are observed in the spaces of MU-modes defined for both muscles combined, FDS + EDC. This is consistent during tasks performed by the dominant (D) and non-dominant (ND) hands and during force production at the middle phalanges (FDS is the prime mover) and at the distal phalanges (FDS plays a supporting role). Reproduced by permission from Madarshahian and Latash (2022b).

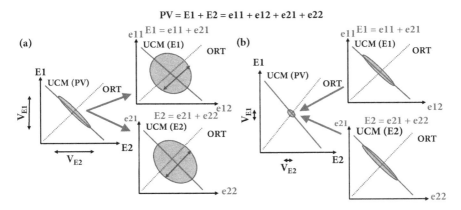

Figure 11.8 Synergies at two levels of a hierarchy are in competition. (a): A strong synergy at the higher level (elements E1 and E2, $V_{UCM} > V_{ORT}$) requires high variability of both E1 and E2, resulting in large V_{ORT} at the lower level (elements, e11, e12, e21, and e22). (b): Strong synergies at the lower level result in small variance of both E1 and E2, thus constraining V_{UCM} at the higher level.

agonist-antagonist muscle pair as a hierarchical system: The task is shared between the two muscles, and then the action of each muscle is shared among its MU-modes. This leads to a trade-off between synergies at the two levels described in Chapter 6 (see also Gorniak et al. 2007, 2009). Indeed, a strong intra-muscle synergy at the single muscle level ($V_{UCM-1} \gg V_{ORT-1}$) makes variance of each muscle's action low (V_{ORT-1} at the single-muscle level in Fig. 11.8). This makes it impossible to have large V_{UCM-2} at the between-muscle level since the projection of V_{UCM-2} on each axis cannot be larger than V_{ORT-1}.

Grouping MUs across muscles into MU-modes is not limited to agonist-antagonist muscle pairs. As shown in a study involving several joints of the human upper extremity, MU-modes (addressed as *synergies* in that study) can unite MUs from muscles crossing different joints if these muscles are involved in a common motor task (Del Vecchio et al. 2023; Hug et al. 2023), for example, if some of the muscles move the hand-held object and other muscles adjust the grip force as needed to ensure grip safety and avoid slippage (explored at the level of referent coordinate [RC] commands in Ambike et al. 2015b).

11.3. Stabilization of reflex-induced force changes

One method to explore possible contributions of spinal vs. supraspinal circuitry to intra-muscle synergies is to quantify them for reflex-induced changes in the salient performance variable. There is, however, a problem of defining *reflex*. Although this term has been used literally for hundreds of years, its explicit, unambiguous definition is elusive, and relatively recently a discussion on the utility of this term in movement studies has not led to a consensus (cf. Prochazka et al. 2000). Commonly, reflexes are defined in contrast to voluntary actions. Reflexes have been assumed to be more stereotypical, tightly linked to the sensory stimulus, and involving relatively simple loops through the central nervous system and limited groups of muscles in the vicinity of the stimulus. In contrast, voluntary movements are more variable, are not always linked to a sensory stimulus, and involve complex neural pathways and large muscle groups distributed throughout the body. All these features are, however, unable to distinguish reflexes from voluntary movements with certainty. There are muscle responses to stimuli that are viewed as reflexes by all researchers, for example, monosynaptic reflexes in response to quick muscle stretch (e.g., as in tendon tap) or electrical stimulation of Ia afferents from muscle spindles. More complex, polysynaptic responses to stimuli, however, are less unambiguous and have been interpreted as both reflexes and quick voluntary actions triggered by the stimulus (reviewed in Pruszynski 2014; Reschechtko and Pruszynski 2020). Some of those reactions have been viewed as involving supraspinal structures, including transcortical loops.

Within the theory of control with shifts in spatial RCs (see Chapters 3 and 7; reviewed in Feldman 2015), any voluntary movement involves important contributions from the stretch reflex. Within this theory, voluntary movements can be defined as those associated with RC shifts for the effectors, and involuntary movements as those produced by changes in external forces in the absence of RC shifts. In human studies, however, it is hard to ascertain that subjects are not responding to a change in external forces, although this ability has been assumed in multiple studies where the subjects were instructed and trained "not to intervene voluntarily" with effector reactions to changes in the external forces. One study directly compared the behavior under two instructions, "not to intervene voluntarily" and "react as quickly as possible" (Latash 1994). It reported more consistent behavior under the former instruction, indirectly corroborating the assumption that movements under this assumption were primarily of a reflex origin.

With all these caveats in mind, one study explored whether covaried contributions of MU-modes stabilized force changes induced by a smooth change in the coordinate of force application, leading to changes in the length of muscles involved in the action by the effector (Madarshahian et al. 2022). This study used a special device called the "inverse piano" (IP; Martin et al. 2011; see also Chapter 7), which could lift and lower fingers in a controlled fashion while recording simultaneously the forces produced by the fingers. Figure 11.9A illustrates the changes in total pressing force produced by the four fingers of a hand pressing in parallel induced by smooth lifting of the fingers by 1 cm over 0.5 s. These parameters were selected to minimize the chance of phasic reflexes and unintentional reactions from subjects. Note the smooth increase in the total finger force during the finger-lifting episode. As shown in earlier studies, such a manipulation was associated with strong linear correlation between the force and coordinate expected from the RC control and the subject's noninterference with the perturbation (e.g., Ambike et al. 2016a; see Fig. 7.6 in Chapter 7).

The study documented MU-mode-based synergies in both the agonist (FDS) and antagonist (EDC) muscles, but not in the space of MU-modes defined across the two muscles—similar to the aforementioned results during voluntary force production. The synergies stabilized force magnitude prior to the IP perturbations and after the

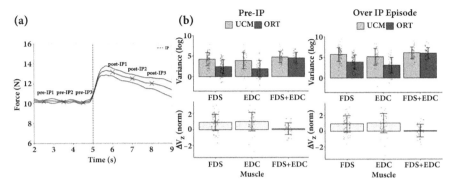

Figure 11.9 (a): Changes in the pressing force (*F*) by a set of fingers produced by a smooth lifting of the fingers by 1 cm over 0.5 s. Note the reflex-induced increase in *F*. (b): Synergies stabilizing steady-state force magnitude ($V_{UCM} > V_{ORT}$; $\Delta V_Z > 0$) before the "inverse piano" episode (Pre-IP) are seen in spaces of MU-modes defined for the agonist (flexor digitorum superficialis [FDS]) and for the antagonist muscle (extensor digitorum communis [EDC]) separately. No such synergies are observed in the spaces of MU-modes defined for both muscles combined, FDS + EDC. The same is true for synergies stabilizing force magnitude change during the "inverse piano" episode (IP episode). Modified by permission from Madarshahian et al. 2022.

perturbations (at the new steady state), and also the change in force produced by the perturbation. The synergy indices are illustrated in Fig. 11.9B.

In contrast, when the same data were analyzed in the space of individual finger forces, synergies were seen over the steady-state phases of the task, but not during the force change induced by the IP. Multi-finger synergies stabilizing force magnitude in isometric conditions have been viewed as mediated by supraspinal subcortical loops, in particular based on observations in neurological patients with disorders of the basal ganglia and cerebellum (reviewed in Latash and Huang 2015; see Chapter 22). The lack of such synergies stabilizing the IP-induced force change indirectly corroborates the spinal origin of these force changes and of the corresponding intra-muscle synergies.

11.4. Spinal vs. supraspinal synergies

Another source of observations supporting the hypothesis of a spinal nature of MU-mode synergies is the lack of effects of hand dominance on indices of those synergies (Madarshahian and Latash 2022a; Madarshahian et al. 2022). These observations stand in contrast to several studies documenting the effects of dominance on multifinger force-stabilizing synergies (Park et al. 2012; de Freitas et al. 2019), including a study that quantified synergy indices at both levels, multi-finger and multi-MU-mode, in a single experiment (Madarshahian et al. 2022). Studies of multi-finger synergies documented higher synergy indices during steady-state force production tasks in the non-dominant hand, consistent with the dynamical dominance hypothesis (Sainburg 2005), which assumes specialization of the non-dominant hand for stabilization of steady states.

Hand dominance is a phenomenon traditionally associated with specialization of the two large hemispheres (see Chapter 10). It is natural to assume, therefore, that phenomena showing effects of dominance involve cortical or transcortical circuitry. This assumption has been supported in a number of studies documenting impaired performance-stabilizing multi-muscle and multi-finger synergies in patients with disorders of the cerebellum and basal ganglia (Park et al. 2012, 2013; Jo et al. 2015). These observations strongly suggest the importance of transcortical loops for multi-effector synergies, in particular those involving a number of fingers within a hand, compatible with the effects of hand dominance on such synergies.

The hierarchical system of control with RCs (see Fig. 6.10 in Chapter 6) theoretically has room for synergies stabilizing the output of any level of the assumed hierarchy. On the other hand, there are inherent trade-offs between synergies at different hierarchical levels (see Chapter 6), suggesting that the neural controller may have to decide stability of what variables is crucial for success at the task. The reviewed findings suggest strongly that spinal mechanisms are involved in stabilization of actions by individual muscles, while supraspinal circuits are involved in stabilization of salient performance variables defined by the task. Whether there are synergies stabilizing action at intermediate levels remains an open question.

Spinal reflexes require numerous repetitions to demonstrate plastic changes (Wolpaw and Carp 1993; Thompson and Wolpaw 2015, 2021). Hence, MU-mode-based synergies in healthy persons may be viewed as robust, acting across tasks and effector sets. Using the language of control with spatial RCs, these synergies stabilize values (time profiles) of RCs for individual muscles, which are equivalent to the stretch reflex threshold λ, by covaried adjustments of RCs to MU-modes. In isometric conditions, these synergies stabilize force magnitude. However, they are expected to stabilize other mechanical variables depending on the external load characteristic. For example, during slow movements in isotonic conditions, stabilization of the λ time profile is expected to result in a stable trajectory of the effector. It is possible to modify the synergic effects of spinal circuitry by acting on spinal interneurons mediating the effects of spinal reflexes. Such effects have not been explored experimentally.

Of course, in cases of major changes in the spinal circuitry, for example, in patients with spasticity, changes in the composition of MU-modes and actions of MU-mode-based synergies may be expected. However, as of now, there are no publications exploring MU-modes and synergies in patients with spasticity (e.g., following mild cortical stroke or incomplete spinal cord injury). Changes in MU-modes and synergies can also be expected in healthy persons under conditions associated with changes in spinal reflexes, for example, under fatigue. So far, only one study explored the effects of fatigue on MU-modes and MU-mode-based synergies (Ricotta et al. 2023a). It did not find consistent changes in the MU-mode composition associated with fatigue, and the synergy index dropped, possibly as a result of gain changes in feedback loops following an increase in presynaptic inhibition caused by fatigue (cf. Bigland-Ritchie et al. 1986a, 1986b; Woods et al. 1987).

11.4. SPINAL VS. SUPRASPINAL SYNERGIES

In contrast, supraspinal circuits are known for widespread plasticity, which creates room for changes in task-specific multi-muscle synergies in cases of neurological injury and also with learning motor skills. These changes in task-specific synergies are not always intuitively predictable. In particular, cortical stroke leads to inconsistent changes in performance-stabilizing synergies: Depending on the site of injury and the task, synergy indices show either no change or a decrease (Reisman and Scholz 2003; Gera et al. 2016a, 2016b; Jo et al. 2016). In contrast, dysfunction of subcortical structures, such as the cerebellum and basal ganglia, leads to consistent impairments of performance-stabilizing synergies across tasks and effector sets (see Chapter 22).

Depending on the stage of skill development, practice can lead to variable effects on synergy indices, which can increase, decrease, or stay unchanged (reviewed in Wu and Latash 2014; see Chapter 20). Practice is expected to lead to more accurate performance, that is, to a drop in V_{ORT}. Changes in V_{UCM}, however, are harder to predict because, by definition of the UCM, they do not affect the salient performance variables. Figure 11.10 illustrates three scenarios using the task of accurate production of the sum of two elemental variables (e.g., total force produced by two fingers). Across the three bottom panels, V_{ORT} is reduced, but V_{UCM} is not changed (A, the synergy index becomes larger), reduced proportionally (B, the synergy index is unchanged), and reduced more than V_{ORT} (C, the synergy index drops). Typically, early stages of practice are associated with emergence and strengthening of synergies—as illustrated in Fig. 11.10A. Later, practice can lead to a decrease in the variance component along

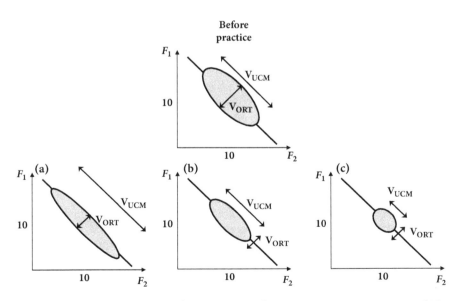

Figure 11.10 Possible changes in the two inter-trial variance components, V_{UCM} and V_{ORT}, with skill learning. V_{ORT} is always reduced, but V_{UCM} may not change or increase (a: the synergy index increases), reduce proportionally (b: the synergy index is unchanged), and reduce more than V_{ORT} (c: the synergy index drops).

the UCM (V$_{UCM}$) without further reduction in the synergy index affecting the performance variable (V$_{ORT}$), possibly due to a floor effect (as in Fig. 11.10C). In other words, actors explore smaller and smaller areas within the UCM, possibly associated with some kind of an optimality criterion. Whether such changes can happen at spinal levels remains unknown. Recent analysis based on the notion of heksors (Wolpaw and Kamesar 2022) suggests that this is indeed possible.

11.5. Missing pieces of the mosaic

One of the least intuitive findings in studies of multi-MU-mode synergies is their presence in spaces of MU-modes defined for a muscle (no matter agonist or antagonist!), but not in spaces of MU-modes defined across the agonist-antagonist muscle pair. Indeed, resultant effector force is defined by the balance of forces produced by the agonist and antagonist. However, a few studies have confirmed the absence of force-stabilizing synergies in the spaces of MU-modes reflecting potentially the two basic commands, the *R*- and *C*-commands. Exploration of the potential role of the two commands led to another unexpected finding. In contrast to earlier studies suggesting that the *R*-command was hierarchically higher than the *C*-command, studies of accurate cyclical isometric force production showed consistent phase modulation of the *C*-command but not the *R*-command. This is a feasible strategy (see Fig. 11.3) but not the most intuitive one. The role of the *C*-command in movements and force production tasks is far from being understood as reflected also in recent studies of force perception and unintentional force drifts (Abolins et al. 2020; Cuadra et al. 2021a, 2021b).

Can MU-based synergies be organized at higher levels within the control hierarchy illustrated in Chapter 6 associated with muscle coordination in larger groups? A number of studies provided evidence for correlated changes in the MU firing rates across muscles involved in a common task (Del Vecchio et al. 2023; Hug et al. 2023). None of those studies, however, tried to link those findings to stability of salient task-specific performance variables, which is a crucial feature of synergies. This obvious knowledge gap is waiting to be filled.

Another method developed recently to explore coordination in groups of MUs involves computing indices of coherence of MU firing frequencies. The coherence analysis quantifies the cross-correlation of MU spike trains in the frequency domain, with distinct frequency bands representing varying neural levels of control. In particular, the alpha band (5 to 12 Hz) has been associated with the spinal reflex action, in particular from muscle spindle endings (Christakos et al. 2006; Erimaki and Christakos 2008). The beta band (15 to 30 Hz) has been assumed to reflect cortical and subcortical processes (Halliday et al. 1999; Lowery et al. 2007). The gamma band (30 to 60 Hz) has also been linked to cortical activity, especially during quickly changing muscle contractions (Farmer et al. 1993; Lowery et al. 2007). So, quantifying the coherence at different frequency bandwidths potentially allows one to distinguish the effects of different origins of the diffused input to alpha-motoneurons. Relations between indices of intra-muscle synergies and indices of coherence have not been explored.

A related question is: Can within-muscle MU-based synergies coexist with similar synergies defined by the action of MUs across muscles? The reviewed studies of agonist-antagonist muscle pairs so far have provided a negative answer to this question. This result is also consistent with the aforementioned trade-off between synergies defined at different levels within a hierarchy. Accepting this answer, however, requires exploration of MU-based synergies across multiple agonists or across muscles crossing different joints but united by the task into units contributing to the same mechanical variables.

The conclusion that MU-based synergies reflect spinal mechanisms while multi-muscle synergies reflect supraspinal circuitry potentially makes exploration of both types of synergies in patients important to identify secondary changes in apparently unaffected neurophysiological structures and circuits. For example, is spinal cord injury associated with secondary changes in supraspinal structures affecting multi-muscle synergies? Also, can a supraspinal disorder (e.g., Parkinson disease) lead to secondary changes at the spinal level affecting intra-muscle synergies? At this time, these questions remain without an answer.

12
The Hand

The human hand is a unique effector, which combines versatility with specialization, and stability of actions with their quickness, and makes even the best artificial grippers look clumsy. It is also involved in several uniquely human actions, such as handwriting and playing musical instruments. These unique abilities of the hand are based on its anatomical design, the elaborate muscular apparatus, and neurophysiological structures and circuits involved in the hand function. In particular, the hand has disproportionally large representations, both motor and sensory, in the cortex of the large hemispheres, reflecting its importance in the everyday functioning of the body and evolutionary success of *Homo sapiens*. The two hands are not identical in their neural control and contributions to the range of everyday hand functions, and the effects of dominance have been discussed in Chapter 10.

In this chapter, we will focus primarily on indices of finger interaction reflecting their nonindependence and the coordination of digits ensuring success in a variety of multi-digit tasks. We will try, as much as possible given the available information, to link features of digit and hand action to the theory of control with spatial referent coordinates (RCs) and the concept of synergies stabilizing salient performance variables (see Chapters 6 and 7). We are not going to cover features of hand action related to selection of specific grasp configurations based on properties of the object, selection of particular locations for the grasp and fingertip contacts, and reactions of hand muscles to changes in the external forces applied to the object. These topics are covered in extensive published reviews (Cesari and Newell 1999; Santello and Soechting 2000; Lukos et al. 2007; Santello et al. 2016).

We are also going to focus primarily on flexion and extension actions of the digits, not so much on their abduction and adduction, although movement and force production along these directions are coupled by muscle action, connective tissues, and, likely, the organization of the neural control of the hand. As a result, the main tasks discussed in this chapter are those of pressing normally to the surface and holding/moving objects using prismatic grasps when only the tips of the digits contact the object and the thumb acts in opposition to the four (or fewer) fingers. We assume non-sticky contact between the fingertip and the object—that is, the fingers can only press, not pull.

12.1. Muscle organization of the hand

Movements of the digits are produced by contractions of two major groups of muscles, intrinsic and extrinsic (Fig. 12.1). The intrinsic muscles have bodies located within the palm. The direct action of an intrinsic muscle is digit specific, producing

12.1. MUSCLE ORGANIZATION OF THE HAND

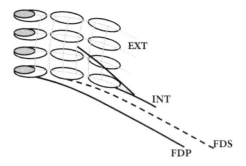

Figure 12.1 Schematic illustration of the intrinsic (INT) and extrinsic hand muscles. FDS, flexor digitorum superficialis; FDP, flexor digitorum profundus; EXT, extensor apparatus.

flexion in the metacarpophalangeal (MCP) joint of the digit. These muscles also contribute to action of the so-called extensor mechanism, a complex passive network of connective tissue on the dorsal side of the hand. As a result, activation of an intrinsic muscle leads to a combination of flexion action in the MCP and extension actions in the interphalangeal (IP) joints.

Extrinsic muscles of the hand produce action by all four fingers. There are two major extrinsic flexors, the flexor digitorum profundus (FDP) and flexor digitorum superficialis (FDS), with the bodies located in the forearm. Each of those muscles has four distal tendons inserted at the four middle phalanges (FDS) and four distal phalanges (FDP). Each of those tendons crosses the MCP joint as well as the wrist joint. As a result, activation of an extrinsic muscle has a multi-joint flexion action.

FDS and FDP are commonly addressed as compartmentalized muscles (Jeneson et al. 1990; Serlin and Schieber 1993; Mariappan et al. 2010). This term implies that each muscle consists of four compartments characterized by a degree of anatomical and neurophysiological independence and consisting of subsets of motor units dedicated primarily to force production by one finger only. In other words, this notion implies that there are effectively four digit-specific muscles in close proximity to each other. In apes, the muscle control of the thumb is similar to that controlling the fingers, and extrinsic flexors consist of five compartments each (Marzke 1992). In humans, this is reflected, in particular, in similar patterns of finger interaction and nonindependence quantified across the four fingers and between the thumb and the fingers (Olafsdottir et al. 2005b). The notion of muscle compartment remains defined imprecisely because of numerous factors including, in particular, force transmission across the compartments.

The extrinsic finger extensor, the extensor digitorum communis, also has four tendons directed at the four fingers. The extensor finger action is also defined by the aforementioned extensor mechanism, which provides strong passive connections across the fingers. The extensor mechanism is a likely contributor to the large indices of finger nonindependence during voluntary extensor action, larger than during similar flexion action (Oliveira et al. 2008; Sanei and Keir 2013).

12.2. Indices of finger interaction

Finger interaction is defined by several factors, which can be classified into peripheral and central (reviewed in Schieber and Santello 2004; Zatsiorsky and Latash 2008). Peripheral factors include connective tissue links between fingers and the presence of multi-finger extrinsic muscles. Central factors range from overlapping cortical representations of fingers, in particular in the primary motor area, to possible reflex projections on subsets of alpha-motoneurons innervating different muscle compartments.

Two phenomena have been observed during single-finger and multi-finger force production tasks, *force deficit* and *enslaving* (Li et al. 1998; Zatsiorsky et al. 2000), illustrated in Fig. 12.2. When a person is asked to press as strongly as possible with a finger, maximal voluntary contraction (MVC) force is significantly larger as compared to a similar task performed by the same finger acting together with other fingers of the hand. Force deficit refers to the apparent loss of force when a finger acts in a group. It has been quantified as a fraction of MVC in a one-finger task lost when the finger is acting in a multi-finger task. When a person is asked to perform the MVC pressing task with one finger, other fingers of the hand produce force in the same direction unintentionally. This phenomenon has been referred to as finger nonindependence or enslaving. The instructed finger is typically addressed as the master finger, while other fingers of the hand are the enslaved ones.

The magnitudes of force deficit and enslaving vary across healthy persons, but there are some consistent patterns in those indices. In particular, force deficit increases as a monotonic nonlinear function with the number of involved fingers. Enslaving indices are typically highest for the ring finger and lowest for the index finger. Note that enslaving can be quantified to reflect how strongly a finger enslaves other fingers or how strongly a finger is enslaved by other fingers. These indices can be computed as the off-diagonal values in Fig. 12.2 averaged over the columns or over the rows, respectively.

A number of studies have provided evidence for the predominantly neural origins of enslaving and force deficit. In particular, similar indices of finger interaction are

	FINGER			
TASK	Index	Middle	Ring	Little
I	**50**	10	6	3
M	9	**45**	15	5
R	8	18	**40**	13
L	6	8	17	**27**
IMRL	33	25	32	17

Figure 12.2 A typical data set of force magnitudes produced by the four fingers during single-finger (I, index; M, middle; R, ring; L, little) and four-finger (IMRL) maximal voluntary contraction (MVC) tests. The master (instructed) finger forces are shown in bold font; enslaved finger forces are those off the main diagonal of the table. Force deficit is the drop in maximal force of all fingers in the four-finger task.

observed when the subjects are pressing not with the fingertips but with the proximal phalanges (Latash et al. 2002b). Note that pressing with the proximal phalanges involves finger-specific intrinsic muscles as the prime movers. So, the presence of multi-finger compartmentalized extrinsic muscles is not expected to define patterns of force production. Enslaving has also been shown to change with specialized training, for example, in musicians (Slobounov et al. 2002), and even following a single session of practice (Wu et al. 2013). More recently, drifts in enslaving toward larger values by 10% to 20% have been reported in trials lasting for 15 to 20 s (see Chapter 16). Clearly, peripheral factors are unlikely to be sufficient to interpret those observations.

Enslaving is not limited to finger pressing. This phenomenon has also been explored and quantified during finger force production in different directions. During extension tasks, the magnitude of enslaving is typically larger than during flexion (pressing) tasks (Oliveira et al. 2008; Sanei and Keir 2013). Abduction-adduction tasks are characterized by complex patterns of unintentional force production, with enslaved force produced in the same direction or in the opposite direction compared to the master finger force (Pataky et al. 2007). Similar patterns of enslaving have also been reported in finger movement tasks and in mixed tasks when the master finger was performing an unopposed movement into flexion and extension and other fingers of the hand were constrained and produced unintentional force changes (Kim et al. 2008).

12.3. Finger modes

The notion of finger modes—hypothetical neural commands reflecting desired involvement of fingers—was originally introduced to describe finger force patterns in MVC tasks when the mode for the master finger was always assumed to be equal to one and the modes for the enslaved fingers equal to zero (Zatsiorsky et al. 1998). Later (Danion et al. 2003), this concept was generalized for submaximal force production and summarized as

$$F = \frac{1}{N^{0.7}}|E|m \qquad (12.1)$$

where F is a four-dimensional finger force vector, $|E|$ is the enslaving matrix, m is a four-dimensional mode vector, and N is the number of instructed (master) fingers. The first term on the right side of this equation reflects the phenomenon of force deficit assumed to produce force scaling not only for the MVC tasks but also for any intermediate force level. In this equation, m can be expressed in fractions of maximal intentional finger involvement. For example, if a person presses with two fingers, index and ring, as strongly as possible, $m = [1\ 0\ 1\ 0]$ (Fig. 12.3). On the other hand, in studies of multi-finger synergies stabilizing resultant mechanical variables (see later), such as total force and total moment of force, it is more convenient to express m in absolute units of force, newtons. For example, when a person tries to produce 10 N of force by pressing with the middle finger only, $m = [0\ 10\ 0\ 0]$.

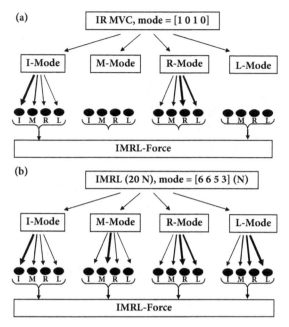

Figure 12.3 Illustrations of the concept of finger mode for the case of two-finger maximal force production task (a) and in the case of four-finger force production of 20 N (b). I, index; M, middle; R, ring; L, little fingers.

Since the enslaving matrix is 4 × 4, it is invertible, and one can compute mode values based on observed combinations of force values assuming that the person-specific E matrix is known, for example, based on a set of earlier observations. For example, the E matrix can be computed based on a set of tasks with ramp force production performed by one instructed finger at a time with all other fingers staying on force sensors and producing force unintentionally (Fig. 12.4). Performing linear regression analysis of the resultant force against the individual finger forces allows estimating entries of the E matrix as regression coefficients $a_{i,j}$ in $F_{i,RES} = a_{i,j}F_j + b_{i,j}$, where i stands for the instructed finger and j for one of the four fingers (i,j = I, M, R, L), and $a_{i,j}$ and $b_{i,j}$ are constants. As described later in this chapter, analysis of multi-finger synergies stabilizing resultant force and moment of forces can be performed in spaces of individual finger forces or modes. There are advantages and disadvantages inherent to each of the two types of analysis.

The concept of enslaving and notion of finger modes assume that patterns of finger nonindependence are person specific and robust. Under this assumption, these concepts have been related to the idea of "cortical piano" introduced by Marc Schieber (2001). Under this concept, undefined brain areas produce inputs into the primary motor cortex organized based on everyday finger use and leading to correlated changes in individual finger force changes within a hand when the actor tries to produce force with one finger only.

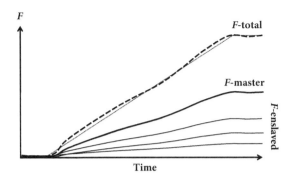

Figure 12.4 A typical single-finger ramp force production task. Note the relatively unchanged proportion of the total force (*F*-total, thick, dashed trace) produced by the individual fingers, master (*F*-master) and enslaved (*F*-enslaved).

Within the theory of movement production with patterns of spatial RCs (reviewed in Feldman 2015), the phenomenon of enslaving suggests correlated RC inputs into muscles and muscle compartments acting on individual fingers. Since any effector is controlled with a pair of commands, reciprocal and coactivation (*R*-command and *C*-command; see Chapters 3 and 7), there is an infinite number of ways a required force magnitude can be produced in isometric conditions. The most intuitive and parsimonious description assumes that the *R*-commands, $R_{i,j}(t)$, to individual fingers are correlated when only one finger *i* produces force intentionally. This assumption is compatible with the idea that the *R*- and *C*-commands are organized hierarchically, with the *R*-command changing in a task-specific way and the *C*-command adjusting, as needed, or staying constant (Levin and Dimov 1997). On the other hand, this description assumes that the *C*-command stays the same for all four fingers, resulting in comparable "apparent stiffness" values, k_j: $F_{i,j} = RC_{i,j} \cdot k_{i,j}$. This is a strong assumption given that the individual fingers vary significantly in their force-producing capabilities (e.g., Li et al. 1998).

An alternative explanation is that changes in the *C*-command define force changes in individual fingers. This explanation has been supported by studies of motor unit firing patterns during sine-like cyclical force production (Madarshahian and Latash 2022a). Since the *C*-command during fingertip force production is defined by commands to compartmentalized extrinsic hand muscles, it is feasible that a change in this command to one of the compartments is associated with proportional changes of the *C*-command to other compartments, leading to correlated changes in k_j. So far, no definitive data-based answer has been obtained for the question of how the correlated changes in finger force are related to changes in the *R*- and *C*-commands.

12.4. Grip force

During a range of object manipulation tasks, the so-called prismatic grip is used with the thumb opposing a set of fingers (Fig. 12.5). If the object is oriented vertically and

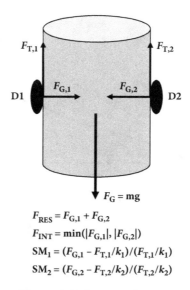

Figure 12.5 An illustration of the contacts between the two opposing digits (D1 and D2) and the vertically oriented object during the grasp. Definitions of the safety margin (SM) for each digit separately, internal force (F_{INT}) and resultant force (F_{RES}).

held motionless, the normal forces of the digits provide necessary friction to allow the application of vertical tangential (shear) forces counteracting the gravity. For a given friction coefficient k, the minimal normal force ($F_{N,min}$) applied by a digit is $F_{N,min} = F_T/k$, where F_T is tangential force. Typically, F_N is larger than $F_{N,min}$ by a fraction addressed as the safety margin: $SM = (F_N - F_{N,min})/F_{N,min}$, which can also be expressed in percent of $F_{N,min}$ (Westling and Johansson 1984). The safety margin varies within a broad range depending on such factors as slipperiness of the surface, possibility of unexpected change in the force applied to the object, fatigue, age, and neurological disorder (Cole and Johansson 1993; Kinoshita et al. 1995, 1997; Gordon et al. 1997; Gorniak and Alberts 2013; Singh et al. 2013). However, object manipulation is commonly accompanied by relatively unchanged safety margin values (Johansson and Westling 1984; Cole and Johansson 1993; Burstedt et al. 1999). Complex changes in the safety margin can be observed when individual digits act on surfaces with different friction coefficients (Aoki et al. 2007).

In a more general situation, when the object can move in a horizontal direction along the line connecting the opposing digits, the horizontal forces produced by the digits on the object can be viewed as contributors to two force components, the resultant force (F_{RES}) and internal force (F_{INT}) (reviewed in Zatsiorsky and Latash 2008). F_{RES} accelerates the object in the horizontal direction. F_{INT} compresses the object with no contribution to its acceleration. Grip force can be defined as F_{INT} or, in the simple case illustrated in Fig. 12.5, as the minimal of the two forces produced by the thumb and by the opposing fingers. Some studies, however, defined grip force as the mean of the absolute values of the two opposing forces (e.g., Emge et al. 2013;

Uygur et al. 2014), which, in our opinion, is confusing and leads to nonzero grip force values even when only one digit (e.g., the thumb) presses on the object—a counterintuitive definition.

Grip force shows feedforward adjustments in conditions when the load can change in a predictable way (reviewed in Flanagan and Wing 1993, 1995; Gordon 2001). This happens, for example, when the hand-held load is lifted quickly by the person or when the person walks with the load in hand. In both examples, predictable variations in the load vertical acceleration modify the baseline gravity contribution to the load. Feedforward grip force adjustments are also observed when a person holds an object with a prismatic grasp and then applies or removes supporting force to the bottom of the object with the other hand (Scholz and Latash 1998). Some of the observations suggest that predicted changes in the load force are only one of the factors defining changes in grip force. In particular, when adjusting grip force to task requirements, the central controller modifies grip force based not only on the expected magnitude of the load force but also on such factors as whether the force is gravitational or inertial and on the contributions of the object mass and acceleration to the inertial force. This was shown in a study of grip force adjustments during cyclical motion of a handle loaded with different weights (Zatsiorsky et al. 2005).

If the external force acting on the hand-held object is changed quickly and unexpectedly, grip force shows adjustments at time delays longer than typical spinal reflex time delays and shorter than the simple reaction time, within the range of 50 to 80 ms (Johansson and Westling 1984). These adjustments represent an example of a large class of phenomena addressed as long-loop reflexes, preprogrammed reactions, or triggered reactions (reviewed in Tatton et al. 1978; Chan and Kearney 1982; Reschechtko and Pruszynski 2020). These reactions can be induced by activity of different types of peripheral sensory endings including those in muscle spindles, Golgi tendon organs, and cutaneous and subcutaneous receptors.

Within the theory of control with spatial RCs, normal force production on a rigid surface is a consequence of the difference between the coordinate of the effector (e.g., fingertip) and RC, which is under the surface. In the case of two digits facing each other, their RCs virtually penetrate the object (Fig. 12.6; cf. Pilon et al. 2007). The spatial difference between the actual coordinates and RCs translates into force units with a coefficient, which can be addressed as apparent stiffness (see Chapter 7), k. Changes in k can be produced by changes in the coactivation of the opposing muscles acting at each of the digits, the C-command, which defines the spatial range within which opposing muscle groups show nonzero activation levels. Anticipatory and reactive changes in grip force are peripheral consequences of changes in the R-command affecting RC and the C-command affecting k.

12.5. Prehension synergies

When a vertically oriented object is held by the hand (as illustrated in Fig. 12.5), conditions for static equilibrium require that the resultant force and moment-of-force vectors acting on the object are zero. Analysis of the neural control of digit forces has frequently assumed a hierarchical scheme based on considerations from anatomy and

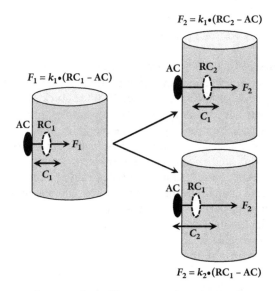

Figure 12.6 The same change in force (from F_1 to F_2) applied to the grasped object can be produced with changes in the referent coordinate (RC) from RC_1 to RC_2 caused by a change in the reciprocal command (R-command, right, top) or the coactivation zone from C_1 to C_2, which translates into changes in the apparent stiffness (k) with the help of the coactivation command (C-command, right, bottom). AC, actual coordinate of the digit.

physiology (Fig. 12.7; Arbib et al. 1985; reviewed in Zatsiorsky and Latash 2008). At the top level of the presumed hierarchy, the task-related magnitudes of the resultant force and moment of force are shared between the thumb and virtual finger (VF)—an imagined digit acting in opposition to the thumb and producing force and moment vectors equal to the summed action of the actual fingers of the hand. At the bottom level of the hierarchy, the VF action is shared among the actual fingers.

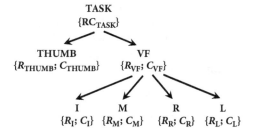

Figure 12.7 Hierarchical control of the hand during prehensile tasks. At the higher level, the task is shared between an imagined digit, virtual finger (VF), and the thumb (TH). At the lower level, action by the VF is distributed across the actual fingers: I, index; M, middle; R, ring; L, little. {R; C}, the reciprocal and coactivation commands.

12.5. PREHENSION SYNERGIES

In the two-dimensional case, this means that the normal forces of the thumb and of all the fingers should sum up to zero, the tangential forces of all the digits should sum up to the weight of the object and be directed against the gravity force, and all the components of the moment of force acting on the object should be balanced. Using the concept of VF, these conditions at the upper level of the presumed hierarchy can be summarized as

$$F_{TH}^N + F_{VF}^N = 0$$
$$F_{TH}^T + F_{VF}^T + L = 0 \qquad (12.2)$$
$$M_{TH}^N + M_{TH}^T + M_{VF}^N + M_{TH}^T + M_{EXT} = 0$$

where the superscripts refer to normal (N) and tangential (T) forces and moments of force and the subscripts stand for the thumb (TH) and virtual finger (VF). M_{EXT} stands for external moment of force. Each of the three equations has an infinite number of solutions in the space of digit forces. Since the object can never be held perfectly motionless (e.g., due to the physiological tremor), a more realistic version of the three constraints should use not = but ≈. In other words, some nonzero variance may be expected for the left-hand expressions in each of the equations if a person repeats holding the same object multiple times. The potential abundance of solutions for each equation allows stabilizing the resultant force and moment acting on the object by covaried contributions of the left-hand summands. In other words, performance-stabilizing synergies may be expected (see Chapter 6).

Such synergies have indeed been observed in studies where subjects were asked to hold statically handles with different combinations of the external load and moment of force. Collecting multiple trials for each combination revealed strong synergies expressed in the inequality $V_{UCM} > V_{ORT}$; that is, intertrial variance along the solution space (uncontrolled manifold [UCM]) for each of the three equations was larger than orthogonal to that space.

Analysis of synergies can also be performed at the level of individual finger forces and moments of force, which sum up to the force and moment of force of the VF. At that level, the equations linking finger-level variables to the VF-level variables are

$$\sum F_i^N = F_{VF}^N$$
$$\sum F_i^T = F_{VF}^T$$
$$\sum M_i^N = M_{VF}^N \qquad (12.3)$$
$$\sum M_i^T = M_{VF}^T$$

In these equations, the subscript i refers to the individual fingers: i = I, M, R, and L. Each of these equations has an infinite number of solutions, which allow synergies to be organized. Note also that the vertical coordinate of the normal finger force application can vary, thus increasing the number of contributing variables in the third equation (the coordinate of the tangential force application is considered fixed by the dimensions of the rigid handle).

180   12. THE HAND

As described in Chapter 6, performance-stabilizing synergies at different levels of a hierarchy show inherent trade-offs. Analysis of synergies at the two levels Task => {TH; VF} and VF => {Fingers} has shown nontrivial results summarized in Fig. 12.8 for the discussed two-dimensional analysis (Gorniak et al. 2009). Note that two of the variables were stabilized at both levels, which is the most nontrivial strategy, whereas the other variables were stabilized at one of the levels only, as expected from the aforementioned trade-off. The observations of stabilization of different variables at one or both levels of analysis were discussed as a window into strategies of the central nervous system.

The concept of prehension synergies has been generalized to three-dimensional tasks (Shim et al. 2005a). This generalization is relatively straightforward but includes more complex equations as compared to those presented earlier. The existence of synergies has also been confirmed in studies where the subjects were required to produce resultant force acting on the handle in a prescribed direction (Gao et al. 2005; see also Kapur et al. 2010a). Those studies documented large intertrial variance in the direction of the force vectors produced by individual digits, which was associated with much lower variance of the resultant force direction. In other words, individual finger forces showed compensation of each other's errors in force direction, a signature of performance-stabilizing synergies.

So far, only a single study explored the control of a hand-held object during its motion within the framework of the theory of control with spatial RCs (Ambike et al. 2015b). This study reconstructed the time patterns of mechanical variables reflecting the R- and C-commands with respect to the vertical motion of the object and with respect to the grip force. This was done with the help of analysis across repetitive cycles under smoothly varying external conditions and the instruction to the subject not to adjust the movement to the changes in the external conditions (similar to the method of reconstruction equilibrium trajectories described in Chapters 3 and 7; see Latash and Gottlieb 1991). The study reported complex coupling within the four-dimensional control space for the two dimensions of analysis, two pairs of {R; C}

VARIABLE	THUMB-VF	FINGERS
F_{GRIP}	$\Delta V > 0$	$\Delta V \approx 0$
F_{LOAD}	$\Delta V > 0$	$\Delta V > 0$
M^T	$\Delta V < 0$	$\Delta V > 0$
M^N	$\Delta V > 0$	$\Delta V < 0$
M_{TOT}	$\Delta V > 0$	$\Delta V > 0$

Figure 12.8 Synergies stabilizing different components of the force and moment-of-force vectors at the two control levels: Level-1, where the task is shared between the thumb (TH) and virtual finger (VF), and Level-2, where action of the VF is distributed across the actual fingers. F_{GRIP}, normal force; F_{LOAD}, tangential, load-resisting force; M^N, moment of normal forces; M^T, moment of tangential forces; M_{TOT}, total moment of force. Cases of synergies stabilizing the variables are shown with bold ($\Delta V > 0$).

commands for the vertical and horizontal (aligned with the grip) dimensions, which produced the experimentally observed coupling between the time-varying vertical force applied to the object and grip force.

12.6. Principle of superposition

Analysis of the variation of mechanical elemental variables with changes in the external load and torque applied to the handle gripped with the prismatic grasp showed patterns suggesting the existence of two control processes (Zatsiorsky et al. 2002a, 2002b, 2004). One of them was related to changes in external torque, and the other one to changes in external load. All the elemental variables (forces and moments of force produced by individual digits) changed with both load and torque, but there was no interaction between those two factors (illustrated for one of such variables in Fig. 12.9). Another group of observations pointed to the so-called chain effects: sequences of relatively straightforward cause-effect links directly related to mechanical constraints leading to nontrivial strong covariation between pairs of elemental variables (reviewed in Zatsiorsky et al. 2004; Zatsiorsky and Latash 2008).

Taken together, these results suggest that human prehension obeys the so-called principle of superposition. In a series of seminal studies, Arimoto and colleagues (Arimoto et al., 2000, 2001) suggested a principle of superposition for the control of robotic hand action. The main idea of this approach was to separate complex motor tasks into subtasks controlled by independent controllers. The output signals of the controllers converged onto the same set of actuators, where they were summed up. This method of control has been shown to lead to a decrease in the computation time as compared to the control of the action as a whole.

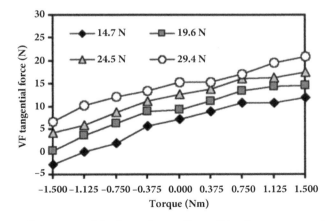

Figure 12.9 An illustration of the dependence of one of the elementary mechanical variables (tangential force by the virtual finger [VF]) on the external load and torque. Note that the variable changed with both load and torque but without an interaction of these factors. Reproduced by permission from Zatsiorsky et al. 2004.

This idea is far from being trivial. For example, if a robotic hand tries to manipulate a grasped object, it has to produce adequate grasping and rotational actions. However, a straightforward change in the grasping force, in general, leads to a change in the total moment produced by the digits on the object. Hence, the grasping and rotational components of the action are not independent, and if a controller responsible for the grasping changes its output, a controller responsible for the rotational action also has to change its output. Having independent controllers for the two action components becomes possible for a mechanically redundant system; the redundancy (abundance at the control level; see Chapter 5) allows decoupling the two action components.

12.7. Force- and moment-stabilizing synergies

Studies of performance-stabilizing synergies in multi-finger tasks have always faced a problem: Finger forces cannot be modified by the neural controller independently from each other because of the well-established phenomenon of enslaving (see earlier and Zatsiorsky et al. 2000). This means that, in the absence of any special control strategy, an inter-trial data point cloud in the finger force space is expected to form not a sphere but an ellipsoid (i.e., be elongated along certain directions). These directions, by pure chance, could be along the UCM for a performance variable or orthogonal to the UCM resulting in spurious inequalities $V_{UCM} > V_{ORT}$ or $V_{UCM} < V_{ORT}$, which cannot be easily interpreted as signatures of synergies or their absence. In a number of studies (Latash et al. 2001; Scholz et al. 2002; reviewed in Latash et al. 2007), this problem was addressed by converting finger forces into modes, hypothetical neural variables that were assumed to be manipulated by the controller one at a time (see earlier in this chapter). This transformation was made with the help of the enslaving matrix computed based on special trials. While this method seems to solve the problem of finger force interdependence, it assumes that the enslaving matrix is robust over time and across conditions, an assumption questioned in a recent series of studies showing drifts in the magnitude of enslaving with time (Abolins et al. 2020; Hirose et al. 2020; reviewed in Abolins and Latash 2021; see also Chapter 16).

A number of early studies explored synergies stabilizing total force (F_{TOT}) and total moment of force (M_{TOT}) in pronation-supination during two-finger, three-finger, and four-finger cyclical F_{TOT} production tasks (Latash et al. 2001; Scholz et al. 2002). In those tasks, visual feedback on F_{TOT} was presented, and the subjects got no instruction with respect to M_{TOT}. Note that, in a two-finger task, the UCM for F_{TOT} is orthogonal to the UCM for M_{TOT} (Fig. 12.10A). So, in principle, it is impossible to stabilize both F_{TOT} and M_{TOT}. Although the instruction specifically mentioned producing an accurate cyclical pattern of F_{TOT} and the visual feedback on F_{TOT} was presented, the subjects consistently showed data clouds elongated along the UCM for M_{TOT}; that is, they stabilized M_{TOT} at the expense of not stabilizing F_{TOT}.

In three-finger tasks, in principle, it is possible to stabilize both performance variables simultaneously as illustrated in Fig. 12.10B with two 2-dimensional UCMs for F_{TOT} and M_{TOT}, which have an intersection shown as the thick straight line. Nevertheless, no F_{TOT} stabilization was seen, while M_{TOT} was stabilized throughout the force cycle. Only in four-finger tasks was stabilization of F_{TOT} observed within a

12.7. FORCE- AND MOMENT-STABILIZING SYNERGIES

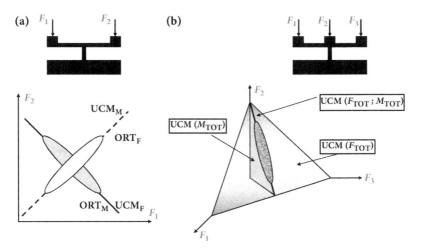

Figure 12.10 (a): In a two-finger pressing task, it is impossible to stabilize both total force (F_{TOT}) and total moment of force (M_{TOT}) because the uncontrolled manifold (UCM) for one of the two variables coincides with the orthogonal space (ORT) for the other variable. Data distributions for F_{TOT} and M_{TOT} stabilization are shown with the gray ellipse and with the white ellipse, respectively. (b): If three fingers take part in the task, both F_{TOT} and M_{TOT} can be stabilized because the two UCMs have an intersection shown as the bold, thick line. A data distribution for $\{F_{TOT}; M_{TOT}\}$ stabilization is shown with the dark gray ellipse.

certain phase range corresponding to relatively large F_{TOT} magnitudes, and M_{TOT} continued to be stabilized throughout the force cycle.

Taken together, these observations lead to a number of nontrivial conclusions. First, the subject's brain can reinterpret the explicit instruction and stabilize a different performance variable, even at the expense of destabilizing the one specified by the instruction. Second, everyday experience conditions the central nervous system to stabilize M_{TOT} in pronation-supination, obviously a very important variable for such actions as taking a sip from a glass, eating soup, writing, and using a variety of hand-held tools. Third, analysis of synergies stabilizing different performance variables can provide information on the relative importance assigned by the brain to those variables.

The two components of variance in multifinger tasks, V_{UCM} and V_{ORT} computed with respect to F_{TOT}, showed very different behaviors within the force cycle (Latash et al. 2002a; Friedman et al. 2009). V_{UCM} changed in-phase with F_{TOT}, whereas V_{ORT} showed a double-peaked profile timed similarly to the force time derivative, dF/dt (Fig. 12.11). These observations led to a hypothesis that V_{ORT} reflected errors in setting a timing parameter defining frequency of force change, and V_{UCM} reflected errors in setting an amplitude parameter. This hypothesis fit an earlier scheme with the level defining timing of an action being hierarchically higher than the level defining metrics of the action (Schöner 1995; Martin et al. 2009). This was formalized in a model

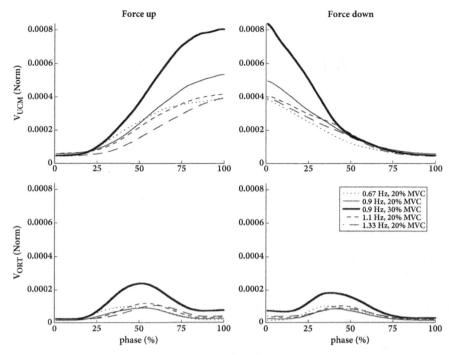

Figure 12.11 Patterns of the two components of variance, along the uncontrolled manifold (UCM) and orthogonal to it (ORT), V_{UCM} and V_{ORT}, within the force cycle during four-finger cyclical force production. Note that V_{UCM} changes in phase with force, whereas V_{ORT} changes with force derivative. The data are means over the subjects. Thin lines show tasks that were performed at various frequencies, from 0.67 to 1.33 Hz. One of the tasks (bold line) was performed over a larger force range. The left plots show the curves during the half-cycle with force increase. The right plots show the curves during the half-cycle with force decrease. Reproduced by permission from Friedman et al. 2009.

assuming independent trial-to-trial variation in setting the timing and amplitude parameters of action (Goodman et al. 2005).

Is it true that synergies can stabilize magnitudes of salient variables but not their timing? In other words, if in a certain trial one of the variables changes too quickly, can other variables compensate for this error by changing a bit slower? So far, the available experimental material suggests a negative answer to this question. In particular, one of the studies (Latash et al. 2004) required producing an accurate F_{TOT} pulse while pressing with the four fingers of a hand, and the target specified both amplitude and timing of peak F_{TOT}. Analysis confirmed multi-finger synergies stabilizing F_{TOT} magnitude and failed to show synergies stabilizing F_{TOT} timing.

Note that force production even by a single finger is abundant at the level of control. As illustrated in Fig. 12.12, the same force can be produced by an infinite number of combinations of the R-command and C-command to the agonist-antagonist muscle pair involved in the task. In a linear approximation, the R-command defines

12.7. FORCE- AND MOMENT-STABILIZING SYNERGIES

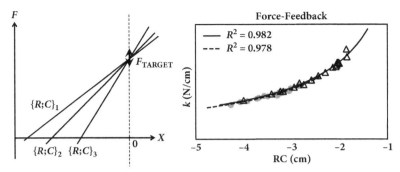

Figure 12.12 An illustration of the abundance at the control level during force production by a single effector (finger) in isometric conditions. A: The task of producing a certain value of force can be performed with various combinations of the reciprocal and coactivation commands translating into pairs of referent coordinate and apparent stiffness, {RC; k}. B: Performing the task multiple times results in a family of variable {RC; k} values, which show strong hyperbolic covariation to satisfy the task constraint: $F = -k \cdot RC$. The figure shows two families of data points, early in the trials and later in the trials. Modified by permission from Reschechtko and Latash 2017.

the RC for the fingertip, and the C-command is reflected in its apparent stiffness, k. Measuring {RC; k} pairs across trials with the production of constant force level confirmed large intertrial variation in both variables, which was primarily constrained to the hyperbolic solution space ($F = -k \cdot RC$, assuming that the actual fingertip coordinate is zero), which is the UCM for this variable (Ambike et al. 2016a). This is not a trivial observation. The target size limited deviations from the UCM (along ORT), but deviations along the UCM were not in any way defined by the task and could be larger than, smaller than, or equal to deviations orthogonal to the UCM. In future chapters, the concept of performance-stabilizing synergies at the {RC; k} level of analysis will be used to analyze a range of unusual behavioral phenomena (see Chapters 16 through 19).

An important phenomenon originally documented in a study of multifinger F_{TOT}-stabilizing synergies is anticipatory synergy adjustment (ASA) discussed earlier in Chapter 8 (Olafsdottir et al. 2005a). To remind, ASA represents graded attenuation of synergies stabilizing a salient variable in preparation of a quick change in that variable. In a later study, ASAs were documented in prehensile tasks with respect to a number of salient performance variables representing components of the resultant force and moment vectors in preparation of self-triggered changes in the external load and torque (Shim et al. 2006).

12.8. Missing pieces of the mosaic

In spite of a number of studies exploring neurophysiological mechanisms involved in the hand motor function, from the cortex to the spinal cord, their role in defining control variables to the hand and individual digits remains largely unknown. The importance of subcortical circuitry for multi-finger synergies has been documented (see Chapters 10 and 24), in addition to synergies stabilizing muscle action in spaces of motor unit groups (MU-modes, see Chapter 11). This information is, however, fragmented, and the relative role of spinal and supraspinal circuitry in ensuring stability of salient performance variables by the hand is still unknown.

So far, studies of the hand function have been performed primarily in spaces of mechanical and muscle activation variables. Only a few studies have explored the neural control of the hand in spaces of hypothetical control variables, such as the R- and C-commands, including the role of synergies in spaces of such variables in ensuring stability of hand actions. This is a major gap in the current knowledge on the processes of the neural control of the hand.

The unique ability of the hand to perform a variety of dexterous actions has been only minimally used to understand the phenomena of motor learning. In particular, the amazing ability to write with various effectors (Bernstein 1947) and in a variety of external conditions taken together with dysfunctions of handwriting in neurological disorders offers a unique opportunity to explore the role of various brain structures in learning, transfer, and generalization of effects of practice. Another unique ability of humans is to modify patterns of handwriting, for example, to produce mirror writing (Bray 1928; Latash 1999). At this time, we are unaware of feasible neurophysiological interpretations of those observations.

13
Reaching Movement

Reaching for a target is arguably one of the most common everyday actions. This action involves all of the most commonly discussed problems in the field of motor control. In particular, reaching involves a redundant number of joints, an even larger number of muscles, and numerous motor units. At the level of mechanics, multi-joint movements are associated with time-varying external loads for the involved joints and muscles due to changes in the joint configuration and the orientation of the extremity with respect to gravity. There are also potentially complex patterns of joint interaction forces. Movements in individual joints are rotations, which makes it non-trivial to produce the typical straight trajectories of the hand in the external space during reaching (e.g., Morasso 1981). All of the aforementioned problems become apparent, not real, within the theory of the neural control of movements with spatial referent coordinates (RCs) coupled to the idea of dynamical stability of salient performance variables organized via synergic mappings of the task-level RC(t) function to such functions at hierarchically lower levels related to the involvement of joints, muscles, and motor units (see Chapters 6 and 7).

Within this chapter, we are not going to repeat all the arguments in favor of the RC control theory as compared to alternative views on motor control, which do not seem to offer productive, physiologically valid hypotheses and have to resort ultimately to assuming magic devices in the brain addressed as "internal models" or invisible links between the brain and effectors that allow the former to prescribe variables produced by the latter. We will rather focus on important experimental observations that shed light on yet unsolved problems within the RC control theory. In particular, we will discuss the potential role of the spinal cord in the control of multi-joint reaching movements, multi-joint synergies and their reflections in the structure of inter-trial variance and in motor equivalence, the issue of equifinality and its violations in different spaces, and the two main views on the control of a particular class of reaching movements, the so-called reach-to-grasp actions. We will also discuss what salient performance variables and in what spaces can be stabilized during multi-joint limb movements. As an exercise, consider a virtuoso playing the violin while moving it in the environment and with respect to the body.

We have to mention, however, in this introductory part a few of the alternative views on the control of multi-joint movements based on considerations from mechanics and engineering. Studies of a variety of movements led to the introduction of the so-called linear covariation principle, which states that the torque time profiles in individual joints follow a single template (Gottlieb 1996; Gottlieb et al. 1996). This principle obviously cannot be viewed as general because, for example, some multi-joint movements are associated with monotonic trajectories in some of the involved joints and with non-monotonic trajectories (those with a reversal of the movement direction) in other joints. A number of studies focused on interaction torques during multi-joint movements and claimed that the ability to predict those torques and compensate for their effects was crucial for healthy movements. The most detailed hypothesis in this group is arguably the so-called leading-joint hypothesis (Dounskaia 2007, 2010), which posits that the central nervous system defines the torque profile in only one joint of the moving limb (the leading joint, commonly, the most proximal joint) and allows the interaction torques to move other joints while introducing only minor adjustments in their trajectories. Let us remind that the length and velocity dependence of muscle forces makes it impossible for the brain to prescribe joint torques (Bernstein 1947 and Chapter 4). Numerous optimization approaches based on mechanical variables have been used to describe multi-joint movements, but their success has been limited (Chapter 9). Overall, attempts to describe the control of reaching via prediction and prescription of performance variables by structures within the central nervous system may safely be viewed as having failed.

13.1. Spinal coordination of multi-joint movements

Articulated extremities can be seen across a range of animals, starting from insects. In some actions, animals with non-articulated extremities, such as the octopus, seem to turn them into articulated ones during grasping and manipulating external objects (Zelman et al. 2013; Levy et al. 2015). Whether there is an evolutionary advantage of reaching and grasping with articulated extremities remains an enticing but speculative issue. We begin analysis of reaching movements from vertebrates when the contribution of the spinal cord can be identified.

In the 19th century, Pflüger described the wiping reflex in decapitated frogs, which was induced by an irritating stimulus (a small piece of paper soaked in a weak acid solution) placed on the back of the frog's body. The reflex involved coordinated action by the ipsilateral to the stimulus hindlimb, which placed the toes close to the stimulus and then wiped it off the body with a quick extension movement. The wiping reflex in spinal frog preparations, that is, after the spinal cord had been cut at a high cervical level, was later studied in much more detail (Fukson et al. 1980; Berkinblit et al. 1986a; Schotland et al. 1989; Giszter et al. 1993). The main observations can be summarized as follows.

A single stimulus placed on the back of the spinal frog induced a sequence of coordinated actions by the ipsilateral hindlimb placing the toes close to the stimulus and wiping it off. A single stimulus produced several wiping actions of the irritated area in different directions. If a stimulus was placed on a forelimb of the frog, the wiping was also accurate even in conditions when the position of the forelimb with respect to the body had been modified across trials. Placing a stimulus on the low portion of the back induced a qualitatively different movement pattern, with the wiping action produced not by the toes but by more proximal portions of the ipsilateral hindlimb. Taken together, these observations show that the spinal cord is able to generate a variety of coordinated multi-joint movements to an irritating stimulus placed on the body. Moreover, a single stimulus could induce variable movements.

Further experiments (reviewed in Latash 1993) have shown that successful wiping movements could be produced with a rather heavy lead bracelet placed on the distal

13.1. SPINAL COORDINATION OF MULTI-JOINT MOVEMENTS

portion of the hindlimb. Moreover, restricting motion in one of the major hindlimb joints did not prevent the spinal frog from reaching the target during the very first wiping movement. The frog was also able to wipe the stimulus off the body if an unexpected brief electrical stimulus was applied to the hindlimb knee extensors, thus producing a major transient disruption of the ongoing movement—an example of equifinality (see later in this chapter). Across all the mentioned manipulations, the most consistent variable was the final location of the toes prior to the wiping extension movement.

These observations lead to a number of conclusions that have already been featured in this book. First, the output of the reflex-producing spinal circuitry did not encode time profiles of muscle forces, joint moments of force, joint trajectories, or muscle activation patterns. Second, the action was driven by a control process targeting the endpoint of the hindlimb and organized to bring it to a spatial location compatible with successful wiping action. This process was redistributed among neural variables to individual joints and muscles in a variable, adaptive way to ensure stability of the hindlimb endpoint coordinate. Some of these conclusions have been supported in experiments with intraspinal electrical stimulation and analysis of force vectors at the hindlimb endpoint (Giszter et al. 1993; Bizzi et al. 1995).

These conclusions have been reflected in a simple kinematic model (Fig. 13.1), where individual joint rotational velocities were defined by efficacy of this joint in moving the endpoint of the limb to the target (Berkinblit et al. 1986b). Of course, as with any kinematic model, this one has to be viewed as a metaphor, not as a realistic reflection of the neural organization of the wiping reflex. Later experiments have falsified some of the predictions of this model when applied to human reaching movements (Karst and Hasan 1987, 1990).

Walking in quadrupeds and in humans (for more detail see Chapter 14) represents another example of movement, primarily controlled by spinal central pattern generators, which shows stabilization of the endpoint trajectory of each limb by covariation

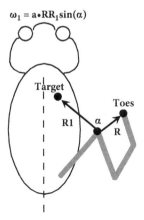

Figure 13.1 A kinematic model of the frog wiping movement. Rotational velocity (ω) in each joint is defined by its efficacy in moving the endpoint (toes) to the target.

of individual joint rotations. Studies of multi-joint synergies stabilizing the foot (paw) trajectory within the framework on the uncontrolled manifold (UCM) hypothesis have documented the existence of such synergies during walking in both humans (young and elderly) and cats (Krishnan et al. 2013; Klishko et al. 2014; Rosenblatt et al. 2014). The presence of such synergies confirms the conclusion on the neural control process organized at the level of the limb endpoint.

13.2. Control of reaching with spatial referent coordinates

Within the theory of control with spatial RCs, reaching movement by a multi-articular limb is associated with setting a time profile, $RC(t)$, for the endpoint of the limb. Further, a sequence of few-to-many mappings generates $RC(t)$ patterns for the involved elements, joints, and muscles (see Fig. 6.10 and Chapter 7). These mappings are not prescriptive but adaptive to the actual conditions of movement execution. In other words, they are associated with synergies among RCs at lower levels of the hierarchy stabilizing salient performance variables such as the endpoint trajectory. Such synergies are reflected in covariation of elemental variables at other levels of analysis, for example, kinematic and electromyographic (EMG), which are typically studied in experiments.

Sequences of equilibrium states of the endpoint in the external force field have been addressed as equilibrium trajectories (Flash 1987; Latash and Gottlieb 1991). Note that equilibrium trajectories cannot be prescribed by the neural process because they emerge in the interaction of $RC(t)$ with the external force field, which is typically not 100% predictable. However, if the external force field remains unchanged, equilibrium trajectories become adequate reflections of $RC(t)$. Note also that actual trajectories during limb movements are not copies of equilibrium trajectories because of a number of time-varying factors. This statement is trivial if one considers unopposed movements to movements against a very heavy inertial load or in conditions when the limb is blocked in the initial position. Less obvious factors include, in particular, the shapes of the endpoint apparent stiffness properties (the so-called stiffness ellipses; Flash 1987) and changes in the inertial properties with changes in the joint configuration. The shapes of the stiffness ellipses have been invoked to account for the differences between the presumed straight equilibrium trajectories and slightly curved actual endpoint trajectories.

The control of reaching movements with $RC(t)$ for the endpoint has been confirmed in studies of reaching by standing persons who had to bend their trunk to reach the target (Tomita et al. 2017). This motion naturally led to hip displacement backward (Fig. 13.2). In a few trials, a stop behind the subject unexpectedly blocked the hip movement, causing a self-generated push from behind and stepping forward. Under such conditions, the trajectory of the endpoint (the hand) in the external space was practically unchanged, while the contributions of the arm and trunk changed at a short time delay. In other words, the movement showed the feature of motor equivalence (see Chapter 5) with respect to the endpoint trajectory, a signature of its stabilization by covaried adjustments of commands to the individual effectors.

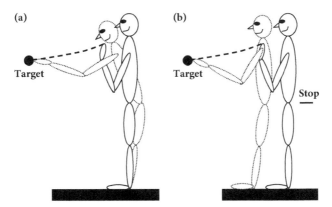

Figure 13.2 A schematic illustration of an experiment involving reaching a spatial target while standing. (a): In control conditions, the hips moved backwards. (b): When the hip movement was blocked unexpectedly, the subject made a step, which led to no major changes in the hand trajectory in external space.

Although mapping from RC(t) at the task level (e.g., related to hand motion in space) to RC(t) for the contributing effectors (e.g., related to individual joint rotations) is not prescriptive, it is also not random and shows preference for certain patterns of effector involvement. These preferences lead to regularities in the mechanical and EMG patterns observed at the level of individual effectors. Examples include the aforementioned linear covariation of joint torques observed across a variety of reaching movements and the parallel scaling of EMG patterns observed in the agonist-antagonist pairs crossing individual joints (Koshland et al. 1991; Hong et al. 1994; Latash et al. 1995b). Similar timing of RC(t) to the wrist and elbow joints during tasks that required a quick movement of only one of the joints has been reported (Latash et al. 1999; Fig. 13.3). The shapes of the two RC(t) were similar, but one of them stopped at a new value encoding the desired final position of the joint and the other one ended close to its initial value encoding a control pattern with the purpose not to move the joint but to avoid its motion under the action of the transient, motion-dependent forces.

13.3. Multi-joint synergies

The presence of multi-joint synergies stabilizing the endpoint trajectory in the joint configuration space has been supported in many studies, starting from the classical study by Bernstein of professional blacksmiths who performed a sequence of well-learned labor movements of hammering the chisel (Bernstein 1930; Chapter 2). Bernstein reported relatively high inter-trial variability of the individual joint trajectories and a relatively low variability of the hammer trajectory. These observations suggest error compensation across the involved joint rotations with respect to the salient performance variable—the hammer trajectory. Error compensation across joint

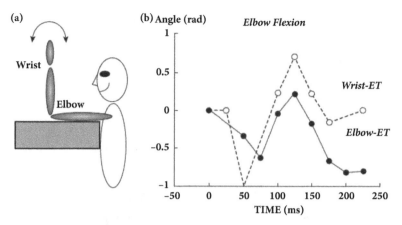

Figure 13.3 Equilibrium trajectories reconstructed for the wrist and elbow joints when the subject performed a very fast movement in the elbow joint only. (a): The schematics of the subject's position. (b): The equilibrium trajectories (ETs). Note the large-magnitude changes in the ET to the wrist. Reproduced by permission from Latash et al. 1999.

rotations was also reported in a study of pointing, which documented larger inter-trial variability in the coordinates of a marker placed over the wrist as compared to the marker placed at the end of the pointer (Jaric and Latash 1998). This is only possible if wrist motion partly compensated for the effects of intertrial variability in proximal joint rotations on the pointer coordinate.

Multi-joint movements show synergies stabilizing salient variables that vary across specific tasks. For example, a study of quick-draw shooting from an infrared pistol to a target (Scholz et al. 2000) compared the indices of inter-trial variance within the UCM and within the orthogonal to the UCM space (ORT) computed with respect to three feasible performance variables. The first one was the angle between the pistol barrel and the direction from the backsight to the target. This angle had to be small at the time of pressing the trigger to ensure accurate shooting. However, it was unconstrained at other phases of the trajectory of the hand with the pistol. The second variable was the coordinate of the pistol in space. It seemed feasible that the subjects selected a comfortable ("optimal") pistol trajectory and tried to reproduce it across trials. The third variable was the trajectory of the center of mass of the arm with the pistol in space. This choice was justified by the consideration of the mechanical interactions between the arm and the trunk: Kinematic effects of these interactions on the hand trajectory could be more effectively reduced if the interactions were predictable and reproducible across trials. Analysis in the joint configuration space has shown that only the first variable was consistently stabilized across subjects and conditions ($V_{UCM} > V_{ORT}$; Fig. 13.4). Note that this inequality was seen throughout the trajectory; that is, the pistol orientation with respect to the target was stabilized even when the pistol was pointing away from the target. The other two variables were stabilized only

13.3. MULTI-JOINT SYNERGIES 193

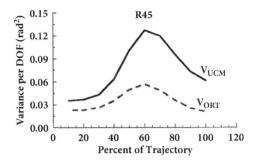

Figure 13.4 Two components of variance, along the uncontrolled manifold and orthogonal to it (V_{UCM} and V_{ORT}), computed across a series of quick-draw shooting trials with respect to the angle between the pistol barrel and direction from the pistol to the target. Note $V_{UCM} > V_{ORT}$ starting early in the movement. Reproduced by permission from Scholz et al. 2000.

at movement initiation but closer to the moment of shooting $V_{UCM} \approx V_{ORT}$, suggesting no specific synergies in the joint configuration space.

The shooting study also used a condition when the elbow joint was blocked by a rubber band in the initial posture. Although the subjects could see the band, they had no practice under this new condition, which was associated with major motion-dependent perturbations to all the joints of the arm. Note that the movement was very fast, typically under 0.5 s, and led to strong joint coupling effects. Nevertheless, most subjects shot the target accurately on the first attempt, demonstrating robustness of the synergy that ensured stability of the salient variable and aligned the pistol barrel with the direction to the target at the moment of shooting.

Multi-joint synergies can stabilize several salient variables simultaneously. This was shown, in particular, in a study of multi-joint pointing to two types of targets, a sphere and a cylinder (Mattos et al. 2011). The former target required only accurate positioning of the pointer tip in space, while the latter also required proper orientation of the pointer with the main axis of the cylinder. Multi-joint synergies stabilizing both pointer orientation and pointer tip coordinate were confirmed ($V_{UCM} > V_{ORT}$). The study also used conditions, which involved constraining the elbow joint with rubber bands of different stiffness, introduced unexpectedly for the subject. The subjects quickly corrected the pointer trajectory and brought it close to the target. Comparing control trials, in the absence of rubber bands, to trials with the rubber bands in the joint configuration space revealed large motor equivalent motion during the trajectory corrections, which was consistently larger than the non–motor equivalent motion. In other words, at the joint configuration level, most of the corrective action had no effect on the pointer tip trajectory, that is, did not correct the ongoing movement (Fig. 13.5). This is an unexpected result if the movement follows some kind of an optimization criterion. The result is expected given the lower stability in the UCM compared to ORT. The quick corrections did not change the stability properties of the action and, in a sense, acted as perturbations for the neural mechanisms,

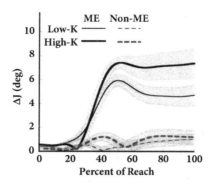

Figure 13.5 Motor equivalent (ME, solid lines) and non–motor equivalent (nME, dashed lines) components of trajectories in the joint configuration space during corrections of the unexpectedly perturbed pointing movements. The target was a cylinder. The elbow joint could be spanned by rubber bands with low and high stiffness (low-K and high-K). Note ME (solid lines) > nME (dashed lines). ΔJ, length of projection of joint difference vector between perturbed and unperturbed conditions on the UCM (ME) and on ORT (nME). Modified by permission from Mattos et al. 2011.

ensuring stability of the salient variables. As a result, the pointer tip moved primarily not in the desired direction but in the less stable directions, that is, within the UCM.

13.4. Equifinality of reaching movements and its violations

The phenomenon of equifinality during reaching movements has been a major source of misunderstanding of the equilibrium-point (EP) hypothesis. A number of studies reported violations of equifinality under the action of transient, motion-dependent forces and interpreted them as evidence falsifying the EP-hypothesis (Lackner and DiZio 1994; Hinder and Milner 2003). Indeed, as discussed in Chapter 3, the EP-hypothesis predicts equifinality under certain conditions but definitely not under any conditions. In particular, equifinality under a transient perturbation requires that the control trajectory, $\lambda(t)$, or, more generally, RC(t), does not change as compared to movements without perturbations. Simply instructing the subject not to react to the perturbation may not be sufficient: As described later in Chapter 16, unintentional drifts in RC(t) are relatively common and can bring about violations of equifinality.

One should also distinguish equifinality in the task-related salient variable, such as hand position in space, and in the space of elemental variables, such as joint rotations. As shown in a series of experiments with hand positioning in space against a constant external force, a transient change in the force, its quick moderate increase and immediate decrease to the initial level, leads to relatively minor violations in the hand position in space but major changes in the joint configuration (Zhou et al. 2014a). In other words, the arm returns to a different joint configuration compatible with about the same hand position; that is, the joint deviations were primarily along the corresponding UCM. These phenomena can be expected based on the relatively low

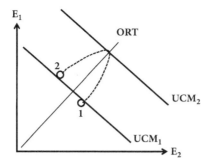

Figure 13.6 The task of producing a certain value of the sum of two elemental variables, E1 and E2. A quick change in (E1 + E2) to a larger value and return close to the initial value results in larger deviations along the uncontrolled manifold (UCM, thick line) as compared to deviations along the orthogonal direction (ORT, thin line). The trajectory is shown with the dashed line.

stability along the UCM, which leads to larger deviations within this space, caused by an external force perturbation, as compared to deviations orthogonal to the UCM. This is illustrated in Fig. 13.6 where a task-specific variable represents the sum of two elemental variables, E_1 and E_2. A transient change in the task-specific variable (E_1 + E_2) brings the system back close to the original UCM, that is, with relatively small non–motor equivalent deviations, accompanied by a relatively large motor equivalent deviation along the UCM.

In a similar experiment, if the force increase and force decrease segments were separated by a dwell time, during which the force was kept at the new level, major violations of equifinality were also seen in the hand coordinate: After the force was reduced to the original magnitudes, the hand moved back toward the initial coordinate, but with a major undershoot (Zhou et al. 2014b, 2015a, 2015b, 2015c, 2016; Fig. 13.7). These violations of equifinality could reach close to 50% of the total hand displacement induced by the force increase when the dwell time was on the order of 2 to 3 s. These phenomena have been interpreted as specific examples of a more general class of unintentional movements associated with drifts in RC at the task level (see Chapter 16).

Overall, violations of equifinality during a variety of motor tasks, including arm reaching movements, are rather common. They can be seen in task-specific salient variables as well as in the space of abundant elemental variables. In the former case, they reflect the common tendency of natural systems to move to states with lower potential energy. In the latter case, they reflect stability properties of the salient variables as seen in the relatively low stability along the corresponding UCM.

13.5. Reach-to-grasp

Reach-to-grasp movements have been traditionally viewed as consisting of two components controlled by two independent visuomotor channels (reviewed in Jeannerod

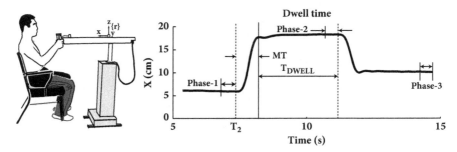

Figure 13.7 A transient force perturbation is applied during the task of holding a handle steady in the external space along the coordinate X (left panel). If the perturbation had a non-zero dwell time, a major violation of the equifinality was observed (right panel, compare the X coordinate at Phase-1 and Phase-3). Modified by permission from Zhou et al. 2016.

1988). The first component involves the grip formation by the hand adjusting to the size, orientation, and other features of the object. The second component involves transport of the hand to the object location in external space. A number of studies provided evidence compatible with this general view. For example, if an object changes its location in space at the onset of a reach-to-grasp movement, both hand transport and grip development are adjusted. In contrast, if an object changes its size without a change in location, only the grip component is adjusted without major effects on the transport component. These findings were interpreted as reflections of a hierarchical relation between the visuomotor channels involved in the transport and grip components.

An alternative description of reach-to-grasp movements views them as the processes of selecting target points on the object and performing pointing movements by opposing digits, commonly the thumb and index finger (Smeets and Brenner 1999). The two views are illustrated schematically in Fig. 13.8. Note that both schemes involve specification of a pair of kinematic variables, hand coordinate and digit aperture vs. coordinates of the two opposing digits.

Similarities between pointing and reach-to-grasp movements have been cited as providing support for the description of reach-to-grasp movements as a combination of two simultaneous pointing movements by the opposing digits (Voudouris et al. 2012, 2013). In particular, the digits show similar trajectories during reach-to-grasp movements and during single-digit pointing movements to the same points on the object. Similar trajectories have also been found during grasping movements by the dominant and non-dominant hands and during bimanual grasping with the two index fingers (Smeets and Brenner 2001).

Another source of data supporting this view comes from studies of the scaling of movement time with the index of difficulty computed within the classical Fitts' law (Fitts 1954; Fitts and Peterson 1964). Similar patterns of scaling were found across many studies during single-digit pointing movements and during grasping (Smeets et al. 2019).

Both views can be incorporated into the theory of control with RC(t), as illustrated in Fig. 13.9. At the task level, the first view implies neural specification of RC

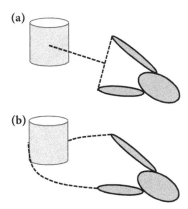

Figure 13.8 Schematic illustrations of the control of reach-to-grasp movements. (a): Two components are controlled separately, hand position in space and digit aperture. (b): Pointing movements are performed by the opposing digits.

trajectories for the hand (RC_{HAND} for transport) and for the digit aperture (RC_{AP} for grasping). Placing RC_{AP} inside the object leads to active grip force production. The second view implies specification of RC trajectories for the opposing digits, for example, RC_{THUMB} and RC_{INDEX}. Both RC_{THUMB} and RC_{INDEX} virtually penetrate the object, leading to the production of opposing forces.

Although the two pairs $\{RC_{HAND}; RC_{AP}\}$ and $\{RC_{THUMB}; RC_{INDEX}\}$ look equivalent, they can lead to different experimentally testable predictions. Imagine, for example, that a person is asked to perform a reach-to-grasp action from an initial hand position far away from the object location and then a similar action is performed when the digits are already very close to the object so that no transport of the hand is needed (Fig. 13.10). In this case, RC_{HAND} does not need to change, and only RC_{AP} is available within the $\{RC_{HAND}; RC_{AP}\}$ pair. Note that changing RC_{AP} produces coupled trajectories of the opposing digits. If, in the initial position, one of the digits is very close to the object surface and the other one is farther away, a change in RC_{AP} is expected to lead to asynchronous force production by the digits: The closer digit would start producing force earlier while the other digit is still moving toward the object. This does not look like an optimal strategy of grasping because, for some time, the resultant horizontal force on the object may be different from zero and lead to its displacement or tilting. The control with $\{RC_{THUMB}; RC_{INDEX}\}$ allows the neural control signals to the digits to be asynchronous and timed to lead to synchronous force production on the object.

13.6. Reaching with the dominant and non-dominant arms

The effects of dominance are seen in various aspects of reaching movements. For example, when a target can be reached by either hand, its relative location with respect to the body defines which hand will perform the action (Coelho et al. 2014). There is typically bias toward reaching with the dominant arm, which is more likely to cross

198 13. REACHING MOVEMENT

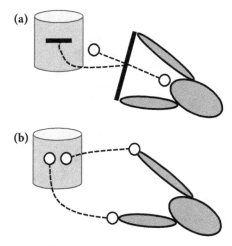

Figure 13.9 The control of reach-to-grasp movements in terms of specifying spatial referent coordinates (RCs). (a): Assuming the control of hand transport (open circles show the RC locations) and aperture formation (the black bars show the initial and final referent aperture). (b): Assuming the control of opposing digits (open circles show the initial and final RC locations). The dashed lines show the RC trajectories. (Gravity is not considered in this illustration.)

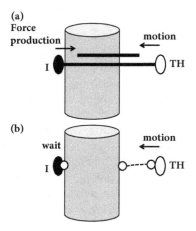

Figure 13.10 The two modes of control illustrated in Fig. 13.9 make different predictions for movements of the opposing digits starting from different initial coordinates. (a): The control with changes in the referent aperture (black bars) predicts an increase in the index finger (I) force simultaneously with movement by the thumb (TH). (b): The control with referent coordinates for the digits (open circles) allow the I finger to wait for the thumb to touch the object before initiating force generation.

the body midline. More subtle aspects of joint coordination in the two arms have been addressed within the dynamic dominance hypothesis, which challenged the common wisdom that humans have a "bad arm" and a "good arm" (reviewed in Sainburg 2002, 2005). According to this hypothesis, both arms are "good" but specialized for different aspects of everyday tasks, which is reflected in different patterns of joint coordination.

Planar arm reaching movements have been explored experimentally using a variety of manipulations affecting inter-joint interactions (Sainburg and Kalakanis 2000; Bagesteiro and Sainburg 2002). Overall, the right arm showed a stronger tendency to keep the hand trajectory straight, while the left hand was more likely to show curved trajectories, which were, nevertheless, more accurate in reaching the target. Additional evidence for specialization of the two hemispheres for different aspects of the control of arm reaching movements came from observations of reaching in persons suffering from mild unilateral stroke, which showed effects in specific aspects of reaching that could be seen in both arms, contralateral and ipsilateral to the stroke site (Schaefer et al. 2007, 2012).

The dynamic dominance hypothesis has been based primarily on the analysis of mechanical variables and links the effects of dominance to the ability of the central nervous system to predict and handle the effects of joint interaction torques. General problems with torque control schemes have been discussed earlier (see Chapter 4). The main insights of the hypothesis can be put into other theoretical frameworks, in particular those that address directly issues of performance stability by abundant systems of elements (see Chapters 6 and 9). In particular, within the UCM hypothesis, observations in the non-dominant hand suggest an advantage in stabilization of salient performance variables during steady-state tasks, while observations in the dominant hand suggest better optimality of reaching action and higher agility during fast actions (Zhang et al. 2006; Park et al. 2012; de Freitas et al. 2019). These insights have been confirmed in studies of multi-finger force production tasks, which are not associated with motion-dependent interaction torques. In particular, the cited studies have confirmed higher indices of stability ensured by interfinger coordination in the non-dominant hand and advantage of the dominant hand in preparation for quick action (as reflected in anticipatory synergy adjustments) and during such actions.

13.7. Missing pieces of the mosaic

Most studies of reaching movements are still performed at the levels of kinematic, kinetic, or electromyographic variables with no attempts to interpret the findings within a feasible theory on the neural control of movement. Sometimes these studies also include various methods of quantifying neural activation patterns in the brain (see Chapter 10). Potentially, such studies produce highly valuable information that has to be deciphered because the brain controls movements indirectly, not by specifying performance variables but by defining changes of parameters within corresponding laws of nature. Movement patterns emerge in the interaction between the neural process and external force field. Unfortunately, frequently, analysis is stopped at claiming that specific brain structures and neuronal circuits are responsible for specific characteristics of movements measured in the study based on the results of correlation analyses between spaces of neural and mechanical variables. There is an uneasy feeling that researchers assume the existence of invisible strings from brain structures to the effectors, leading to observations of correlated changes in indices of brain activity and peripheral movement patterns. Such assumptions are obviously not productive.

Quantifying neural control variables is complicated (see Chapter 24). As of now, we have no reliable tools to measure time changes in $RC(t)$ or $\lambda(t)$, although there has been progress in this direction. However, most frequently, we have to rely on indirect measurements to understand salient features of the indirect control of movements. This refers to reaching movements, which, by their nature, involve multiple elements such as joints, muscles, and motor units. Reconstructing the time patterns of the reciprocal and coactivation commands during such movements would represent a major step in understanding the basics of inter-joint coordination and the coordination between components of reach-to-grasp actions. This would also be a major step in advancing the current understanding of the differences between movements by the dominant and non-dominant arms.

14
Posture and Whole-Body Actions

Most everyday actions, including whole-body actions, rely on maintenance of vertical posture in the field of gravity. The notion of posture, however, is not very well defined, although it is intuitively appealing and used frequently. Many studies of keeping vertical orientation of the body in the field of gravity imply under posture an ability not to fall down while the standing person may be performing various whole-body movements. Imagine a person unexpectedly slipping while walking on ice. Typically, complex movements of all the limbs and trunk are seen that are successful in preventing a fall in a high percentage of cases (although not always!). In such situations, it is hard to identify a salient geometric characteristic corresponding to maintenance of vertical posture.

In biomechanics, posture is traditionally defined as joint configuration (Zatsiorsky 1998). This definition is appealing because, in certain situations, joint configuration has to be kept unchanged during a whole-body movement (e.g., during diving or figure skating). On the other hand, when clinicians describe a patient with postural disorders, they do not mean an inability to keep joint configurations unchanged but rather problems with keeping vertical posture (not falling down) during a variety of actions. This term may also address an impaired ability to keep the orientation of a glass with water close to the vertical that does not lead to spilling the contents of the glass, or to maintain a certain grip of a writing implement, or to keep the orientation of a violin and bow while playing the instrument, and so forth. All these examples suggest a much broader understanding of the concept of posture as compared to the aforementioned biomechanical definition.

Within this chapter, we will define posture as a geometrical characteristic of the body or of an object manipulated by the body that is kept dynamically stable during performance of movements. Keeping posture is nontrivial given the unavoidable spontaneous changes in body states, including excitability of neural elements, and action of external perturbations (i.e., unpredictable changes in external forces). Abilities not to fall down, not to spill water from the glass, and to keep proper orientation of the violin and bow are particular examples fitting this definition. This definition links the notion of posture directly to a number of earlier discussed concepts, such as equilibrium, dynamical stability, synergy, and a few others (see Chapters 3, 6, and 9).

Whole-body contributions to functional movements are not limited to postural control. The most obvious example is movements leading to changes in the location of the body in external space, that is, various types of locomotion. Other examples include reaching involving trunk movement or making a step, hitting a soccer ball with a foot, and lifting a heavy object. To discuss such movements, we will accept the theory of movement control with changes in spatial referent coordinates (RCs; Chapter 7) and the idea of synergic control as the means of ensuring dynamical stability of actions (Chapter 6).

14.1. Postural sway and its components

When a person is asked to stand quietly, the body shows spontaneous deviations from the vertical—the so-called postural sway (reviewed in Collins and DeLuca 1993; Zatsiorsky and Duarte 1999; Duarte and Freitas 2010). Traditionally, postural sway has been viewed as a reflection of the mechanical properties of the body modeled as an inverted pendulum moving primarily about the ankle joint (Fig. 14.1; Winter et al. 1996, 1998). Neural control has been associated with specifying agonist-antagonist muscle activation levels translating into time changes in the net moment of force about the ankle joint and its apparent stiffness. The effects of neural noise (not a well-defined concept) have also been incorporated. Overall, within this framework, sway has been viewed as a reflection of imperfections in the design of the human body and/or an unavoidable byproduct of the bipedal stance. Sway has been quantified using characteristics of the horizontal migration of the center of mass (COM), center of pressure (COP), and sometimes other kinematic characteristics (e.g., head trajectory). Increased sway, quantified with different variables such as its mean velocity, root mean square deviations, and so forth, has been viewed as a reflection of poor postural stability.

In spite of its popularity, this view is obviously inadequate. It suffers from a number of flaws. The first and most obvious one is the assumption that the central nervous system prescribes electromyographic (EMG) or mechanical peripheral variables, which obviously depend on the action of reflex feedback loops. Note that changes in the ankle joint angle are associated with changes in many local mechanical variables,

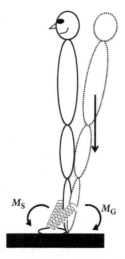

Figure 14.1 An illustration of the control of vertical posture with spring-like properties of the agonist-antagonist muscles acting about the ankle joint. Any changes in the moment of gravity force (M_G) caused by deviation of the posture (dashed image) are counteracted by changes in the "ankle spring" length, causing changes in its force and moment (M_S).

such as muscle length, velocity, and force, that affect activity levels of numerous sensory endings, which exert reflex effects on activation of alpha-motoneurons (see also Chapter 3). The assumption that changes in the apparent stiffness of the ankle joint are sufficient to equilibrate the body has also been falsified in experiments (Morasso and Sanguineti 2002; Casadio et al. 2005; Loram et al. 2005). Equating large postural sway with poor postural stability has a very strong counterexample: Patients at later stages of Parkinson disease show very little postural sway, but their vertical posture is very unstable as reflected in their inability to keep it under the action of even moderate perturbations (Horak et al. 1992).

A number of features of sway suggest that it represents not "noise" but a purposeful process controlled by the brain and possibly linked to exploration of the borders of postural stability during standing. In particular, sway increases when a person is asked to stand on a board with a narrow beam fitted to its bottom (Mochizuki et al. 2006). Note that, if the beam is not too narrow (e.g., 4 to 5 cm wide), its smaller dimension is still much larger than typical peak-to-peak deviations of COP and COM during standing on a flat surface (typically, under 1 cm). So, the subjects could ignore the beam and stand naturally without any danger of losing balance. Nevertheless, the subjects show a rather large increase in the sway, which sometimes leads to losing balance. An opposite effect—a large sway reduction—is seen when standing subjects touch an external object with a fingertip (Jeka and Lackner 1994; Jeka et al. 1998). Mechanical effects of the touch in such experiments are very small and cannot contribute to body balance. So, these effects have been interpreted as reflections of a sensory cue, which informs the subject on body orientation with respect to the vertical and alleviates the need to scan the borders of stability.

Postural sway has been discussed as superposition of at least two processes. One of them reflects descending control of the body and is associated with migration of the instantaneous equilibrium coordinate of the body in the field of gravity. This component has been addressed as *rambling*. The other component reflects mainly peripheral factors such as body mechanics, including the mechanical properties of muscles, tendons, and ligaments, as well as the action of spinal reflex loops. This component has been addressed as *trembling*. Zatsiorsky and Duarte (1999, 2000) suggested a method for sway decomposition into these two components along two main coordinates, anterior-posterior and mediolateral, traditionally used in studies of postural control. The method is based on identifying points in time when the resultant force along one of the horizontal coordinates is zero and then extrapolating those points into a smooth trajectory—rambling. Note that these points occur relatively frequently because the force has to change direction to avoid movement of the body incompatible with standing. The difference between the COP trajectory and rambling has been termed trembling.

Figure 14.2 illustrates typical trajectories of the sway, rambling, and trembling during quiet standing. Note that most of the sway power is in the rambling component, which is also characterized by relatively slow processes. In contrast, trembling is typically faster, but its power is relatively modest. Associating rambling with the trajectory of the instantaneous equilibrium coordinate provides direct links between this component of sway and the concept of control with spatial RCs for the body (see Chapter 7). This link will be discussed in more detail later in the chapter.

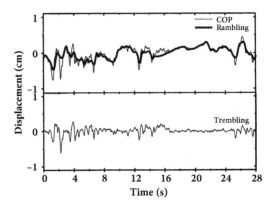

Figure 14.2 Typical trajectories of the center of pressure (COP), rambling, and trembling during quiet standing. Reproduced by permission from Mochizuki et al. 2006.

Analysis of the relations between the sway components and force magnitude in the same direction revealed no such correlations for rambling and strong, close-to-linear correlations with trembling expected from pendulum-like behavior. The lack of such correlations with rambling suggests that this component of sway (which dominates the total process in terms of power!) is unrelated to pendulum-like mechanics, in line with the view that sway is primarily reflecting processes within the central nervous system.

Rambling-trembling sway decomposition has been used in a number of studies exploring both natural sway in young healthy persons and changes in sway under special instructions and in populations with mildly impaired postural control (Sarabon et al. 2013; Bolbecker et al. 2018; Costa et al. 2022). In particular, when young healthy persons stand with purposefully increased coactivation of muscles acting at the ankle joint, both sway components become faster (Yamagata et al. 2019a). The direct effects of muscle coactivation on the apparent stiffness of the ankle joint may be expected to lead to higher speed of pendulum-like processes, that is, to faster trembling. An increase in the speed of rambling, which is unrelated to peripheral pendulum-like processes, is not straightforward. If rambling is indeed a reflection of the search process of the limits of stability, this increase may reflect the perceived difficulty of maintaining balance with additional muscle coactivation.

A few studies have suggested that there may be a third, very slow component of postural sway with typical frequencies of under 0.1 Hz (Duarte and Zatsiorsky 1999; Yamagata et al. 2019b). These phenomena have been interpreted as reflections of slow motion of the referent orientation (RO) of the body toward its actual orientation (see Chapter 16).

Figure 14.3 illustrates the idea of the control of vertical posture with two basic commands, reciprocal (*R*-command) and coactivation (*C*-command), acting primarily at the ankle joint (see Chapters 3 and 7). The difference between the body RO (a version of the RC in angular units) defined by the *R*-command translates into moment of force with the help of the *C*-command, which defines apparent rotational stiffness.

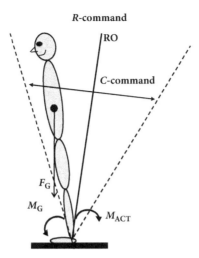

Figure 14.3 A schematic illustration of the control of vertical posture with spatial referent coordinates. The reciprocal command (R-command) defines the referent body orientation (RO) different from the actual orientation (AO). The coactivation command (C-command) translates the difference between RO and AO into moment-of-force units (M_{ACT}), which counteracts the moment (M_G) by the gravity force (F_G).

Rambling is associated with purposeful time changes in the R-command (and possibly also the C-command). There is, however, a back-coupling process of unintentional changes in the R-command, which tends to move the RO toward the actual body orientation. These phenomena are discussed in more detail in Chapter 16. They are typically slow, with characteristic times of 10 to 20 s, compatible with the very slow sway component. According to the scheme in Fig. 14.3, if the drift of RO toward actual orientation (AO) is allowed to accumulate, it can lead to a decrease in the active moment of force, which is needed to counteract the effects of gravity. Ultimately, it may lead to loss of balance and necessity to take a protective step. The slow drift component becomes particularly pronounced if the subject's attention is distracted and they stop correcting the effects of the hypothetical back-coupling process.

14.2. Posture-stabilizing mechanisms

Posture is stabilized by a number of mechanisms acting at different characteristic latencies. If a person performs an action associated with perturbation of posture, changes in activation levels of muscles involved in the postural task component are seen prior to the action initiation (Belen'kii et al. 1967). Such *anticipatory postural adjustments* (APAs) can be seen, in particular, in leg and trunk muscles when a standing person performs a quick arm movement or lifts/catches/drops a load (reviewed in Massion 1992). APAs show scaling with the expected magnitude and direction of the postural perturbation as well as with other factors such as the magnitude of action (even if the associated postural perturbation remains unchanged) and postural stability. In particular, APAs are smaller if the action triggering a standard perturbation is smaller (Aruin and Latash 1995). APAs can be reduced when the initial posture is more stable (e.g., when an additional support is available during standing) and also when it is more unstable (e.g., when standing on a surface with reduced support area) (Nardone and Schieppati 1988; Nouillot et al. 1992; Aruin et al. 1998). Traditionally, APAs have been quantified and interpreted in spaces of mechanical variables, such as COP shifts, and muscle activation indices. They have also been interpreted as shifts in the referent configuration of the body. Figure 14.4 illustrates possible shifts of the body RC in a sagittal plane resulting in typical muscle activation patterns during APAs, which have been associated with the so-called ankle and hip strategies (Horak and Nashner 1986).

If a change in external forces comes unexpectedly, APAs obviously cannot be seen, and the first line of defense is peripheral reactions of muscles and tendons to changes in muscle length, the so-called *preflexes* (cf. Loeb 1999), which act against external perturbations nearly instantaneously. The next quickest responses are spinal reflexes, which typically represent negative feedback loops and act at relatively short latencies on the order of 30 to 50 ms. Both preflexes and reflexes are insufficient to compensate for the posture-destabilizing effects of typical perturbations, but they provide partial compensation and gain time for later, more sophisticated posture-stabilizing mechanisms.

The next group of responses comes at somewhat longer latencies, about 50 to 90 ms, which are still under the latency of simple reaction time. These responses have been addressed with various names including long-loop reflexes, triggered reactions, and preprogrammed reactions (reviewed in Marsden et al. 1976; Chan and Kearney 1982). Their pattern and magnitude show strong dependence on the instruction to the subject. In particular, unlike spinal reflexes, they are reduced when the subject is instructed not to resist an upcoming perturbation as compared to the instruction "resist as quickly as possible." A number of studies provided evidence for at least two different loops involved in such preprogrammed reactions, a shorter-latency loop via the brainstem and a longer-latency transcortical loop (Duchateau et al. 2002; Reschechtko and Pruszynski 2020). Patterns of muscle activation during

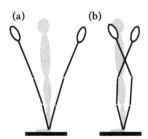

Figure 14.4 A schematic illustration of changes in the referent configuration of the body associated with the ankle (a) and hip (b) strategies. The actual body orientation is shown with gray lines. Referent configurations for body movements in opposite directions are shown with black lines.

preprogrammed responses are highly sensitive to conditions when a perturbation emerges and to neurological disorders. In particular, typical healthy subjects most commonly demonstrate reciprocal patterns of activation in agonist-antagonist muscle pairs (i.e., changes in the R-command; see Chapter 3). These patterns turn into coactivation (i.e., changes in the C-command) in conditions of postural instability and/or uncertainty with respect to the direction of the upcoming perturbation (Slijper and Latash 2000; Piscitelli et al. 2017). Coactivation patterns are also seen in populations with impaired postural control, from the healthy elderly to persons with atypical development to neurological patients (Aruin and Almeida 1996; Nagai et al. 2011; Hirai et al. 2015; Lee et al. 2015).

14.3. Whole-body voluntary movements

One of the central problems in motor control has been formulated by von Holst as the *posture-movement paradox*. If a person maintains a certain posture, changes in external forces lead to postural deviations, which are counteracted by involuntary posture-stabilizing mechanisms described in the previous section. Now imagine that a person performs a voluntary movement away from the initial posture (Fig. 14.5). The posture-stabilizing mechanisms are expected to oppose the movement. The person can apply extra effort to overcome these mechanisms, but he or she will not be

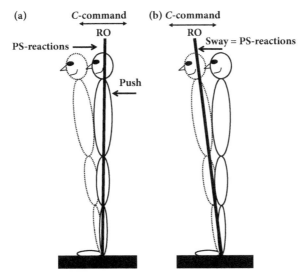

Figure 14.5 An illustration of the posture-movement paradox. (a): When an external force deviates the body from an equilibrium, involuntary posture-stabilizing mechanisms (PS-reactions) act at a short time delay and tend to bring the body back to its original state. (b): When the same body deviation is produced voluntarily, the same posture-stabilizing mechanisms contribute to body sway to a new equilibrium. RO, body referent orientation; C-command, coactivation command.

able to relax at the new position after the movement because, otherwise, the involuntary posture-stabilizing mechanisms would bring the effector back to the initial posture. It is well known, however, that animals are able to relax their muscles following voluntary movements in a variety of the initial and final postures. The ultimate question is: How can movements be produced without triggering resistance of posture-stabilizing mechanisms?

Traditional views on this problem have included suppression of posture-stabilizing mechanisms during voluntary movements and applying extra effort to counteract those mechanisms. The latter argument has already been considered and shown to contradict the ability to relax at any position following voluntary movement. The former argument contradicts results of numerous experiments showing that involuntary posture-stabilizing mechanisms are not turned off during voluntary movements and can be observed if the trajectory is unexpectedly perturbed (reviewed in Shemmell et al. 2010; Fautrelle and Bonnetblanc 2012; Reschechtko and Pruszynski 2020). A solution to the seeming paradox is given by the idea of indirect control of movement mechanics with spatial RCs. Indeed, as illustrated in Fig. 14.5B, a change in RC (RO for body orientation) turns the previous posture into a deviation from the one compatible with the current RC value and leads to the generation of active forces moving the effector to the new posture. In other words, this type of control turns posture-stabilizing mechanisms into movement-producing ones.

Consider once again a person standing in the field of gravity (Fig. 14.6). During quiet standing, the COM projects in front of the ankle joints, and the gravity force generates a moment of force about the ankle joints. This moment has to be counteracted by the active moment of force, which is produced indirectly, by establishing the trunk RO (a particular example of RCs in angular coordinates) different from its actual orientation (AO in Fig. 14.6). The angular difference (RO—AO) is translated into moment-of-force units with the help of the ankle joint apparent stiffness, k. RO is established by the R-command, and k is a mechanical consequence of the C-command (see Chapter 3).

If this person wants to perform a quick voluntary body sway, this scheme of control makes different predictions with respect to patterns of muscle activation during backward sway and forward sway. Backward sway is initiated by moving the RO backwards, resulting in a typical triphasic pattern of muscle activation starting from an agonist EMG burst (see Chapter 4). In contrast, moving the RO forward to produce forward sway is not expected to produce agonist EMG changes until the RO moves ahead of the AO. During the initial time interval, body movement is produced by the gravity force, and its acceleration is naturally limited by this force. As a result, the movement is not started with an expected agonist EMG burst but with a drop in the background activation level of the antagonist muscles, and its speed is smaller than that of the backward sway (Mullick et al. 2018; Nardini et al. 2019). Taken together, these observations confirm that whole-body movements are not controlled with patterns of muscle activation prescribed by the central controller but with patterns of shifts in referent body coordinates, such as its RO. A similar conclusion was reached in a study of the sit-to-stand action controlled by time patterns of the referent body configuration (reviewed in Feldman 2015).

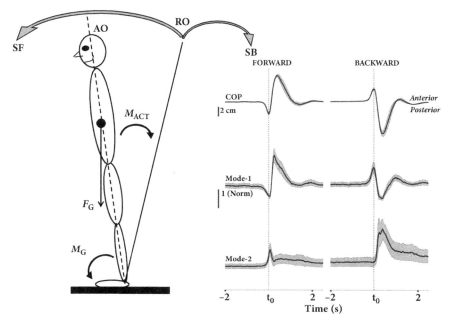

Figure 14.6 To perform voluntary body sway, the person shifts the referent orientation (RO) in the direction of movement. Sway backward (SB) is associated with a burst of muscle activation in the agonists. Sway forward (SF) shows suppression of the antagonist activity and no agonist burst until RO reaches the body actual orientation (AO). M_G, moment of the gravity force (F_G); M_{ACT}, active moment of force. Reproduced by permission from Nardini et al. 2019.

14.4. Whole-body synergies

The importance of stability of the vertical posture suggests the existence of synergies stabilizing salient variables at the level of control reflected in synergies in other spaces of analysis, kinematic, kinetic, and EMG. Synergies at the level of muscle activation patterns have been analyzed during a variety of actions including voluntary body sway, releasing the load from extended arms, and initiating a step (Krishnamoorthy et al. 2003b; Wang et al. 2005; Danna-dos-Santos et al. 2007). These studies involved three major steps (see also Chapter 6). First, stable muscle groups (muscle modes or M-modes) with parallel scaling of EMGs were identified using various matrix factorization techniques including principal component analysis. Second, mapping small changes in the M-mode magnitudes onto changes in salient performance variables was identified using multiple linear regression techniques. This step produced a Jacobian matrix, and the null-space of this matrix was taken as a linear approximation of the uncontrolled manifold (UCM). Third, inter-trial (or inter-cycle, during cyclical sway tasks; Danna-dos-Santos et al. 2007) variance in the M-mode space was quantified within the estimated UCM and orthogonal to it. Taken together, these studies confirmed the existence of synergies stabilizing the trajectory of the COP in the spaces of M-modes across a variety of whole-body tasks.

These studies have shown also that the composition of M-modes was robust within variations of a task (e.g., across cyclical voluntary sway tasks performed at different frequencies), but this composition could change when similar tasks were performed under challenging conditions (e.g., while standing on a board with a reduced supporting surface or in the presence of vibration applied to the Achilles tendons) (Danna-dos-Santos et al. 2008; see also Chapter 15). Major changes in the M-mode composition were seen during unusual whole-body tasks (e.g., when the subjects were asked to produce an accurate shear force pulse in the anterior-posterior direction) (Robert et al. 2008). Although analysis of M-modes and M-mode-based synergies was performed in muscle activation spaces, its results have to be interpreted in terms of changes in the body referent configuration. An earlier figure (Fig. 14.4) illustrated referent body configurations corresponding to the two typical sets of M-modes. The first set corresponds to typical tasks such as swaying in the anterior-posterior direction. Note that the referent body configuration resembles the so-called ankle strategy. The second set corresponds to the unusual task of shear force pulse production. It resembles the so-called hip strategy (cf. Horak and Nashner 1986).

Analysis of synergies during whole-body tasks has also been performed in spaces of kinematic and kinetic variables. In particular, analysis in the joint configuration space was applied to both spontaneous postural sway during quiet stance and voluntary cyclical actions performed while standing (Hsu et al. 2007; Freitas et al. 2010). Note that the idea of postural sway reflecting pendulum-like action about the ankle joint has no room for multi-joint synergies stabilizing such global variables as the trajectory of the COM or of the head in space: Any deviation in the ankle joint angle is expected to lead to deviations of those variables without a possibility of correcting those deviations by motion in other parts of the body. Analysis of joint deviations along the body has shown, however, that strong inter-joint synergies stabilize both the COM and head trajectories. Hence, spontaneous postural sway represents a multi-joint movement reflecting the hierarchical control with spatial RCs (see Chapter 7) organized in a synergic way.

When a person is asked to perform accurate cyclical movement of the COP or of one of the leg joint angles while standing, coordinated motion of all the major joints is observed. Analysis of multijoint coordination within the framework of the UCM hypothesis confirmed the existence of synergies stabilizing the COM trajectory in the anterior-posterior direction and of the trunk orientation (Freitas et al. 2010).

A number of studies performed analysis of synergies in spaces of kinetic variables, such as individual joint moments of force, during cyclical actions, such as hopping (Auyang et al. 2009; Yen et al. 2009). These studies provided evidence for covariation of the joint moments stabilizing the resultant force components acting on the supporting surface. These results, however, could be consequences of varying joint configurations across cycles, that is, reflections of a kinematic synergy seen in kinetic variables. Indeed, if the configuration of a serial kinematic chain remains unchanged, the endpoint force/moment vector defines unambiguously moments of force in all the intermediate joints independently of their number. Covariation of joint moments to produce the same endpoint force/moment vector becomes possible

14.4. WHOLE-BODY SYNERGIES 211

if joint configuration is allowed to vary. There is another possibility of kinetic synergies in a serial kinematic chain: if only some components of the force/moment endpoint vector are prescribed by the task and other components are allowed to vary (Xu et al. 2012). Overall, the results of studies of kinematic and kinetic synergies have to be interpreted in spaces of control variables such as body RO defined by the R-command and apparent stiffness (k) reflecting the C-command (see Chapters 3 and 7).

A study exploring postural control during standing in the {RO; k} spaces was performed using a smooth unloading technique (Nardon et al. 2022) similar to those used in the very early studies forming the foundation of the equilibrium-point hypothesis (EP-hypothesis) (Feldman 1966) and also in studies using the "inverse piano" (see Martin et al. 2011 and Chapter 7). The subject was asked to stand within a spatial target for the COP in the anterior-posterior direction against a load acting backwards (Fig. 14.7A). The load was smoothly reduced and then returned to the initial value unexpectedly for the subject, who was instructed not to react to the (small) body deviations induced by the load manipulation. Further, linear regression between the kinetic and kinematic variables, the COM coordinate and load-related force or body orientation and load-related moment of force, was used to compute RO/RC and k values (Fig. 14.7B, top). Repeating this procedure multiple times led to a family of data points, which typically were aligned along a hyperbolic curve reflecting a synergy in the {RO; k} space stabilizing the initial moment of force counteracting the

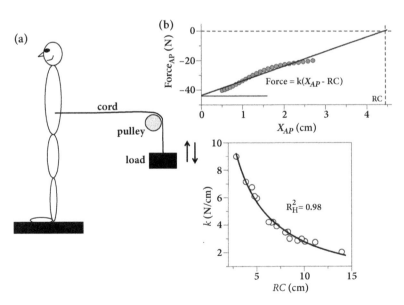

Figure 14.7 An illustration of a study of posture-stabilizing synergies at the level of the reciprocal and coactivation commands (R- and C-commands). (a): A load was smoothly removed and reapplied to follow the force-coordinate characteristic. (b): The intercept and slope of the characteristic were measured in each trial (top right). Across trials, they showed strong hyperbolic relations (bottom right). Reproduced by permission from Nardon et al. 2022.

load (Fig. 14.7B, bottom). Using hyperbolic regression confirmed the covaried adjustments in RO and k interpreted as synergies in the space of the R- and C-commands. The indices of these synergies were smaller in persons who stood with higher muscle coactivation, as reflected in higher average k values, confirming the earlier conclusion that increasing muscle coactivation leads to worse stability of the vertical posture (Yamagata et al. 2019a, 2021).

The conclusion on posture-stabilizing synergies in the spaces of the R- and C-commands has also been supported by analysis of motor equivalence (see Chapter 18). The subjects were asked to occupy an initial posture and then to perform a quick voluntary sway to another target and back to the initial position. The mentioned unloading procedure was used and {RO; k} pairs were computed twice, before and after the sway. Analysis of the {RO; k} pairs confirmed larger deviations along the UCM as compared to deviations orthogonal to the UCM as expected from a synergy stabilizing the salient performance variable.

When a standing person prepares to perform a fast action or counteract a predictable perturbation, anticipatory synergy adjustments (ASAs; see Chapter 8) are seen about 200 to 250 ms prior to the action initiation (Klous et al. 2011; Krishnan et al. 2011). During ASAs, the synergy index shows a smooth drop reflecting gradual destabilization of the salient variable, such as the COP coordinate in the anterior-posterior direction, to facilitate its planned change. ASAs and APAs show some similarities. For example, both are reduced in the healthy elderly (Woollacott et al. 1988; Olafsdottir et al. 2008), and both shift toward the action initiation time under the simple reaction time instruction (De Wolf et al. 1998; Olafsdottir et al. 2005a). On the other hand, there are significant differences in characteristics of ASAs and APAs, reflecting their different nature. In particular, APAs are specific to the direction of an expected perturbation. As a result, if the subject knows only the timing of a self-triggered perturbation but not its direction, APAs are delayed and reorganized: They show larger changes in the muscle coactivation level as compared to the patterns observed when the direction of the perturbation is known in advance (Piscitelli et al. 2017). In contrast, ASAs show no differences between the conditions with known and unknown direction of a self-triggered perturbation. Indeed, destabilizing the salient performance variable helps to change it quickly, independently of the direction of its change. In other words, ASAs are more universal and APAs are more situation specific.

14.5. Locomotion and central pattern generators

To start walking, one has to make the first step. Step preparation has been studied at the level of COP trajectory, muscle activation patterns, and changes in COP-stabilizing synergies (ASAs). During forward step initiation, the COP trajectory shows a consistent pattern over 0.5 to 1 second prior to lifting the stepping foot (Fig. 14.8). These patterns may be viewed as mechanically necessary. They include a transient shift of the COP in the mediolateral direction toward the stepping foot, which then reverses and moves toward the supporting foot. At the same time, the COP moves backwards. At the level of body mechanics, this allows unloading the stepping foot and creating a moment of force rotating the trunk forward. Synergies

14.5. LOCOMOTION AND CENTRAL PATTERN GENERATORS 213

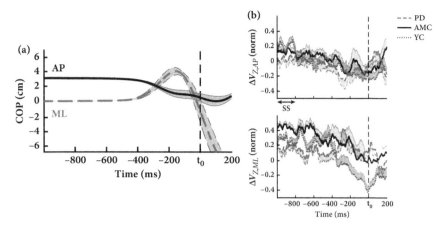

Figure 14.8 (a): Typical shifts of the center of pressure (COP) in preparation of making a step. AP, anterior-posterior; ML, mediolateral. (b): Changes in the COP-stabilizing synergies in the muscle activation (M-mode) space for the anterior-posterior (AP, top) and mediolateral (ML, bottom) directions. The three curves show the results for young controls (YCs), age-matched controls (AMCs), and patients with Parkinson disease (PD). Reproduced by permission from Falaki et al. 2023.

stabilizing the COP coordinates are seen at the level of muscle groups (M-modes), and they tend to become weaker about 200 to 300 ms prior to the take-off (Fig. 14.8B; Wang et al. 2005; Falaki et al. 2023). These patterns can be interpreted at the level of neural control with RCs as consequences of the mapping from the task-level RCs, which encode desired changes in the body position in space, to the RCs at the level of individual extremities and joints.

Steady-state locomotion is viewed as produced by central pattern generators (CPGs) in the spinal cord, which are under both descending control and action of reflex feedback loops from proprioceptors. Although the notion of CPGs is relatively old (e.g., Graham Brown 1914), the neural circuitry of CPGs has been deciphered only for relatively simple animals such as lampreys (reviewed in Orlovsky et al. 1999). The existence of CPGs has been demonstrated in other animals, including quadrupeds, such as cats and dogs, in experiments showing locomotor-like activity in animals whose spinal cord had been surgically separated from the brain (spinal preparations). This activity could be induced by moving the treadmill under the animal's feet and by injecting certain neurotransmitters into the spinal cord (reviewed in Grillner 1975; Grillner and Wallen 1985). Moving the treadmill at different speeds could produce, in spinal animals, different gaits such as walking, trotting, and galloping.

In decerebrate preparations, when the brainstem structures have access to the spinal cord, locomotion can be induced by electrical stimulation of the so-called mesencephalic locomotor region in the reticular formation (reviewed in Shik and Orlovsky 1976). Changing the strength of the stimulation resulted in different gaits. Note that neither decerebrate nor spinal animals could maintain posture on the treadmill and had to be supported by belts during these experiments.

Important observations were made in studies of so-called *fictive locomotion* when the neuromuscular transmission was blocked by curare and, as a result, no overt movement happened. Recording in the ventral roots of the segments innervating the limbs showed, in such animal preparations, rhythmical patterns of activation of alpha-motoneuronal pools resembling those seen during locomotion. These experiments proved that reflex feedback modulated during locomotion was not obligatory for the functioning of spinal CPGs. Of course, this does not mean that reflex feedback is not important. In particular, it participates in modulating the step cycle and also forms the basis of the stretch reflex—the crucial component of the neural control of movement according to the EP-hypothesis and control with spatial RCs.

A series of experiments, primarily on patients with spinal cord injury, provided evidence for the existence of CPGs in humans. In those studies, electrical stimulation of the lumbar enlargement could produce cyclical leg movement resembling running (reviewed in Shapkova 2004; Gerasimenko et al. 2008). Note that the patients were supine, and their legs were suspended in the air with a system of belts. Using different sites of stimulation could lead to bilateral cyclical leg movement, unilateral cyclical movement of one of the legs, and even cyclical movement of only one of the major leg joints. These observations suggest the existence of a hierarchy of CPGs at the spinal level, possibly organized in a synergic way. This conclusion has been supported by the observations of various effects of constraining motion of a leg or of a joint during the stimulation-induced "air stepping." In particular, blocking one leg induced more vigorous cyclical motion of the other leg, and blocking a joint led to larger amplitudes of motion in other joints of the leg.

A number of studies reported that air stepping could be induced in healthy human subjects by muscle vibration applied to one of the major leg muscle groups (Gurfinkel et al. 1998; Selionov et al. 2009; Solopova et al. 2016; see also Chapter 15). The subjects in those studies were lying with their legs suspended in the air. They were instructed to ignore the vibration. In about 50% of the subjects, vibration induced air stepping, which could be facilitated by various factors such as the Jendrassik maneuver, electrical stimulation of a cutaneous nerve, and rhythmical motion of an arm produced by the experimenter. Taken together, these observations suggest that CPGs for locomotion can be facilitated by a variety of relatively nonspecific spinal circuits and descending pathways.

Locomotion has been discussed within the idea of hierarchical control with spatial RCs (Feldman et al. 2021). This analysis, in particular, has shown the inadequacy of the idea that stepping is associated with loss of postural control and "catching the falling body" with the leading leg. The bipedal posture shows high dynamical stability during locomotion, as shown in many experiments with unexpected mechanical perturbations. Indeed, even if a person unexpectedly steps on a slippery surface (e.g., a patch of ice), commonly, a complex, seemingly unique, sequence of movements by all parts of the body can be observed, which prevents falling down in a large percentage of cases (cf. Akulin et al. 2019). Such unique corrective actions can be described as consequences of using a single, simple rule for RC changes resulting in different RCs to effectors due to the different initial conditions when the slip occurs.

Synergies stabilizing salient performance variables during locomotion have been studied at the kinematic and EMG levels. In particular, synergies in the joint configuration space have been shown to stabilize the mediolateral foot trajectory during the swing phase (Krishnan et al. 2013; Rosenblatt et al. 2014). This is of particular importance because dynamical stability in the mediolateral direction is crucial: Passive walking bipedal devices typically lose balance in the mediolateral direction, not in the anterior-posterior direction (Kuo 2002). Such synergies also have been documented in quadrupeds (cats) during natural locomotion (Klishko et al. 2014). The pattern of the synergy index was, however, qualitatively different. The synergy index was higher in the mid-swing in humans, and it showed a drop close to the touch-down. In cats, the synergy index was the highest at the take-off and touch-down and showed a drop in the mid-swing. This can be related to several factors. In particular, in humans, there is a possibility of the swing leg touching the supporting leg in mid-swing, which is impossible in cats. Cats frequently navigate complex environments with high-precision demands for foot placement, while humans most commonly walk on less demanding surfaces.

Studies of muscle activation patterns in young adults have shown a relatively small number of muscle groups (M-modes or factors) observed across subjects and velocities of walking (reviewed in Ivanenko et al. 2004, 2006). Stepping in newborns shows only two of those M-modes, and the number of factors and their relative timing during the step cycle emerge and are refined over the first few years of life (Dominici et al. 2011). The characteristic patterns of muscle activation for individual M-modes have been mapped onto activation patterns within the spinal cord (Ivanenko et al. 2008). So far, these patterns have not been explored in relation to the second basic feature of synergies, dynamical stability of salient performance variables. They are likely to reflect mapping of RCs at the level of whole-body motion to RCs at the level of individual joints and muscles.

14.6. Missing pieces of the mosaic

The current limitations in understanding the mechanisms of the neural control of whole-body movements are directly related to the availability of tools such as force platforms, motion analysis systems, and EMG systems, which are frequently seen as sources of direct information on neural processes. To put it mildly, this is very far from reality. As emphasized by many researchers (see also Chapter 4), peripheral variables recorded in typical studies carry mixed information on a variety of factors, including the frequently unpredictable external forces and action of reflex feedback loops. Such information becomes useful if it is interpreted within a realistic physiological theory on the neural control of movements. As of now, there seems to be no competition to the theory of control with spatial RCs, which readily incorporates the idea of hierarchical control and performance-stabilizing synergies. Assuming computational processes in the central nervous system that would likely be placed there by an engineer is not a viable alternative.

An encouraging example of an analysis of mechanical variables offering new insights into the nature of spontaneous whole-body postural sway is the rambling-trembling decomposition. It puts sway within the context of control with RC time profiles and offers new interpretations of changes in sway characteristics across tasks and populations.

There has been no comparable progress in studies of locomotion, although the idea that CPGs generate time profiles of RCs to individual limbs, effectors, and muscles has been formulated (Feldman 2015). Patterns of natural locomotion represent superposition of multiple factors including descending control, spinal CPGs, action of reflex and reflex-like feedback loops, and interactions with the environment. So far, deciphering the contributions of these factors to observed patterns of mechanical and EMG variables has been unsuccessful.

Another missing piece of the puzzle is the transition between standing and walking. Step initiation has been studied in detail. Likely, step initiation is dominated by descending control as exemplified, for example, by problems with step initiation (e.g., episodes of "freezing of gait") in patients with Parkinson disease (reviewed in Watts and Koller 2004; Fahn and Jankovic 2007; also see Chapter 24). If we accept the dominant view that steady walking is produced with major contributions from spinal CPGs, an issue emerges of how sharing of the control responsibilities changes when step initiation turns into steady walking.

15
Kinesthetic Perception

In most situations, even with closed eyes, we are aware of the orientation of the body in space, the configuration of the limbs and other effectors (e.g., the head, trunk, and digits), and the contact forces with the environment. Taken together, this awareness is commonly referred to as kinesthetic perception. Traditionally, kinesthetic perception has been viewed as a major contributor to the creation and updating of an undefined neural representation of the body and its interaction with the environment necessary for planning and performing successful actions. An alternative idea of direct perception has been developed within the field of ecological psychology (Gibson 1979). Within this idea, neural signals reflecting perceptual processes can produce changes in ongoing actions without updating the neural representation of the body in the world. The most obvious examples of such perception-action coupling are reflexes and reflex-like adjustments of actions based on changes in afferent information from peripheral sensory endings. There are also examples of more complex action adjustments to changes in sensory information that apparently bypass the step of updating the neural representation of the body.

The importance of the sensory signals contributing to kinesthetic perception for the neural control of movements is illustrated by major problems with the production of functional movements in persons with large-fiber peripheral neuropathy, a rare disorder leading to interruption of transmission of action potentials along large afferent fibers (Rothwell et al. 1982a; Sainburg et al. 1995). In addition to nearly complete lack of kinesthetic perception, this disorder leads to elimination of typical spinal reflexes, including the stretch reflex, which forms the foundation of the neural control of movements in healthy persons (see Chapter 3). Persons affected by this disorder cannot perform functional movements in darkness, and even in the presence of vision, their movements are slow and deliberate and show major impairment in joint coordination.

Kinesthetic perception obeys regularities described for other perceptual modalities such as visual and auditory. In particular, across all modalities, the minimally distinguishable difference between two stimuli changes with the average magnitude of the stimuli, leading to a logarithmic relation between intensity of sensation and magnitude of the sensed physical variable (the Weber-Fechner law). It is also likely that perception across modalities can be used for two different purposes, to guide an action and to report percepts (including reporting to oneself). The presence of two different neurophysiological streams for these two purposes has been documented and explored in detail for visual perception (Goodale et al. 1991; Goodale and Milner 1992). Recent studies of kinesthetic perception, described in one of the following chapters (Chapter 17), suggest that perception-to-report can differ from perception-to-act for this perceptual modality as well, although the two corresponding neurophysiological streams have not been identified yet.

We will view perception as a process of measurement of salient physical variables such as joint angles, coordinates of parts of the body, and forces between effectors and the environment. As with any measurement, one has to have a proper tool and a system of coordinates. We assume that signals from sensory endings represent elementary tools, which can be combined to reflect more ecologically valid variables, while processes related to the generation of actions define systems of coordinates, which are used to estimate sensory signals and convert them into relevant units. Since most of the relevant neural signals cannot be recorded in humans during natural actions, we will need to use indirect evidence, including errors in perception and action, to formulate and test hypotheses on how the two main contributors to kinesthetic perception interact to produce adequate and inadequate percepts.

15.1. Ambiguity of sensory information

Traditionally, kinesthetic perception is viewed as based primarily on signals from proprioceptors, which are specialized neurons with the body located in spinal ganglia and a very long T-shaped axon. The peripheral branch of the axon ends with a special structure, a sensory ending, which converts physical signals into sequences of action potentials. Most sensory endings are sensitive to deformation, and their responses to physical variables, such as forces and displacements, depend on their location in the muscles, tendons, skin, and so forth. Some of the endings also respond to chemicals (such as products of muscle metabolism), temperature, inflammation, and other factors unrelated to kinesthetic perception.

The main types of proprioceptors seem to provide ample information on salient mechanical variables. Sensory endings located in muscle spindles modulate their firing rate with muscle length (primary and secondary endings) and velocity (primary endings only). Sensitivity of spindle endings to muscle length and velocity is modulated by small motoneurons (gamma-motoneurons), which prevents, in particular, complete silence of these sensory endings during voluntary movements when the muscle is shortened quickly. Golgi tendon organs located at the junction between the muscle fibers and tendon modulate their firing rate with force. Articular receptors are sensitive to joint angle. In addition, any movement is associated with skin deformation, which is sensed by various cutaneous and subcutaneous sensory endings.

However, considering each of these sources of information in more detail suggests that deriving salient mechanical variables (e.g., joint angle and moment of force) from these signals may not be trivial. For example, joint movement is linked anatomically to changes in the length of the "muscle + tendon" complex, which may differ from changes in muscle fiber length sensed by spindle endings. In particular, during muscle contraction in isometric conditions, muscle fibers shorten, the tendon stretches, and the length of the "muscle + tendon" complex remains unchanged (Fig. 15.1). These differences may even be qualitative, for example, when the "muscle + tendon" complex stretches under the influence of external forces while muscle fibers contract actively (e.g., Loram et al. 2005). Changes in activity of gamma-motoneurons can also lead to changes in activity of spindle endings, unrelated to changes in muscle length.

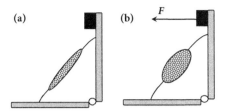

Figure 15.1 During muscle contractions, changes in muscle fiber length may not reflect joint motion, which is linked to changes in the "muscle plus tendon" complex. (a): Relaxed muscle. (b): The muscle produces force F in isometric conditions.

Golgi tendon organs inform the central nervous system on tendon force magnitude. However, salient variables for joint movements are rotational. The effects of tendon force on joint moment of force depend on the lever arm of the force vector, which changes with joint position. This means that information on joint angle is needed to convert force values reported by signals from the Golgi tendon organs into moment-of-force values.

Articular receptors also show features that prevent them from being considered reliable sources of information on joint angle. In particular, these receptors show nonmonotonic changes in their firing rate with joint angle, they are primarily sensitive to joint angle values close to the anatomical limits of rotation, and they also modulate their response with joint capsule tension and a few other factors, for example, joint inflammation (Burgess and Clark 1969; Clark and Burgess 1975).

We come to the conclusion that each type of proprioceptor generates ambiguous information, which requires nontrivial deciphering to extract values of salient mechanical variables such as joint angle, its velocity, and moment of force. Of course, it is possible to offer computational means of extracting these values, in particular given that spindle and Golgi tendon organ signals come from all muscles crossing a joint. However, we are trying to avoid assuming computational processes within the central nervous system and would like to account for stable kinesthetic perception based on physiological processes.

15.2. Perception of muscle length and force

The idea that percepts emerged with an important role played by neural commands for movement was suggested in the 19th century by Helmholtz. He reached this conclusion by observing the differences in visual percepts during natural eye movements and during eyeball displacements produced by pressing on the eyeball with a finger. In both cases, images from the environment move over the retina, but in the latter case only, a percept of the external world moving with respect to the person emerges. Later this insight was formalized using the concept of efference copy (a.k.a. efferent copy, corollary discharge), a copy of neural commands for action participating in perceptual processes (von Holst and Mittelstaedt 1950/1973; Sperry 1950). This concept was developed later (for reviews see Feldman 2009, 2016; Latash 2021c) and is still very much under discussion (for more detail see Chapter 17).

Within the theory of control with spatial referent coordinates (RCs; λ for a single muscle; see Chapters 3 and 7), neural command for an action is represented as a time-varying RC(t) function for the involved effector(s). Figure 15.2 illustrates the dependence of effector force F_X along a spatial coordinate X for a fixed value of RC. Note that steady states of the effector are constrained to the curve $F_X(X)$ and depend on the external load characteristic as a function of X. The problem of perceiving a particular combination of $\{F_X; X\}$ can be viewed as a problem of identifying a point from the two-dimensional set limited by biomechanically possible values of X and F_X. Setting a value of RC limits the number of possible solutions to points along the $F_X(X)$ curve, that is, turns the problem of selecting a point from a two-dimensional space into one of selecting a point along a single dimension. Sensory signals can be used to identify a point along this dimension and create percepts of both force and coordinate.

Note that, for a single muscle, signals from a number of sources increase monotonically with deviation of the point from RC (λ). When muscle length moves away from λ to larger length values, this causes an increase in the firing rate of spindle endings. At the same time, muscle force increases, leading to an increase in the firing of Golgi tendon organs. There is also an increase in the firing of alpha-motoneurons due to reflex effects on muscle activation (Fig. 15.2). Theoretically, any of these sources of information can be used to identify a point along the curve and lead to perceiving muscle force and length. A number of recent studies have provided evidence that spindle endings sometimes change their firing rate with muscle force and that Golgi tendon organs change their firing rate with muscle length, indirectly corroborating this scheme (Watson et al. 1984; Luu et al. 2011; Proske and Gandevia 2012). Interestingly, according to this scheme, as long as the stretch reflex mechanism is intact, a degree of rudimentary kinesthetic perception can be based on signals from the Renshaw cells excited by the alpha-motoneurons, which contribute to ascending spinocerebellar pathways.

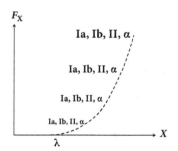

Figure 15.2 The control of a muscle is associated with setting its spatial referent coordinate—the stretch reflex threshold, λ. Setting a value of λ defines a one-dimensional space of possible combinations of muscle length (coordinate, X) and force (F_X) at steady states constrained to the thick, dashed line. A number of neurons show increased levels of activity with deviation from λ along the $F_X(X)$ line as shown with fonts of different size.

Within the scheme in Fig. 15.2, perception of muscle force is intimately linked to perception of its length by the $F(L)$ dependence defined by the stretch reflex. This conclusion leads, in particular, to predictions of interactions between kinematic and kinetic perceptual illusions and errors, a prediction confirmed experimentally in studies of the effects of high-frequency muscle vibration (see later and Reschechtko et al. 2018).

Similar schemes can be offered for different effectors and effector sets. Indeed, the control of any effector acting along a single dimension is associated with setting RC values for the agonist and antagonist muscle groups or, equivalently, values of the reciprocal and coactivation commands (R- and C-commands; see Chapters 3 and 7). The number of sources of relevant sensory signals increases with the number of muscles, tendons, and joints housing respective sensory endings. Figure 15.3 illustrates the afferent and efferent components of kinesthetic perception for different levels of analysis, from single muscles to single joints to limbs to the whole body. It is important to note that the number of sources of information is always larger than the number of salient mechanical variables, which may be perceived by the actor. For example, the steady state of a joint with one kinematic degree of freedom spanned by two muscles is characterized by two mechanical variables, joint angle and moment of force (possibly, a third variable can be considered reflecting internal joint capsule tension). The number of sources of sensory information is at least five: spindle endings and Golgi tendon organs in each of the muscles and articular receptors. In addition, there are two efferent commands, the R- and C-commands.

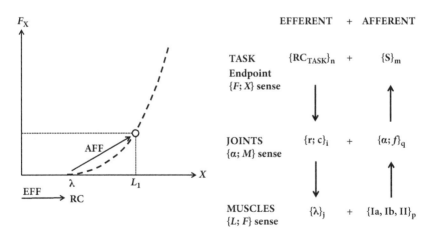

Figure 15.3 Left: The afferent (AFF) and efferent (EFF) components of kinesthetic perception. Right: The sources of information at different levels of analysis. Note that the number of elemental variables is always larger than the number of variables describing the coordinate and force variable to be perceived. F, force; M, moment of force; α, angle; X, coordinate; RC, referent coordinate; r, R-command; c, C-command; S, relevant sensory sources; f, individual muscle forces; Ia, Ib, and II, afferents; the subscripts refer to the dimensionality of the signals.

The illustration in Fig. 15.3 can be viewed as a case of redundancy: A relatively low-dimensional percept has to be defined based on a higher-dimensional set of elementary variables (firing patterns from specific sensory sources and neural commands to the effector). Alternatively, such a set of signals can be viewed as abundant, that is, not as a source of computational problems for structures within the central nervous system involved in perceptual processes but as a useful feature that allows ensuring stability of percepts of salient variables. Within the latter approach, all the problems and hypotheses described for motor abundance (Chapter 5) are equally applicable to perceptual abundance.

15.3. Stability of percepts: Iso-perceptual manifold

One of the striking and poorly understood features of percepts is their stability. There are examples of bistable percepts and other cases of compromised stability of perception (Hupé and Rubin 2004; van Ee et al. 2005; Naber et al. 2010) including unstable kinesthetic illusions (Feldman and Latash 1982c), but they are not typical during behavior in the everyday environment. It seems obvious that evolutionary success requires stability of percepts reflecting salient variables describing the environment and the body. In particular, these percepts have to remain stable in the presence of unavoidable small changes in afferent and efferent signals to allow quick and unambiguous identification of predators and prey and planning of successful actions. Stability of percepts is not a trivial feature. Indeed, consider the following example: You coactivate strongly muscles acting about the elbow joint without moving it and without looking at the arm. This is easy to do, and I encourage the reader to try it. You will have a veridical percept of the joint angle being unchanged with changes in the muscle coactivation level. Where does this steady percept come from? This is not a trivial question.

Indeed, the efferent commands to the muscles crossing the joint are changed. Signals from all the major sensory endings are changed too. Articular receptors are sensitive to joint capsule tension, which is going to change with muscle coactivation. Golgi tendon organs will change their firing with tendon force. Muscle spindle endings will change their firing rate with changes in the activity of gamma-motoneurons expected from the well-known phenomenon of alpha-gamma coactivation. In addition, in the absence of joint movement, muscle contraction leads to changes in the muscle fiber length (see Fig. 15.1): They shorten with activation and stretch the tendon such that the length of the "muscle + tendon" complex remains unchanged. It seems that each of the main signals contributing to perception of joint angle changes without a change in the respective percept. How can this be?

This observation suggests that changes in all the mentioned signals from the contributing sources (elemental variables) are associated with their covariation keeping a particular function of those signals unchanged, or, in other words, these elemental variables are constrained to a subspace associated with perceiving an unchanged joint angle. This conclusion can be viewed as a definition of percept, one that allows drawing predictions and performing quantitative analysis (see later). The subspace corresponding to a fixed value of a percept has been addressed as iso-perceptual manifold (IPM; Latash 2018a).

Figure 15.4 illustrates schematically an IPM in a cartoon situation when one efferent variable (RC) and two afferent variables (A1 and A2) contribute to perception of a variable: V = RC + A1 + A2. The gray triangle shows the IPM for this percept. Note that all the variables can change along the IPM without affecting the percept, which is a schematic illustration of what happens in the earlier example of perceiving an unchanged joint angle under variations in muscle coactivation. Motion along the IPM can be termed perceptually equivalent (similar to the concept of motor equivalence; see Chapter 5). Motion away from the IPM is non–perceptually equivalent, and it leads to changes in the percept.

The IPM concept and the related concept of perceptual equivalence lead to a nontrivial prediction that perception of a variable produced by an abundant set of contributing elements should be more accurate compared to perception of elemental variables produced by each of the elements. Indeed, stable and accurate perception of the task-related variable is compatible with variation of the state of the system along the respective IPM. This variation, however, is associated with variable percepts of elemental variables. In other words, if an actor is confident in residing within an IPM for the task variable, this does not mean that he or she is equally confident in the exact location of the system within the IPM.

This prediction was tested in a series of studies with single-finger and multi-finger accurate force production (Cuadra and Latash 2019; Cuadra et al. 2021b). The method of matching force magnitude with the symmetrical, contralateral effector(s) was used to estimate force perception. In both multi-finger and single-finger tasks, the subjects were rather accurate in matching the force by pressing with the same finger(s) of the other hand. In multi-finger tasks, when the subject was asked to match the force of only one of the involved fingers (an elemental variable), both constant and

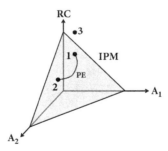

Figure 15.4 An illustration of the concept of iso-perceptual manifold (IPM) for a system with one efferent variable (referent coordinate [RC]) and two afferent variables (A1 and A2). They contribute to perception of a variable V = RC + A1 + A2. The IPM is shown with the gray triangle. Motion along the IPM, from point 1 to point 2 (perceptually equivalent [PE]), is associated with stable percept of V. Moving outside the IPM (e.g., to point 3) changes the percept.

variable errors were significantly larger than those during matching the force of the same finger in its single-finger tasks.

15.4. Vibration-induced illusions

Low-amplitude, high-frequency muscle vibration is known as a very powerful stimulus for the primary endings in muscle spindles (Matthews and Stein 1969). Vibration at about 1 mm peak-to-peak amplitude and the frequency of 60 to 100 Hz can produce driving of all the primary spindle endings in the muscle, which means that each spindle ending generates at least one action potential to each cycle of vibration. Such an unusually high level of activation of the spindle endings leads to a number of motor and sensory phenomena. Golgi tendon organs also show sensitivity to muscle vibration in muscles with non-zero baseline activation (Fallon and Macefield 2007), but their role in the vibration-induced sensory and motor phenomena remains unknown.

Vibration can induce reflex contraction in muscles subjected to the vibration as well as other muscles of the limb—known as the tonic vibration reflex (TVR; Eklund and Hagbarth 1966, 1967). TVR patterns in different muscles depend on the body configuration as well as on some other mechanical factors, such as pressure on the foot (Latash and Gurfinkel 1976; Gurfinkel and Latash 1978). These reflex contractions are typically tonic and characterized by muscle activation patterns resembling those during voluntary muscle contractions. TVR is unique among typical reflexes in the ability of healthy persons to suppress it with mental effort. In other words, humans have an ability to allow TVR to develop or to prevent it from leading to muscle activation. In some persons (about half of the healthy population), vibration applied to a major leg muscle group (e.g., with a vibrator placed on the Achilles tendon) can induce cyclical patterns of muscle activation resembling those during running as long as the leg is suspended in the air (Gurfinkel et al. 1998; Selionov et al. 2009; Solopova et al. 2016). These observations have suggested that vibration-induced afferent inflow can engage the central pattern generator for locomotion, likely located in the lumbar enlargement.

The sensory effects of muscle vibration have been explored for about 60 years (Goodwin et al. 1972; Lackner and Taublieb 1984; Roll and Vedel 1982). Most commonly, these effects represent an illusion of movement in a joint spanned by the muscle with the vibrator corresponding to elongation of that muscle. These vibration-induced illusions do not emerge in all healthy persons, but when they do, their magnitude can be very large, leading to perception of anatomically impossible joint positions (Craske 1977). Illusions of velocity of joint rotation induced by vibration can be much stronger than illusions of joint angle (Sittig et al. 1985). These observations point to the velocity-sensitive primary spindle endings as a major contributor to such illusions. A surprising phenomenon is that vibration-induced illusions can reverse their direction under the influence of seemingly irrelevant stimuli such as visual and auditory stimuli and also with mental effort (Feldman and Latash 1982c). A specific consequence of vibration-induced illusions is the so-called vibration-induced fallings (VIFs) seen as deviations of the body from the natural orientation, close to the vertical, during standing under vibration of leg muscles (Eklund and Hagbarth

15.4. VIBRATION-INDUCED ILLUSIONS

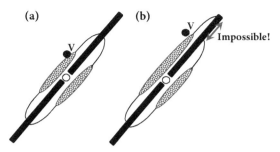

Figure 15.5 Possible effects of muscle vibration. Vibration (V) of one of the opposing muscle groups is expected to lead to perception of elongation of that muscle group without a change in the length of the antagonist (compare a and b). This is impossible without violation of body integrity.

1967; Hayashi et al. 1981) and some other muscles (Lund 1980; Roll et al. 1989). VIFs during vibration of the Achilles tendon are interpreted as consequences of perceiving the ankle joint as moving into flexion, corresponding to stretching the triceps surae muscle, and correcting the corresponding illusory motion of the ankle joint with its actual deviation into extension.

The most straightforward interpretation of the perceptual effects of vibration is that the unusually high activity level of spindle endings is interpreted by the brain as the muscle being stretched. This interpretation, however, cannot account for the aforementioned reversals of illusory movements. Upon closer inspection, it also fails to account for more traditional patterns of vibration-induced illusions. Consider Fig. 15.5. If one of the muscles crossing a joint is subjected to vibration, the increased level of activation of its spindle endings can lead to overestimation of its length. However, spindle endings in the antagonist muscle are unaffected and supply veridical information. If these patterns of signals from the spindle endings in both muscles are interpreted at face value, the illusion should correspond to a change in the point of attachment of one of the muscles and to its elongation without joint motion (as shown in panel B of Fig. 15.5). Such illusions have never been reported, however.

Earlier, we noted that kinesthetic perception is based on interpreting sensory signals within a reference frame defined by the efferent process (RC; see Fig. 15.2). For a single muscle, this illustration suggests that illusions of muscle length have to be coupled to illusions of muscle force. Vibration effects on force perception have indeed been described in support of this scheme (Cafarelli and Kostka 1981; Reschechtko et al. 2018). On the other hand, the presence of motor effects of vibration such as the TVR and locomotion-like leg movements suggests that muscle vibration can also lead to a change in the RC to the muscle as well as to other muscles of the body. As a result, actual RC values can differ from the ones issued by the brain, leading to additional effects on kinesthetic perception (see Feldman and Latash 1982a, 1982b). Figure 15.6 illustrates possible effects of vibration on RCs for the muscle and on the sensory component of perception, suggesting that, depending on the relative magnitudes of these effects, vibration can lead to kinematic and kinetic illusions in different directions.

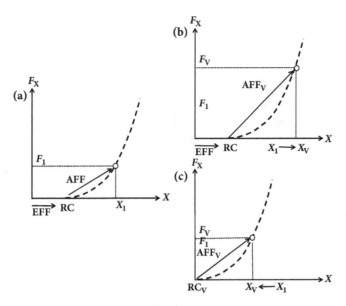

Figure 15.6 Possible effects of muscle vibration (V) on the two components of kinesthetic perception: referent coordinate (RC) and afferent signals (AFF). Vibration can lead to illusions of both coordinate (X) and force (F_x) in different directions (shown with arrows in b and c) depending on its effects on the efferent component (RC). Subscript 1 refers to the initial state of the muscle. Subscript V refers to variables under vibration.

Indeed, the patterns of force illusions described in one of the studies suggest that vibration-induced changes in RCs could play a dominant role (Cuadra et al. 2021a).

15.5. Interpreting impossible sensory signals

As suggested by Fig. 15.5, direct effects of vibration on signals from spindle sensory endings originating in the agonist and antagonist muscles can lead to perception of anatomically impossible consequences violating body integrity. Such percepts are disallowed by the central nervous system, which tries to come up with an interpretation for the conflicting sensory inflow from the opposing muscles. The observations of illusions that correspond to impossible joint positions (Craske 1977) suggest that this rule is not without exceptions. Apparently, the central nervous system permits percepts that go beyond anatomical limits of joint rotation but not those violating anatomical integrity of the body.

Figure 15.7 illustrates expected correlated changes from length-sensitive sensory endings in the agonist-antagonist muscle pair during natural movements of the joint spanned by the muscles. To keep the anatomical integrity of the body, the signals should show a degree of negative correlation corresponding to the rule that an

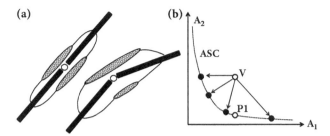

Figure 15.7 (a): Joint motion is associated with opposite changes in the length of the opposing muscle groups, agonist and antagonist. (b): As a result, only certain combinations of sensory signals from length-sensitive endings in the opposing muscles, A_1 and A_2, are accepted as the basis of valid percepts (shown by the line labeled ASC). A point (V) deviating from the ASC line is projected on that line before being interpreted as a percept. Note that various projections are possible, resulting in different illusions.

increase in the length of one of the muscles has to be accompanied by shortening of its antagonist. Such a rule is shown schematically as a line containing all the allowed sensory combinations (ASC line in Fig. 15.7). If an anatomically impossible sensory combination emerges, for example, as a result of vibration applied to one of the muscles (illustrated by the point marked "V"), it has to be projected on the ASC line to allow interpretation in terms of joint motion. Note that such a projection can lead to different points on the ASC line, corresponding to a variety of joint positions. In particular, some of the resulting points may correspond to the joint moving in a direction corresponding to stretching of the muscle with the vibrator while other points may correspond to the joint moving in the opposite direction. Rules defining such projections on the ASC line may depend on various undefined factors, including sensory inputs of other modalities, leading to possible reversals of kinesthetic illusions caused by muscle vibration (cf. Feldman and Latash 1982c).

Of course, this analysis has to be taken with a grain of salt. Indeed, muscle spindles are not the only sensory endings responding to vibration. Golgi tendon organs respond to vibration in the presence of baseline muscle activation. Responses of cutaneous and subcutaneous endings could also potentially contribute to percepts of both kinetic and kinematic variables. In fact, recent studies have suggested that the effects of muscle vibration on kinesthetic perception may be related not to local effects on the activity of sensory endings but to more general effects on the stability of the efferent contribution to perception (i.e., the RC). These studies and their implications will be discussed in more detail in Chapter 17.

15.6. Missing pieces of the mosaic

The concept of IPM suggests that synergies at the level of elemental afferent and efferent signals may exist, contributing to the stability of salient percepts. Such synergies remain a theoretical possibility. So far, there have been no studies exploring whether covaried changes in efferent and afferent contributors to perception can result in stable percepts. Such synergies are likely to exist given that changing commands to muscles acting in isometric conditions do not lead to (illusory) percepts of motion of the effector.

A major problem in the exploration of percepts and their stability is the lack of reliable instruments for measurement of both percepts and the two hypothetical contributing signals, afferent and efferent (see also Chapter 24). Both of the commonly used methods, matching with the same or a homologous contralateral effector and using psychophysical scales, can be criticized as producing outcomes affected by other factors such as imperfect memory, asymmetry of effectors, and the subjectivity of psychophysical scales and their changes with time and experience. This obviously limits experimental verification of the introduced hypotheses.

Can signals from only one of the two main sources, afferent and efferent, lead to kinesthetic percepts? Within the suggested scheme, this seems unlikely. Indeed, if perception is equivalent to measurement of salient variables, one has to have both a measuring tool (sensory signals) and a point from which measurement is performed (reference frame, RC). If one of those is absent, adequate perception becomes impossible. For example, in patients with advanced amyotrophic lateral sclerosis, sometimes no body movement is possible while signals from sensory endings continue to be generated in peripheral structures and transmitted into the brain (e.g., Birbaumer et al. 2012). It is possible, however, that the lack of movement does not imply a lack of associated RC shifts even if alpha-motoneurons do not show signs of activation. Such RC shifts can be used to create reference frames to estimate sensory signals. Patients after amputation commonly describe perception of missing effectors, including perception of their motion. It is possible that signals from neurons within the central nervous system involved in the transmission of sensory signals to the brain remain active after the sensory signals stop arriving from the amputated effector (see Feldman 2009, 2016). So, the presence of such percepts does not necessarily mean that they are based on perceiving the RC component of perception in the complete absence of the afferent component or on perceiving the afferent component in the absence of the RC component.

Vibration effects on kinesthetic perception are far from being completely understood. Presently, we have a scheme for analysis of such effects, but exploration of this scheme and checking its predictions are at a very early stage. In Chapter 17, more material will be presented related to the perceptual effects of muscle vibration, which offers new insights into the possible effects of vibration on the stability of percepts.

PART IV
SURPRISING PHENOMENA

16
Drifts in Action

Unintentional drifts in performance in young, healthy persons have been described relatively recently, possibly due to the fact that such drifts are typically slow and require prolonged observation times. Besides, they are seen reproducibly in isometric force production tasks but require special arrangements to be observed in kinematic tasks, such as reaching, or in whole-body tasks performed while standing, such as whole-body sway.

These phenomena differ from the well-known effects of prior practice on performance, for example, those of body deviations from the vertical observed following prolonged standing on a slanted surface or podokinetic effects following a period of walking on a rotating circle (Gordon et al. 1995; Weber et al. 1998; Earhart et al. 2002; Knapen and van Ee 2006; Higashiyama and Yamazaki 2022). The latter practice leads to walking along a curved line when the person is instructed to walk straight ahead with eyes closed. One more phenomenon in this group is the so-called Kohnstamm phenomenon (reviewed in De Havas et al. 2017). To observe this phenomenon, the subject is asked to perform a very strong voluntary muscle contraction against a stop for about 10 s, typically trying to raise the arms laterally by pressing on immovable obstacles with the dorsal sides of the palms. Then, after the obstacles are removed and the subject is asked to relax, the arms show an unintentional motion upward. We should also mention the so-called catch phenomenon consisting in an increase in the ability of a muscle to produce force to a standard incoming excitation following a brief period of vigorous activation of the muscle (Burke et al. 1970, 1976). In this section, we consider drifts in action that can be observed without any special practice and without changes in the ability of the actor to perform the task, in particular in the absence of fatigue or other factors that can have detrimental effects on performance.

There is another group of well-known drift phenomena observed immediately after the termination of a very quick action. For example, a quick single-joint movement ends with an elevated level of coactivation of the agonist-antagonist muscle pair (Gottlieb et al. 1989b). This level of coactivation subsides slowly over a period of several seconds, which may be viewed as an example of unintentional change in performance. Another example: When a person lifts quickly an object gripped by the fingers using the prismatic grip (see Chapter 11), the grip force shows a quick transient increase and remains somewhat elevated following the movement when the object is kept at the new location. It takes a few seconds for the grip force to return to the pre-action level. These phenomena are likely related to those discussed in this chapter, and we will consider them later.

As in any chapter, we will analyze the phenomena at different levels, behavioral, physiological, and physical. Under the physical level, we mean analysis based on the laws of nature, that is, based on variables and parameters describing physical and physiological processes involved in the neural control of actions (see Chapters 1 and 3). We will use the only theory on the neural control of movement known to us at this time that formulates such variables and parameters explicitly, that is, the theory of control with spatial referent coordinates (RCs; Chapters 3 and 7). We will try, as much as possible, to avoid invoking computational processes within the central nervous system and cognitive phenomena such as motivation, memory, and so forth, which belong to the upper level of the Merleau-Ponty pyramid (see Chapter 1).

16.1. Spontaneous force drifts

To observe unintentional force drifts in a typical experiment, a young, healthy person is asked to press with an effector, for example, a finger, and produce a certain force level using visual feedback on the force magnitude. Then the visual feedback is turned off, and the person is asked to keep the force unchanged. Within a few seconds, the force magnitude starts to change, typically to lower values, and the magnitude of the force drop can reach up to 30% to 40% of the initial force level over 15 to 20 s (Fig. 16.1) (Vaillancourt and Russell 2002; Ambike et al. 2015a). Modeling the force time changes with exponential functions produced time exponents on the order of 8 to 15 s (Ambike et al. 2015a; Solnik et al. 2017). These phenomena are observed at moderate initial force magnitudes, for example, 15% to 30% of the maximal voluntary contraction (MVC) level. At very low initial force levels, the force drop becomes small and can even reverse, leading to small drifts toward higher force magnitudes. Across all these conditions, the actor is unaware of the force changes and reports that he or she followed the instruction to keep the force magnitude constant. Note that

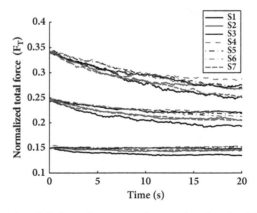

Figure 16.1 Unintentional drifts in force seen after turning the visual feedback on the force magnitude off. The magnitude of the drift could reach about 30% of the initial force level. The initial force produced by the index fingers of the right and left hands was 0.15, 0.25, and 0.35 of maximal voluntary contraction (MVC) level. Different colors show the force drifts corresponding to different initial sharing of total force (F_T) between the fingers, from 80:20 to 20:80. Reproduced by permission from Ambike et al. 2015a.

the mentioned force drifts take place despite the continuous availability of the undistorted somatosensory information.

The first publications on unintentional force drifts linked those phenomena to problems with working memory (Slifkin et al. 2000; Vaillancourt et al. 2003). This interpretation has been supported indirectly by studies of patients with Parkinson disease and brain imaging studies of healthy persons (Vaillancourt et al. 2001; Poon et al. 2012). In particular, patients with Parkinson disease showed larger magnitudes of force drift (Vaillancourt et al. 2001; Jo et al. 2016a), which was interpreted as a consequence of their documented problems with working memory (Vaillancourt et al. 2001). In addition, turning the visual feedback off was associated with early activity changes in the left ventral premotor cortex and ventral prefrontal cortex, areas associated with working memory (Jonides et al. 1993; Owen et al. 2005). This interpretation is not readily compatible with the predominant force drop during the drift—rarely force increase—and uses a concept of working memory that is external to the neural control of movement and is not well defined in terms of the laws of nature.

Two more factors have been invoked in the analysis of force drifts. The first is fatigue. The drifts, however, have been described for modest initial force levels and over modest time intervals. Such contractions, for example, 20% of MVC force over 20 s, do not lead to fatigue in a healthy person. The other factor is adaptation of peripheral receptors sensitive to pressure on the surface of contact (e.g., on the fingertip), which is expected to lead to lower levels of force-related sensory signals. Note that slowly adapting receptors (such as Merkel discs and Ruffini corpuscles) show a slow exponential drop in their firing rate during an ongoing, constant deformation (Iggo and Muir 1969). Such phenomena, however, are expected to lead to perception of lower force levels and, therefore, a corrective drift toward higher actual force magnitudes, not observed in typical studies.

Force drifts during finger and hand force production show relatively unexpected effects of dominance and different drift magnitudes across the fingers. In particular, the drifts are larger in the dominant hand (Fig. 16.2), which has traditionally been viewed as the better-controlled one. The dynamic dominance hypothesis (reviewed in Sainburg 2005, 2014) challenges the notions of "good hand" and "bad hand" and suggests that the two hands are both good and specialized for different classes of tasks or aspects of tasks (see Chapter 10). In particular, it suggests that the non-dominant hand is specialized for tasks that require stability of action, such as holding a nail during its hammering into a board and holding a loaf of bread during its slicing. The smaller drifts in the non-dominant hand suggest direct links between the phenomenon of unintentional force drifts and the concept of action stability discussed in more detail in Chapters 5 and 6.

Along similar lines, larger force drifts have been observed during force production by the index finger, which has been traditionally viewed as the most independent and best-controlled one (e.g., Zatsiorsky et al. 2000). In contrast, smaller drifts have been reported during similar tasks performed by the ring finger, traditionally viewed as the least independently controlled one (Chapter 12). Digit specialization has been suggested earlier in studies of prehensile tasks: The ring and middle fingers show larger involvement in grip force production, whereas the index and little fingers show

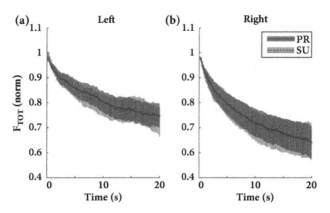

Figure 16.2 Unintentional drifts in total force (F_{TOT}) in force-moment production trials performed by the dominant (right) and non-dominant (left) hands. Note the larger drifts in the right hand for both pronation (PR) and supination (SU) moments. The force scale is in fraction of the initial force level. Reproduced by permission from Parsa et al. 2016.

specialization for the moment-of-force production facilitated by the larger lever arms of their normal forces (Zatsiorsky et al. 2002a, 2002b, 2003). Delegating the gripping component to the ring and middle fingers illustrates their specialization for task components that require high stability. This interpretation links force drifts to the concept of action stability, similar to the difference between the dominant and non-dominant hands.

When subjects are informed of the phenomenon of force drifts, they naturally try to avoid or minimize these drifts given the explicit instruction to keep the force magnitude constant (Abolins and Latash 2022a). This frequently leads, however, to overcompensation and drifts in the opposite direction. It takes multiple information sessions following blocks of trials without visual feedback with information on the average force time profiles to reach relatively consistent behavior. After six such sessions, all the fingers except the index finger were able to avoid force drifts; the index finger was still unable to avoid force drift to lower force magnitudes, although the drift magnitude was reduced as compared to the initial trials performed by naïve subjects.

Force drifts are not limited to studies of force production by the hands and fingers. A recent study confirmed this phenomenon for isometric force production tasks performed by knee extension (Rannama et al. 2023; Fig. 16.3). This is not a trivial generalization because knee extensors are commonly involved in steady force production tasks (e.g., during standing), and a force drift in such conditions could potentially lead to loss of balance. Force drift magnitude in the lower extremities was about half that observed in hand force production tasks, on the order of 10% of the initial force level. Those drifts, however, were faster, with characteristic times on the order of 3 to 5 s.

16.2. DRIFTS IN NEURAL CONTROL VARIABLES   235

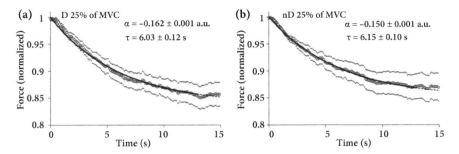

Figure 16.3 Unintentional drifts in force in force production trials performed by the legs. Note the relatively fast drifts caused by turning off the visual feedback on force magnitude. The data are for the initial force level of 25% of maximal voluntary contraction (MVC) level. D, dominant; nD, non-dominant; α, magnitude parameter in the exponential regression; τ, timing parameter in the exponential regression. Reproduced by permission from Rannama et al. 2023.

16.2. Drifts in neural control variables

Unintentional drifts in performance in the absence of salient sensory information can be viewed as reflections of natural behavior of the neuromotor system, in particular as a reflection of the general tendency of all natural objects to move to states with minimal potential energy. This interpretation can be illustrated using the framework of the neural control of movement with RCs for the muscle groups involved in the task. Figure 16.4 illustrates the control of an effector involved in the task of force

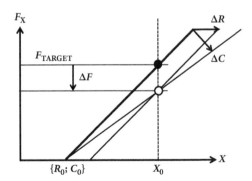

Figure 16.4 Force production in isometric conditions (at a coordinate X_0) by an effector (e.g., a finger) is associated with setting values of the reciprocal (R-command) and coactivation (C-command) commands, $\{R_0, C_0\}$. A force drop (ΔF) can be produced by a change in the R-command (ΔR) and/or in the C-command (ΔC). The initial force-coordinate characteristic is shown with a thick line. Two possible characteristics leading to the force drop are shown with the thinner lines.

production along a single spatial coordinate X with two control variables, RC_{AG} and RC_{ANT}, which define RC values for the agonist and antagonist muscle groups, respectively. Note that the overall mechanical characteristic of the effector, $F_X(X)$, represents the algebraic sum of the two muscle characteristics and is illustrated in Fig. 16.4 as a straight line.

As described in more detail in Chapters 3 and 7, the neural control of the agonist and antagonist muscle groups can be described using the notions of two basic commands, reciprocal and coactivation, $R(t)$ and $C(t)$. The former defines the coordinate where the net muscle force is zero, that is, the intercept of the effector $F_X(X)$ characteristic with the X-axis. This point may be viewed as an RC for the effector because, in the absence of external force, the effector would tend to move to this coordinate. The latter defines the spatial range where both muscle groups are activated simultaneously. In terms of mechanics, it defines the slope of the $F_X(X)$ characteristic, that is, the apparent stiffness of the effector, k. The availability of two control variables makes the neural control of this task abundant, even if it is not redundant at the level of mechanics (see Chapter 7 and Ambike et al. 2016a).

Force drifts can be associated with drifts in the R-command, C-command, or both. Indeed, as shown in Fig. 16.4, a force drop can be produced by different changes in the effector $F_X(X)$ characteristic induced by changes in the two basic commands. This interpretation suggests that the magnitude of the force drift should be sensitive to the compliance of the interface between the effector and the external object. Indeed, as illustrated in Fig. 16.5A, a drift in either command is expected to produce smaller magnitudes of force drift if the surface is not perfectly rigid. This prediction has been confirmed experimentally (Abolins and Latash 2022b; Fig. 16.5B). So, force drift is not a consequence of encoding the wrong force magnitude by the brain based on imperfect cognitive processes (e.g., imperfect working memory) but a consequence of a hidden natural process of drifts in the neural commands, which lead to different consequences at the level of mechanics depending on features of the interface with the environment.

Note that a drift in the R-command and/or C-command illustrated in Fig. 16.4 leads to smaller absolute differences between the actual effector coordinate ($X = 0$) and RC for the agonist and antagonist muscles. A decrease in the difference $|RC - X|$ may be viewed as a drop in the potential energy available for the muscle action. If the RC for either muscle drifts to X, its activation level drops to zero and no further drift has effects on the effector mechanics. This can explain why the amount of force drop in typical experiments is never exceeding a certain magnitude, typically 30% to 40% of the initial force level. We will discuss this issue in more detail in Chapter 19.

Studies of synergies stabilizing force magnitude during pressing tasks revealed broadly varying combinations of R- and C-commands compatible with the required force (Ambike et al. 2015a; Reschechtko and Latash 2017; see Chapter 7). Unintentional spontaneous force drifts may be viewed as (partial) loss of stability of force organized by synergies in various abundant spaces (Figs. 16.6 and 16.7). For a single effector (e.g., a single finger or a hand viewed as a single effector), there is indeed a drop in the synergy index in the space of control variables, mechanical variables RC and k, reflecting the R- and C-commands observed over typical force drifts (Fig. 16.6; Ambike et al. 2016b; Reschechtko and Latash 2017, 2018). This analysis

16.2. DRIFTS IN NEURAL CONTROL VARIABLES   237

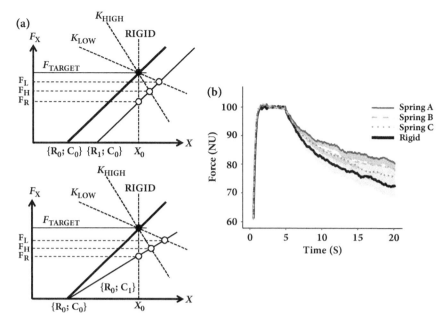

Figure 16.5 (a): A drift in either R-command or C-command is expected to produce smaller force drifts if acting against a compliant spring (as compared to acting against a rigid surface). Two spring characteristics are shown with slanted dashed lines corresponding to low and high stiffness, k_{LOW} and k_{HIGH}. (b): Experimental data support the prediction that force drift magnitude changes consistently with surface compliance. The springs are arranged from the most compliant one (A) to the least compliant one (C). Reproduced by permission from Abolins and Latash 2022b.

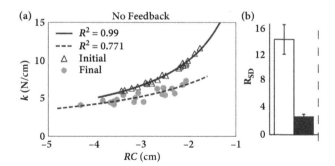

Figure 16.6 Changes in force-stabilizing synergies associated with unintentional force drift in the space of referent coordinate (RC) and apparent stiffness, k. (a): Typical distributions of RC and k under visual feedback (open triangles) and after the force drift (gray circles). Hyperbolic regression lines and equations are shown. (b): An index of covariation (R_{SD}) used with randomization of RC and k values across trials. Open bar, with visual feedback; black bar, after the force drift. Reproduced by permission from Reschechtko and Latash 2017.

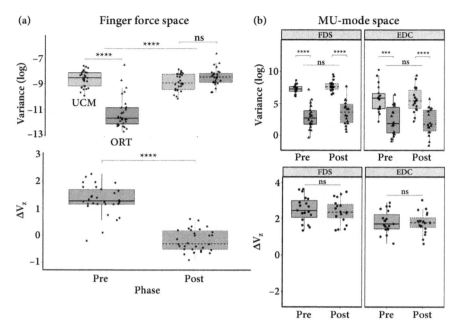

Figure 16.7 Changes in force-stabilizing synergies quantified using inter-trial variance indices within the uncontrolled manifold (UCM) and orthogonal to the UCM (ORT) associated with unintentional force drift. A signature of synergy is $V_{UCM} > V_{ORT}$; $\Delta V_Z > 0$. Note the loss of synergies in the finger force space (a), and no changes in synergies in the MU-mode (groups of motor units) space (b). Reproduced by permission from De et al. 2024.

involved the creation of surrogate data sets with RC and k taken from different trials (see Chapter 5; Müller and Sternad 2003) and computing variability indices across actual trials and those created from the same distributions of RC and k, but without covariation. The ratio between these indices (R_{SD}) was used as a measure of covariation in the original data set.

The hand action can also be viewed as the sum of the actions by the fingers, and force-stabilizing synergies can be analyzed in spaces of control variables to individual fingers. Such analysis also reveals a drop in the synergy index, that is, partial loss of hand force stability quantified in the space of individual finger forces (Figure 16.7; De et al. 2024). Panel B of Fig. 16.7 shows that intra-muscle synergies (see Chapter 11) show no changes during the force drifts, which provides indirect support for the hypothesis that intra-muscle synergies reflect spinal circuitry, which may be immune to changes in RCs defined by the brain. We will return to Fig. 16.7B in the analysis of the control of antagonist muscles and phenomena of muscle coactivation (Chapter 19).

Overall, this interpretation of unintentional force drifts suggests that drifts in basic neural command occur at all times, not only in isometric conditions and not only in one-hand tasks (e.g., Zhou et al. 2015a; Rasouli et al. 2017; Abolins et al. 2023). When accurate sensory information is available on salient performance variables, the drifts are effectively corrected and do not lead to error accumulation over prolonged time intervals. Obviously, somatosensory information on contact forces, by itself, may be insufficient to provide sensitive enough feedback to avoid force drifts.

16.3. Faster drifts triggered by quick force changes

Faster force drifts can be seen during isometric force production tasks when the task requires force changes and under positional perturbations of the effector. For example, when a person is instructed to produce cyclical force changes in isometric conditions at a comfortable frequency (e.g., 1 Hz) between two targets set within a comfortable force range (e.g., between 15% and 25% of MVC force), turning the visual feedback off leads to two effects (Ambike et al. 2016c). First, the midpoint of the force cycle drifts slowly to lower magnitudes, similar to the observations during constant force production described in the previous section. Second, the peak-to-peak force changes increase at a much faster rate, within 1 to 2 s, and this increase can be twofold (Fig. 16.8).

The presence of two unintentional processes at very different time scales suggests involvement of two different mechanisms. Since all human movements, even mechanically non-redundant ones, involve multiple muscles, the control of such movements is always abundant and can be analyzed within the framework of the uncontrolled manifold (UCM) hypothesis (see Chapter 5). According to this hypothesis, two subspaces within the space of elemental variables differ in their stability properties: The UCM is less stable, and the orthogonal to the UCM space (ORT) is more stable (see Figs. 9.1 and 9.2 in Chapter 9). Note that characteristic times of processes are related

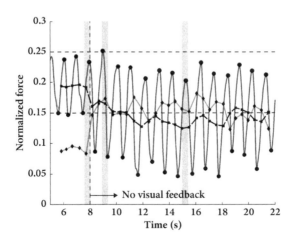

Figure 16.8 During cyclical force production tasks, turning the visual feedback on force magnitude off leads to two effects: a slow drift in the midpoint of the force magnitude within the cycle and a much faster increase in the peak-to-peak force changes. Reproduced by permission from Ambike et al. 2016c.

to their stability: In highly stable directions, processes are expected to be faster, and in less stable directions, they are expected to be slower (imagine, as an illustration, an object on a very stiff spring [very stable and moving fast] and on a compliant spring [less stable and moving slower]). The observation of two processes at markedly different characteristic times suggests that one of them originates from processes within the UCM (slow drifts) and the other one from processes within the ORT (e.g., fast change in force peak-to-peak amplitude). Since force drifts are always observed in the ORT, this assumption implies that the UCM and ORT may be coupled. Note that during constant force production tasks, no major quick deviations are expected in the ORT and, therefore, all force drifts are characterized by times typical of processes within the UCM.

In the mentioned study of cyclical force production (Ambike et al. 2016c), the observation of fast drifts in the peak-to-peak force amplitude was interpreted as a drift toward a preferred force amplitude. A follow-up study, however, failed to confirm this prediction (Ambike et al. 2018). In that study, subjects performed the cyclical force production task starting from various midpoints and various force amplitudes. The results were surprising: The increase in the peak-to-peak force amplitude was observed across all conditions, and the magnitude of the increase, expressed in percentage of the MVC force, was about the same, independent of the initial task parameters. There were no signs of a particular preferred range of force changes. This result remains poorly understood.

Comparably fast force drifts, with the characteristic times of about 1 to 3 s, were also observed in experiments when the effector (e.g., a fingertip) was lifted smoothly and quickly, leading to a quick force increase. This was done using the "inverse piano," described earlier (Chapter 7; Martin et al. 2011). Note that lifting the fingertip perturbs the system in the ORT direction and, therefore, may be expected to lead to faster force drift as compared to steady-state force production tasks.

Force drifts during constant force production tasks are clearly incompatible with holding an object with the prismatic grasp for a prolonged time. Indeed, the safety margin (see Chapter 12) may not be sufficient to allow a 30% to 40% drop in the grip force. It is possible that sensory signals related to mechanical variables other than the grip force normal to the surface contribute to corrections of the expected force drifts. These signals may be related, for example, to tangential skin deformation and partial slip of the object (see Barrea et al. 2018; Delhaye et al. 2021).

Unintentional drifts in the grip force were documented in experiments using a handle that was instrumented to change its aperture, that is, to expand and contract while being held in the air by a person (Ambike et al. 2014). In that study, the actor was asked to allow the hand to behave naturally and not to correct force changes induced by changes in the handle aperture. During slow changes in the handle aperture (by 1 cm over 5.5 s), the digit force increased and decreased with the aperture increase and decrease, respectively. However, when the handle expanded by 1 cm and then contracted slowly returning to the initial aperture, the grip force showed an overall drop by a considerable magnitude, about 20% to 25% of its initial value (see Fig. 16.8). So, grip force is not immune to force drifts, but observing such drifts may require perturbing the opposing digits.

16.4. Unintentional kinematic drifts

Comparable drifts in kinematic tasks have not been reported. However, application of a smooth transient perturbation during a position-holding task did lead to violations of equifinality, suggesting a drift in the RC for the hand (Zhou et al. 2014b, 2015a, 2015b, 2015c). In those experiments, the subjects were holding a handle against an external constant force produced by a programmable robot. The instruction was not to react to possible change in the robot force but to allow the hand to move naturally under those force changes. A smooth, transient force increase (over 1 second) and return to the initial force magnitude led to hand motion away from the initial position and then back to that position. The final position along the direction of force action did not differ from the initial one; that is, this transient perturbation led to equifinality (see Chapter 18). Analysis in the joint configuration space showed, however, that there were substantial differences between the initial and final joint configuration, both compatible with about the same hand position, that is, motion within the corresponding UCM, which can be seen as an example of motor equivalent motion.

The behavior changed when the force increase and decrease phases were separated by an interval of dwell time and the force was kept constant at the new level. Under those conditions, the final hand position was significantly shifted in the direction of force increase as compared to the initial hand position: The return movement of the hand covered only about half of the distance to the initial position. This result was interpreted as a consequence of an unintentional shift in the RC for the hand triggered by the perturbation moving the hand away from the original RC (Fig. 16.9). This interpretation received support in further studies exploring possible changes in the RC with the help of additional small perturbations applied during the dwell time when the hand was motionless at the new position (Zhou et al. 2015a).

These results suggest that RC drifts are common across effectors and types of tasks. During the dwell time, RC shifts led to relatively weak effects on the hand location

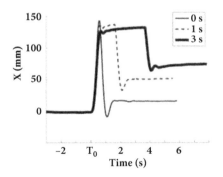

Figure 16.9 When the hand holds a position in space against an external force, a transient force increase-decrease sequence leads to about the same hand coordinate (thin trace). When there is a dwell time at the peak force level (two cases are illustrated, with the dwell times of 1 second and 3 s), the hand stops short of the initial coordinate; that is, it shows violations of equifinality. Reproduced by permission from Zhou et al. 2014b.

(due to associated changes in muscle coactivation defined by the C-command), and as a result, they were not corrected. Estimates of the time profiles of the RC drifts suggested that they evolved over 1 to 2 s, similar to the mentioned fast drifts in force production tasks. Such relatively fast drifts are compatible with the idea that they originated from processes within the ORT triggered by the hand displacement, which obviously happened in the ORT space.

A related effect commonly observed following fast movements is the aforementioned drop in muscle coactivation. The coactivation level is typically increased following a fast movement, and it takes a few seconds or even tens of seconds for the coactivation to subside to the level seen prior to the movement (e.g., Gottlieb et al. 1989b). These phenomena suggest a slow drift in the C-command, resembling the drifts observed during force production tasks. Note that the R- and C-commands have been viewed as forming a hierarchy, with the R-command being hierarchically higher and defining movement pattern, and the C-command adjusting and transferring in space to new locations defined by the R-command (of course, in its interaction with the environment) (Levin and Dimov 1997; reviewed in Feldman 2015). Within this idea, a change in the C-command following a movement against zero external load is expected to produce no drift in the final effector coordinate defined by the R-command. We will return to the role of the C-command and its possible drifts in Chapter 19.

16.5. Drifts in whole-body tasks

Vertical posture is strongly protected against possible failures (see Chapter 14). It is of no major surprise, therefore, that no obvious drifts are observed aside from the well-known phenomenon of postural sway, which, however, does not lead to accumulation of the body deviation from the vertical. Drifts in the body orientation can be observed, however, when the task is not to keep body position but to change it rhythmically. In such a study, subjects were asked to perform cyclical movement of the center of pressure (COP) at a comfortable rate prescribed by the metronome in various directions and with various midpoints shifted with respect to the preferred COP coordinates during quiet standing (Rasouli et al. 2017). The trials started with visual feedback on the COP coordinates and then the feedback was turned off while the subjects were asked to continue producing the same COP trajectories.

The idea of unintentional RC drift to the actual body configuration corresponding to minimal muscle activation (i.e., close to the vertical during quiet standing) leads to an unexpected prediction. Indeed (Fig. 16.10), if a person stands quietly with the body deviating slightly forward from the vertical, this requires setting the body referent orientation (RO, equivalent to RC in angular units), which differs from the actual orientation in the direction opposite to that of the body lean. If the RO drifts slowly to the actual body orientation, the difference between the two decreases, leading to a drop in the active moment of force counteracting the moment generated by the force of gravity. As a result, the body is expected to show an increase in its deviation from the vertical, ultimately resulting in a fall or protective step. Of course, this is a very counterintuitive prediction. It is more natural to expect a body drift toward the natural vertical orientation, that is, toward safer states.

16.6. DRIFTS IN INDICES OF FINGER INTERACTION 243

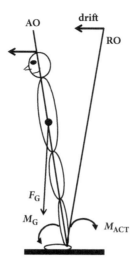

Figure 16.10 Quiet standing is associated with the referent orientation (RO) of the body behind its actual orientation (AO). If RO drifts toward AO, the moment (M_G) of gravity force (F_G) becomes larger than the counteracting active moment of force (M_{ACT}). As a result, the person is expected to lean forward until losing balance.

Experiments with quiet standing and cyclical body sway showed consistent slow drifts in the midpoint of the sway in individual subjects, but the drift could be in different directions across subjects (Fig. 16.11). The amplitude of the sway increased similarly to the observations during cyclical hand force production. Overall, the results suggested competition between two factors, the hypothesized RO drift toward the actual body orientation (which can potentially lead to loss of balance) and an opposite drift toward a safer body orientation.

16.6. Drifts in indices of finger interaction

Slow drifts in movement-related characteristics are not limited to performance variables explicitly used in the task formulation. In particular, drifts in mechanical outputs of effectors not involved in the task have been documented in studies of finger force production. To remind, when one finger of a hand produces force or moves intentionally, other fingers show unintentional force production or movement. This phenomenon, addressed as enslaving (Li et al. 1998; Zatsiorsky et al. 2000; see Chapter 12), has been explored across tasks and populations and quantified using a matrix of finger interaction—the enslaving matrix. Enslaving has been viewed as a robust individual-specific characteristic of finger interaction, which remains unchanged across tasks but can change with specialized practice (Slobounov et al. 2002). The assumed robustness of enslaving led to the introduction of the concept of finger mode, a hypothetical command to a finger leading to proportionally scaled actions by all the fingers of the hand (Danion et al. 2003).

244   16. DRIFTS IN ACTION

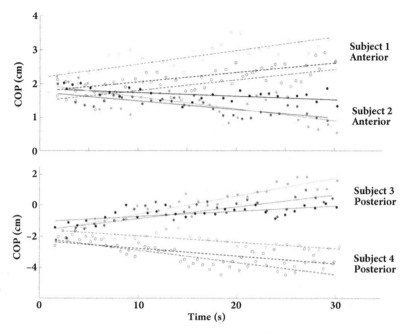

Figure 16.11 When a person performs a whole-body cyclical sway task, turning the feedback on the center of pressure (COP) off leads to a consistent drift in the midpoint of the cycle. Directions of the drift, however, vary across subjects. Reproduced by permission from Rasouli et al. 2017.

A number of studies have shown that enslaving increases smoothly in the course of typical trials. All such studies instructed the subjects to perform a force production task with a subset of fingers of a hand and recorded the forces produced by all the fingers, instructed and non-instructed (enslaved). Figure 16.12 illustrates typical time profiles of the instructed (master) and non-instructed (enslaved) finger forces in two-finger force production tasks (Abolins et al. 2020; Hirose et al. 2020). Subjects in those tasks received continuous visual feedback on the master finger force (F_{MAS}) or enslaved finger force (F_{ENS}), but they were unaware of the nature of the feedback signal (the visual feedback gain was adjusted to make the conditions look similar) and thought that it always reflected the force by the instructed fingers. Consistently, when the feedback showed F_{MAS}, F_{ENS} increased smoothly during the task. In contrast, when the feedback showed F_{ENS}, F_{MAS} dropped smoothly during the task. Of course, there was no drift in the variable used for the feedback. In both situations, the relative amount of F_{ENS} in the total force produced by all the fingers of the hand increased by about 15% over 10 to 15 s.

The relative increase in enslaving has also been confirmed when the feedback showed total force produced by the master and enslaved fingers or when no feedback was presented at all (Abolins et al. 2023). Under the total force feedback, F_{MAS} drifted to lower values, while F_{ENS} increased in a balanced way to keep their sum constant.

16.6. DRIFTS IN INDICES OF FINGER INTERACTION

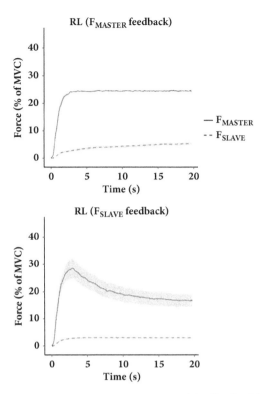

Figure 16.12 When a person receives continuous visual feedback of the force produced by the instructed fingers (master fingers, ring and little [RL]), the force by the enslaved fingers (middle and index) increases. When the visual feedback shows the force of the enslaved fingers, the force by the master fingers drops. Reproduced by permission from Abolins et al. 2020.

When no feedback was available, F_{MAS} drifted to lower values, while F_{ENS} showed inconsistent behavior with variable drifts across subjects.

These observations can be summarized as $F_{ENS}(t) = \eta(t) \cdot \varepsilon \cdot F_{MAS}(t)$, where ε is the initial, subject-specific level of enslaving, and $\eta(t)$ is a monotonically increasing function. Note that a typical drop in $F_{MAS}(t)$ in the absence of visual feedback can lead to different effects on $F_{ENS}(t)$ depending on the competition between the relative increase in $\eta(t)$ and relative drop in $F_{MAS}(t)$, reflecting the hypothesized drifts in RC for the involved muscles and muscle compartments.

Drifts in enslaving over relatively short time intervals are unlikely to reflect changes in the peripheral muscular apparatus and much more likely to represent consequences of neural processes. They can be interpreted using the concept of "cortical piano" introduced by Schieber (2001). According to this concept, hypothetical neuronal pools at a pre-M1 level generate signals corresponding to intentional involvement of individual fingers (i.e., precursors of the hypothetical finger modes). These signals project on M1 representations of all four fingers, leading to observed patterns of enslaving. Such "neuronal chords" are based on personal lifetime experience, leading to individual-specific characteristics of enslaving.

The slow increase in enslaving suggests spread of excitation over neighboring M1 neurons (e.g., due to the "Mexican hat" excitability pattern among cortical columns; Lin et al. 1998; Kang et al. 2003; reviewed in Kandel et al. 2012), leading to slowly increasing excitation received by the slave finger representations. Such spread of excitation during steady-state force production may be related to the phenomenon of beta-band cortical activity typical of steady states (Gilbertson et al. 2005; Engel and Fries 2010) that shows a tendency to spread over time (Rektor et al. 2006; Tan et al. 2016).

Drifts in enslaving have also been documented in two-hand tasks involving one master finger and one enslaved finger per hand (Abolins et al. 2023). In those studies, presenting feedback on master, enslaved, or total force computed over the corresponding fingers in both hands led to very similar observations to those described for one-hand tasks. Somewhat surprisingly, when the feedback was computed using the forces produced by the fingers of one hand only, both hands showed very similar patterns of force drifts despite the fact that changes in the finger forces of one of the hands had no effect on the feedback. One could expect that the "no feedback" hand would always show force drift patterns similar to those seen in the no-feedback condition, but its force drifts depended on the feedback type used for the other hand. Of course, as in all such studies, the subjects were unaware of the feedback manipulations and always thought that the feedback reflected the force produced by the instructed fingers. These observations suggest strong coupling of neural variables in two-hand tasks as illustrated in Fig. 16.13. The observed drift patterns suggest that the sharing process of RC_{TASK} between the two hands (RC_{H1} and RC_{H2}) is very robust and leads to parallel adjustments of the control signals to the two hands even if visual feedback reflects the forces produced by the fingers of one hand only.

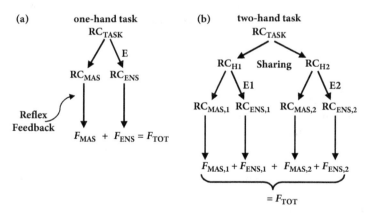

Figure 16.13 In one-hand tasks (a), referent coordinate (RC) to the instructed fingers (RC_{MAS}) is accompanied by scaled RC to enslaved fingers (RC_{ENS}) via the process of enslaving (E). In two-hand tasks (b), when visual feedback was provided on the finger force produced by one of the hands only, the patterns of force drifts were similar between the hands. This suggests strong coupling between the neural networks involved in the force production by the two hands. H1 and H2 stand for two hands; MAS, master fingers; ENS, enslaved fingers. Reproduced by permission from Abolins et al. 2023.

The findings of rather large drifts in enslaving have implications for typical tasks exploring enslaving across populations. Commonly, to quantify enslaving, subjects were asked to press with one finger at a time and produce a slow ramp of that finger's force over a certain range with visual feedback provided on the force magnitude (e.g., Park et al. 2012). Other fingers of the hand stayed on the force sensors, and the subjects were asked not to pay attention to possible force production by those fingers. Ramps of different duration have been used, typically within the range of 6 to 10 s. Given the described drifts in enslaving, it would be much more prudent to use shorter ramps, maybe 3 s or so, to ensure that enslaving stays relatively unchanged in the course of such a trial.

16.7. Classification of movements

The reviewed observations of unintentional movements associated with changes in RCs suggest a modification of the earlier classification of movements. To remind, within the theory of control with RCs, there are two classes of movements, involuntary and voluntary (Fig. 16.14). Involuntary movements are produced by changes in external forces without a change in the neural control variable to the effector, its RC. Voluntary movements are produced by changes in the RC, which may or may not be associated with changes in external forces.

Now, we have to add another level to this classification because RC shifts (i.e., voluntary movements) can be intentional and unintentional. In addition, there are two classes of unintentional movements (changes in performance), slow and fast. Slow drifts are seen in steady-state tasks in the absence of external perturbations of salient

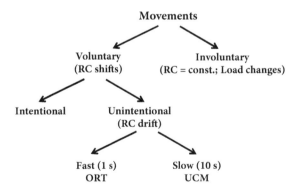

Figure 16.14 Movements can be classified into voluntary (produced by changes in the referent coordinates [RCs]) and involuntary (produced by changes in external forces). Further, voluntary movements can be intentional and unintentional. The latter are seen as drifts in performance typically not perceived by the subjects. Unintentional movements can be relatively slow (typical times of 10 s) or fast (typical times of 1 second), possibly related to processes within the uncontrolled manifold (UCM) and orthogonal space (ORT), respectively.

variables. Fast drifts are seen in action characteristics related to changes in the salient performance variables, produced either voluntarily (as during cyclical force production) or by an external perturbation (as during experiments with the "inverse piano"). Hypothetically, the different characteristic times of the two types of drifts reflect characteristic times of processes within the UCM and ORT spaces.

The similarity of the time profiles of drifts in finger forces and in the index of enslaving tentatively suggests their shared origins. So far, there have not been studies of drifts in finger enslaving during fast force changes produced voluntarily or by external perturbations. It would be important to explore such changes to confirm or disprove the hypothesis on shared origins of the reviewed drifts.

16.8. Missing pieces of the mosaic

As in many other studies, interpretations based on the concept of control with spatial RCs remain speculative because of the lacking methods of continuous recording of RC changes (see Chapter 24). Indirect methods are typically limited to measurements at a specific moment of time, for example, those using the "inverse piano." These methods have been applied during typical force drift studies early in the trial, when the force was produced with the help of visual feedback, and late in the trial, when the drift magnitude was relatively large (e.g., Reschechtko and Latash 2017, 2018). Note, however, that applying the inverse piano perturbation can by itself modify (typically, accelerate) the force drift process (cf. Wilhelm et al. 2013; Reschechtko et al. 2014, 2015). We seem to face one of the examples of the "principle of indeterminacy" in studies of biological objects, which react to attempts to measure their characteristics by changes in the variables the experimenter tries to quantify (more in Chapter 24).

The hypothesis on two types of drift processes originating in the UCM and ORT is appealing and has to be tested rigorously. The most nontrivial step in this hypothesis is selecting the space of elemental variables where the UCM and ORT are computed (see Chapters 5 through 7). It is natural to analyze neural control processes in spaces of control variables such as λ, RC, and R- and C-commands. However, given that most actions involve many levels at which RCs can be defined and analyzed, choosing the right level seems to be highly nontrivial.

One of the least understood phenomena of unintentional actions is the drift in enslaving. Are these phenomena specific to finger action? Can they be seen in other motor phenomena, across tasks and effector sets? What are the hypothetical origins of the finger modes ("pre-M1 levels"), and can processes at those levels also contribute to unintentional actions? Should researchers continue to perform analysis of finger coordination in finger mode spaces in spite of the likely drifts in enslaving? These questions so far have no good answers.

17
Efference Copy

The concept of efference copy (EC; a.k.a. efferent copy, corollary discharge; von Holst and Mittelstaedt 1950/1973; Sperry 1950), as a contributor to perceptual processes, has been frequently used in the literature without an explicit definition. This expression is based on a number of assumptions. The first is that some of the neural streams within the central nervous system can be rather unambiguously classified as efferent (i.e., related to the process of movement generation), while others can be classified as afferent (i.e., related to sensory transmission). The second is that the efferent signals have a copy playing an important role in perceptual processes. Both assumptions seem to be nontrivial and in urgent need of an explicit definition for the involved concepts.

A number of recent studies have associated EC with signals in a variety of pathways and structures within the central nervous system traditionally associated with movement production (Andersen et al. 1997; Fautrelle and Bonnetblanc 2012; Fee 2014; Shadmehr 2017; Brooks and Cullen 2019). Most commonly the invoked structures have included the cortex of the large hemispheres, the cerebellum, the basal ganglia, and associated subcortical loops via the thalamus. All these structures and loops, however, have also been implicated in sensory processing. So, it is hard to claim that these signals are unambiguously "efferent." The situation becomes more complicated by the fact that the notion of EC has been applied to analysis of movements and perceptual processes across species, from insects (Webb 2004; Sheeran and Ahmed 2020) to tadpoles (Straka and Chagnaud 2017) and fish (Pichler and Lagnado 2020) and up to monkeys (Wurtz 2018). Given the broad spectrum of the species, it is hard to link EC to any specific anatomic structure or neural loop.

An overwhelming majority of studies using the concept of EC have assumed that neural structures within the animal's body perform computational operations with neural signals, including EC, predicting both action mechanics and its sensory consequences. This assumption is far from being trivial, in particular because it is not based on any known laws of nature. Nikolai Bernstein (1947, 1967) emphasized that the brain could not in principle predict mechanical consequences of neural efferent signals it generated because of the unpredictable external forces, mechanical coupling among body segments, and spontaneous variations in the state of intermediate neural structures, including those in the spinal cord (see Chapter 2). This conclusion has not been challenged seriously (for a comprehensive review see Feldman 2015). It also means that the brain cannot predict sensory consequences of efferent signals, in particular signals from proprioceptors sensitive to unpredictable mechanical variables such as forces and displacements. The line of thinking traditionally associated with the concept of EC seems to struggle with the famous sarcastic statement by Erwin Schrödinger (1949): "If you suspend laws of nature in the human body, you can explain anything."

To make the EC concept useful for analysis of perceptual processes, one has to define explicitly what efferent signals encode. We will start with the classical attempt by von Holst and Mittelstaedt to offer such a definition. Then, after demonstrating the inadequacy of this definition, we will offer an alternative definition based on the idea of control with spatial referent coordinates (RCs; see Chapter 7). This will allow interpreting a variety of perceptual phenomena including illusions and errors. However, this approach also struggles with behaviors where EC seems to be distorted as compared to the efferent process, that is, when EC(t) differs from the corresponding function RC(t). We will have to admit that EC is not always a copy of signals associated with the efferent process, which makes it a misnomer. Taken together, the current knowledge cannot offer an acceptable account for the role of processes associated with the neural control of movement in perceptual phenomena. Maybe the concept of EC has to be dropped altogether and replaced by a more clearly defined and less biased term.

17.1. Von Holst's concept of efference copy

As mentioned in Chapter 15, arguably the first scientist who paid attention to the effects of action on perception was von Helmholtz. He drew this conclusion based on different perceptual effects of eye motion between the conditions when the eye moves naturally and when it moves in a less usual way, for example, by pressing on the eye with a finger. In the middle of the 20th century, von Holst and Mittelstaedt (1950/1973) and Sperry (1950) made an important step in offering a hypothesis on the role of neural efferent processes in perception. They suggested that a copy of efferent signals was used to estimate afferent signals from sensory endings and to create a percept. This copy was termed *efference copy*.

Von Holst and Mittelstaedt discussed the potential role of EC in both the creation of percepts and corrections of ongoing motion in cases of motor errors, and associated EC with a copy of the signals from alpha-motoneuronal pools to the innervated muscles (Fig. 17.1). They classified signals from proprioceptors into *exafference* and *reafference*. The former term reflected changes in afferent signals produced by action of external forces. They assumed that exafference always led to reflex-induced changes in muscle activation and corrections of the deviations produced by the external forces. Exafference was also always assumed to lead to changes in kinesthetic perception. The term *reafference* described changes in signals from proprioceptors produced by one's own actions.

According to the original scheme, a copy of the signals from alpha-motoneurons was used to predict changes in sensory signals from proprioceptors in the target muscles expected from the planned action (i.e., reafference). The signals carried by collaterals of axons of alpha-motoneurons were termed *efference copy*. They were assumed to be somehow transformed into units that made them commensurable with signals from proprioceptors—a very strong assumption with no obvious neurophysiological mechanism, particularly given that afferent signals from proprioceptors are many and varied. If the prediction was accurate, that is, if a copy of the output of the alpha-motoneuronal pool was equal to the reafferent signals (whatever this

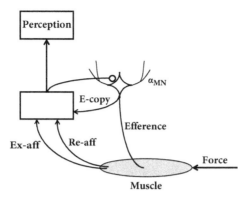

Figure 17.1 An illustration of the concepts of exafference (Ex-aff, changes in afferent signals due to an external influence, e.g., a force perturbation) and reafference (Re-aff, changes in afferent signals due to voluntary movement). The latter are compared to efference copy (E-copy) to generate percepts and corrective actions.

statement means), no correction was issued. If, however, the prediction was wrong, an error signal was sent to the alpha-motoneurons, leading to correction of the action. Reafference affected perception only if it differed from the prediction based on EC.

The scheme in Fig. 17.1 has a number of arbitrary assumptions on processing of neural signals and also contradicts some everyday observations (Feldman 2009, 2016). In particular, it is unable to account for the ability of animals to relax after moving an effector from some initial position to a different final position. Indeed, consider a pair of agonist-antagonist muscles crossing a joint, which moved from one steady state to another steady state against an unchanged external load (Fig. 17.2). If both muscles are relaxed, the output of their alpha-motoneurons is zero, and EC is also zero. This is true independently of joint position. However, length-sensitive sensory endings, in particular those in muscle spindles, are expected to show a change in their level of activity, increased in the longer muscle and decreased in the shorter muscle. This change is reafference according to the introduced classification because

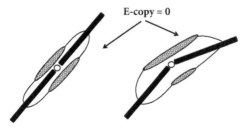

Figure 17.2 If an agonist-antagonist pair of muscles is relaxed in two different joint positions, changes in the length-sensitive afferent signals cannot be compensated by changes in a copy (E-copy) of signals to the muscles because the latter are zero in both positions.

it was produced by active movement. So, a change in reafference can happen without a change in EC, and it has to lead to reflex-mediated changes in muscle activation and produce motion of the effector, in clear contradiction to the everyday experience that animals, including humans, can relax muscles at various joint configurations. So, we can conclude that EC is not a copy of the output of alpha-motoneuronal pools. Note also that a study of the Kohnstamm phenomenon—unintentional facilitation of movement following a period of high voluntary muscle activation in isometric conditions—led the authors to conclude that this phenomenon represented movement without EC (De Havas et al. 2017). Clearly, activation of alpha-motoneurons happens during the Kohnstamm phenomenon, once again leading us to the same conclusion: EC is not a copy of the motoneuronal output. But what is it a copy of?

17.2. Efference copy as a referent coordinate

In Chapter 15, we described a scheme of kinesthetic perception based on the theory of control with spatial RCs (see also Chapter 7 and Feldman and Latash 1982a; Feldman 2009, 2015). Within this scheme, perception of the coordinate, linear or angular, and the force (moment of force) component along that coordinate are linked by superposition of the stretch reflex characteristics of the contributing muscles. The neural control of the effector defines these characteristics for both agonists (producing force in a certain direction along the coordinate) and antagonists (producing force in the opposite direction). Their superposition (Fig. 17.3) defines the force-coordinate characteristic of the effector. Only points along these characteristics are possible as equilibrium states of the effector acting against various external loads.

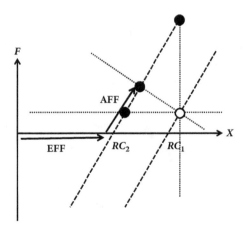

Figure 17.3 An action caused by a change in the referent coordinate (RC) produces different peripheral outcomes depending on the external load (shown with three filled circles). Perception of both coordinate and force is a result of an interaction between the efferent signals (EFF) and afferent signals (AFF) changing with the deviation of the effector state from RC.

As described in more detail in Chapters 3 and 7, the control of any effector can be described with the help of two basic commands, the reciprocal command (*R*-command) and the coactivation command (*C*-command). Hence, identifying values of the *R*- and *C*-commands directly contributes to solving the problem of perception for both kinematic and kinetic variables. Sensory signals are needed to identify a point along the force-coordinate characteristic defined by the neural commands. As long as these signals are sufficient to identify such a point, their exact source is not crucial. Hence, signals from coordinate-sensitive sensory endings may be sufficient to perceive force, and signals from force-sensitive sensory endings may be sufficient to identify the coordinate. These conclusions have been supported by a number of recent studies providing evidence that spindle endings may play the role of force sensors and Golgi tendon organs may play the role of muscle length sensors (Watson et al. 1984; Luu et al. 2011; Proske and Gandevia 2012).

The scheme shown in Fig. 17.3 unites the production and perception of movements with the help of the same variable, RC. Action is defined by an interaction of the force-coordinate characteristic defined by the RC with the external load characteristic (three load characteristics and their corresponding equilibrium points are shown in Fig. 17.3), and perception is defined by an interaction of the RC with sensory signals (EFF and AFF in Fig. 17.3). In other words, perception of the force and coordinate values at the reached steady state emerges as a result of estimating afferent signals within the reference frame provided by the RC.

This scheme allows interpreting a variety of observations in the field of kinesthetic perception (reviewed in Feldman 2009, 2016; Latash 2018a, 2021c). In particular, RC shifts in the absence of sensory signals can lead to perception of motion in persons after amputation of a limb (phantom limb motion). Artificial changes in signals to a muscle group (e.g., by electrical stimulation of those muscles) can produce consistent errors in estimation of the magnitude of the produced motion as a result of the mismatch between the current RC and actual muscle activation level. Artificial changes in sensory signals from a muscle group (e.g., induced by high-frequency muscle vibration) can lead to vibration-induced illusions and related phenomena reviewed in Chapter 15.

17.3. Is efference copy a copy of efference?

A number of studies have suggested, however, that processes of movement production and perception may use different values of the RC variables (Abolins et al. 2020; Cuadra et al. 2021a). In other words, RCs participating in perceptual processes (RC_{PER}) may be not identical to RCs participating in action processes (RC_{ACT}). Or, to put it even more bluntly, EC participating in perception may not be an exact copy of the efference. The last statement makes the term "efference copy" a misnomer, and we are going to use it in quotation marks from now on to emphasize that this is a neural process related to action production, which is used for purposes of kinesthetic perception, but not necessarily an exact copy of that neural process.

It is generally accepted that visual signals use two pathways through the brain circuitry, the so-called *ventral stream* and *dorsal stream*, which lead to conscious perception of the environment with all the visual stimuli (including those from one's own body) and supports acting toward visual targets, respectively (Goodale et al. 1991; Goodale and Milner 1992). In particular, after a stroke affecting parietal brain areas, patients commonly experience severe visual neglect or even denial: a state of not recognizing objects in the contralesional visual field, including one's own body parts. Some of these patients, however, are able to catch a ball thrown into the contralesional visual field. After the catch, they may deny performing the catching movement and claim, for example, that the experimenter placed the ball in their hand.

Several studies have suggested that proprioceptive signals can also have diverse effects on perception of one's effectors and reporting these percepts to another person or to oneself vs. using these signals and percepts to guide actions. In particular, one of the studies reported qualitatively different findings in subjects when they were asked to report finger-pressing force using an earlier-learned psychophysical scale and when they were asked to match the force magnitudes with the contralateral finger (Cuadra et al. 2021b).

Arguably, the most dramatic difference between perception-to-act and perception-to-report with respect to force was observed in a study where young healthy persons were asked to press with a finger and produce a steady force level with the help of visual feedback (Cuadra et al. 2020). The design of the experiment is shown schematically in Fig. 17.4A. Then the feedback disappeared, and the subjects were asked to coactivate the finger and hand muscles without changing the force level. They were given practice of muscle coactivation without net motion in the absence of pressing,

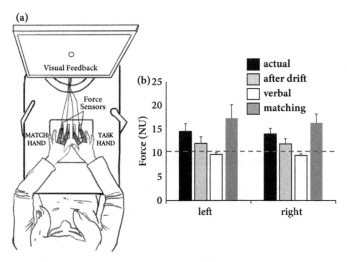

Figure 17.4 (a): The subject is asked to coactivate hand muscles without changing the net pressing force by the fingers. Further, the subject is asked to report the force magnitude verbally and to match it with the contralateral hand. (b): Coactivation results in a major increase in the pressing force, which drifts to lower magnitudes. The subjects report a small force drop. Matching results is a major overshoot of the force. Reproduced by permission from Cuadra et al. 2020.

that is, with the hand kept in the air. During coactivation, the subjects showed a substantial increase in the pressing force, on average by about 50%. Then, they were asked to report the pressing force level assuming that the initial level was 10 units. The subjects reported the opposite effect: a significant (although modest in magnitude) drop in force. When the subjects were asked to match the force with the contralateral finger, they showed a significant overshoot. These results are illustrated in Fig. 17.4B.

Figure 17.5 provides an explanation for at least some of these findings within the scheme of control with RCs. Assume that in the initial state, the required finger force level was produced by a combination of the R- and C-commands, $\{R_0; C_0\}$. The instruction to coactivate muscles implied an increase in the C-command (to C_1), which can be expected to lead to an increase in the slope of the fingertip force-coordinate characteristic (k). If the R-command remains unchanged, an increase in k is expected to lead to an increase in the force level (as observed in the experiment). Likely, however, the R-command was adjusted in the direction of the actual effector coordinate, resulting in a new combination $\{RC_1; k_1\}$, as illustrated in Fig. 17.5, partially correcting for the force increase. This shift of the R-command led to the reported perception of a drop in the force level; that is, when asked to report the force change, the subjects reported values expected from the change in the R-command alone and ignored the fact that the C-command and sensory signals from force-sensitive endings changed, pointing at an increase in the force level. We come to an intermediate conclusion that perception-to-report is dominated by changes in one of the two basic commands

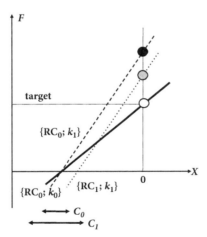

Figure 17.5 Force generation in isometric conditions is accompanied by a combination of the R- and C-commands, $\{R_0, C_0\}$. Coactivation leads to an increase from C_0 to C_1. It has to lead to force increase if the R-command does not change due to an increase in the slope (k) of the force-coordinate characteristic. An adjustment of the R-command (referent coordinate [RC] moves closer to the actual finger coordinate) mitigates the force increase and is interpreted by the subjects as a small force drop. The font size reflects increased or decreased variables as compared to their values in the initial state. Reproduced by permission from Cuadra et al. 2020.

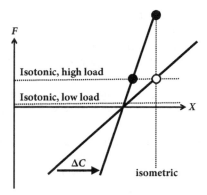

Figure 17.6 In isometric conditions, a change in the coactivation command leads to a major change in the state of the effector. This is not the case in isotonic conditions when acting against a close-to-zero external load. Black circles show final equilibrium states; open circle shows the initial state.

only, namely the R-command. In other words, the "efference copy" differs from the efferent command.

When asked to match force magnitude with the contralateral finger, which had been relaxed throughout the first part of the trial, the subjects were not biased by the initial force level and had to match the final force magnitude. They did this with a tendency to overshoot the actual force level, a phenomenon reported in earlier studies of force matching (reviewed in Proske and Gandevia 2012; Proske and Allen 2019), but without significant differences from the actual, increased force magnitude. Interestingly, misperception of the state of an effector under voluntary muscle coactivation is not observed during movements (i.e., in non-isometric conditions). This may be related to the fact that, when the external load on an effector is close to zero, coactivation may not lead to visible changes in force and coordinate. Such changes happen, however, when the effector is acting against a substantial external load or in isometric conditions when the external load is matched to the voluntary force produced against the external stop (Fig. 17.6).

17.4. Perception and production of force

Within the theory of control with RCs, force production is associated with setting the RC for the effector (e.g., a fingertip) below the interface surface; that is, the effector virtually penetrates the surface (Pilon et al. 2007; Fig. 17.7). This is done with the help of the R-command. The difference between the RC and the actual effector position on the surface in spatial units is converted into force units with the help of the C-command, which can be viewed, in a linear approximation at the level of mechanics, as setting apparent stiffness of the effector, k. Signals from sensory endings sensitive to force (pressure), such as the cutaneous and subcutaneous endings and Golgi tendon organs, by themselves do not ensure accurate perception of force, as

17.4. PERCEPTION AND PRODUCTION OF FORCE

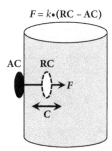

Figure 17.7 Force production as a combination of the R-command and C-command setting the referent coordinate (RC) and apparent stiffness (k) of the effector, respectively.

shown, in particular, by large unintentional force drifts, which are not perceived by the actor (see later and Chapter 16). These signals are estimated within the reference frame defined by the two basic commands.

As mentioned, if a person is asked to match force produced by an effector with a homologous contralateral effector, the matching is relatively accurate with a tendency to overshoot the target force level. The situation changes when the initial force level produced by the first effector drifts unintentionally in the absence of visual feedback (see Chapter 16). In such a situation, the actor is unaware of the force drift, even when it reaches relatively large magnitudes, on the order of 30% to 40% of the initial force level within 15 to 20 s. When asked to match the force level after a drift with the contralateral effector, the actor typically overshoots the actual force level by about the same magnitude independently of the force drift magnitude. One of the studies explored force matching following episodes of force drift across a range of initial force magnitudes, from 10% of maximal voluntary contraction (MVC) force to 30% of MVC (Cuadra et al. 2021a). As shown in Fig. 17.8, when the initial force was 10% of MVC, there was very little force drift after turning the visual feedback off, and the subjects overshot this force with the contralateral effector during the matching episodes. As the initial force increased, the drift magnitude increased and was rather large for the initial force of 30% of MVC. Although the subject was unaware of the drift, the amount of the overshoot of the actual force level remained relatively constant, suggesting that matching was performed taking into consideration the force drift. This is another example of the difference between perception-to-report (the subject reported no drift) and perception-to-act (the matching force depended on the drift) in the field of kinesthetic perception.

Changes in the R- and C-commands during the force drift were quantified using the "inverse piano" device (see Chapter 7; Martin et al. 2011), which uses smooth perturbation of the effector (a finger or a set of fingers) to compute mechanical variables, RC and k, reflecting the two commands (Ambike et al. 2016b; Reschechtko and Latash 2017; Abolins et al. 2020). These studies documented significantly larger absolute magnitudes of RC and smaller magnitudes of k in the matching hand. When RC and k were measured in the original effector (the task hand) prior to and following

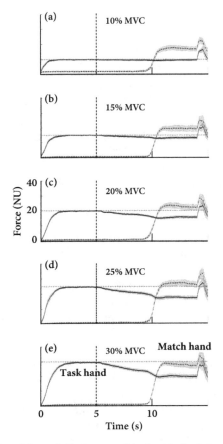

Figure 17.8 Unintentional force drifts in one of the hands (task hand, solid traces) caused by turning visual feedback on force magnitude off. Note the lack of drift for the lowest initial force level (10% of maximal voluntary contraction [MVC]) and significant drift for higher initial force levels. Note that the matching hand (dashed traces) overshot the task hand force by about the same magnitude over the conditions. Reproduced by permission from Cuadra et al. 2021a.

the force drift, they showed a similar pattern: an increase in the absolute magnitudes of RC and smaller magnitudes of k that combined to produce lower force magnitudes.

Analysis of these findings suggests that force drifts are primarily produced by large drifts of the RC for the antagonist muscle group (RC_{ANT}) toward the actual effector coordinate, leading to a drop in the C-command, with a relatively minor change in the R-command associated with a smaller drift of the RC to agonist muscles (RC_{AG}), also toward the actual effector coordinate (Fig. 17.9; Cuadra et al. 2021a). Note that RC_{ANT} in the initial state defines the initial level of muscle coactivation, and its drift magnitude is naturally limited by the difference between RC_{ANT} and the effector coordinate. This may be the factor explaining the limited magnitudes of force drift observed in

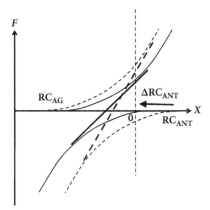

Figure 17.9 Unintentional drop in force associated with a drop in the C-command and an increase in the absolute magnitude of the R-command can be interpreted as a consequence of a drift in the referent coordinate to the antagonist muscles (RC_{ANT}) toward the actual effector coordinate ($X = 0$) and a smaller drift of the RC to agonist muscles (RC_{AG}). The initial force-coordinate characteristics for the opposing muscles and effector are shown with dashed lines; the characteristics after the drift are shown with solid lines. Modified by permission from Cuadra et al. 2021a.

young healthy subjects (rarely more than 35% of MVC) and larger force drift magnitudes in populations with elevated levels of muscle coactivation, such as patients with Parkinson disease (Vaillancourt et al. 2001; Jo et al. 2016a).

We come to a rather unexpected set of conclusions: (1) control signals to antagonist muscle groups are prone to unintentional drifts, and (2) these drifts are not taken into account in the force perception-to-report. These conclusions have also been supported in studies of force perception under muscle vibration (Cuadra et al. 2021a; see also Chapter 15).

17.5. Muscle vibration and stability of percepts

High-frequency, low-amplitude muscle vibration can have effects on both components of perception, afferent and efferent. The effects of vibration on sensory endings, in particular the velocity-sensitive primary spindle endings and some of the cutaneous receptors, have been known for a long time and used to interpret both motor and sensory effects of vibration (see Chapter 15). The effects of vibration on the efferent component of kinesthetic perception ("efference copy") were suggested a long time ago (Feldman and Latash 1982a) to interpret specific patterns of vibration-induced kinesthetic illusions. Such effects have been confirmed in recent studies of the effects of vibration on force perception (Reschechtko et al. 2018). Indeed, an increase in the activity of spindle endings and, possibly, Golgi tendon organs was by itself unable to explain the effects of vibration as estimated using force matching with the homologous contralateral effector. To account for the observed force-matching errors, one had to assume that vibration led to a drift in the RC to the effector.

Another study explored the effects of vibration applied over the extrinsic finger/wrist flexors and finger/wrist extensors on force matching (Cuadra et al. 2021a). In spite of the likely vibration spread through the soft tissues of the forearm (cf. Eklund and Hagbarth 1966), the vibration was expected to be a much stronger stimulus for sensory endings in the muscle groups directly under the vibrator. So, if the effects of vibration were caused by responses of peripheral sensory endings, one could expect the opposite effects of vibration applied to the agonists (flexors) and antagonists (extensors) during finger pressing tasks. However, the patterns of matching errors under both conditions were very similar, suggesting that they were caused not by local effects of vibration but by its likely effects on the "efference copy."

Measurement of the effects of vibration on the RC and k (reflections of the R- and C-command, respectively) in the task and match hand showed patterns suggesting once again a mismatch between the commands to the antagonist muscle group, RC_{ANT} (Cuadra et al. 2021a; cf. Abolins et al. 2020). Figure 17.10 illustrates the effects of vibration applied to the task hand (panel A) and to the match hand (panel B). In both cases, RC_{ANT} in the hand with the vibrator is perceived as being closer to the actual effector coordinate, leading to matching errors in opposite directions. The pattern of the matching errors, however, was the same during vibration applied to the agonists and to the antagonists.

Note that correlations between the antagonist muscle response and kinesthetic illusions produced by vibration have been documented (Calvin-Figuiere et al. 1999) in support of the idea that antagonist muscle reaction to vibration can be associated with perceptual effects. These results, taken together with those described in the previous section, suggest that RC_{ANT} is an inherently unstable signal, which shows spontaneous drifts in the absence of visual feedback. Such drifts can also be triggered by other stimuli, including smooth transient perturbations of the effector (see Chapter 16) and muscle vibration. The fact that actors are unaware of the drifts is most surprising. It suggests that RC_{ANT} is not readily incorporated into kinesthetic perception or, at

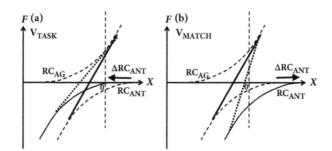

Figure 17.10 Schematic illustrations of the effects of vibration applied to the task hand (a: V_{TASK}) and to the match hand (b: V_{MATCH}). The observed effects can be interpreted as misperception on the command to the antagonist muscles (RC_{ANT}) in the hand with the vibrator: RC_{ANT} is perceived as being closer to the actual effector coordinate ($X = 0$). Reproduced by permission from Cuadra et al. 2021a.

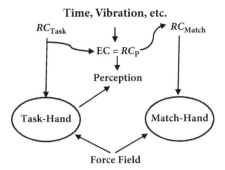

Figure 17.11 Schematic interpretation of the force matching task. Note that the neural command to the task hand (RC_{TASK}) may be distorted before it is used for perception ("efference copy," $EC = RC_P$) and as a command to the matching hand (RC_{MATCH}). Such distortions can happen spontaneously and/or be provoked by muscle vibration.

least, force perception. This conclusion leads to the question: Do we perceive what our antagonist muscles are doing? For example, can one match the activation level of a muscle group when this groups acts as the antagonist in a task performed by an effector? Introspection suggests a negative answer. For example, during a very strong contraction of a muscle group (e.g., an MVC contraction of the elbow flexors), antagonist muscles (triceps brachii) commonly show substantial activation and force levels, possibly on the order of 20% of their own MVC force. However, when asked to match the antagonist force level by a homologous contralateral effector, a naïve actor is confused and reports zero coactivation level.

Figure 17.11 illustrates the main conclusions schematically. According to this figure, the "efference copy" (RC_{PER}) can differ from the efferent commands (RC_{ACT}) to the effector, leading to perceptual errors that can be revealed differently in verbal reports or matching tasks. These differences can happen spontaneously, as in studies of force drifts, or be triggered by external stimuli, in particular muscle vibration.

17.6. The place for sense of effort

Unlike force, effort is not a well-defined physical variable. As a result, it is not easy to handle this concept using the language of natural science. The introduced framework, however, allows addressing the sense of effort and offering a hypothesis on its origin within the same scheme of control with RCs. The first step is to define this notion. In dictionaries, *effort* is defined as conscious exertion of power, exertion of physical or mental power, physical or mental activity needed to achieve something, and use of physical or mental energy. On the one hand, these definitions link effort to physical variables such as power and energy; on the other hand, effort is associated with less precisely defined processes such as mental activity. The vast literature on sense of effort (reviewed in Proske and Gandevia 2012; Proske and Allen 2019) offers a few definitions for this elusive construct, including the following: "signals of motor command ..., which arise at levels rostral to spinal motoneurons" (Gandevia 1982, p. 151) and "centrally generated sensations arising from the motor command" (Jones 1995, p. 305). According to these definitions, sense of effort is viewed as being of a primarily central origin. However, the possibility of modulation from sensory signals has also been emphasized. A number of studies presented examples of dissociation between senses of effort and force, in particular under such manipulations as fatigue, paralysis, or sensory deprivation (Jones 1995; Gandevia 1982; Proske and Allen 2019). On the other hand, a few studies have suggested that, when subjects are asked to match forces by two effectors (e.g., the left and right hands), they tend to match efforts (i.e., measures of central neural commands), leading to force mismatch in conditions when the external force fields for the two hands differed (van Doren 1995, 1998).

In more lay terms, effort is an internal measure of how hard one tries to perform the task as reflected in commands to motoneurons, which makes it compatible with the introduced definition of RC and suggests that sense of effort may originate from the "efference copy." The aforementioned experiments exploring the effects of voluntary coactivation on force perception suggest that effort during an action by an effector equals its RC or, more precisely, its difference from the current effector coordinate (i.e., it reflects the R-command). The naturally emerging changes in the C-command may have no direct effects on the sense of effort. Although this hypothesis may look strange, it is readily compatible with the idea that the R- and C-commands form a hierarchy: The R-command changes in a task-specific way to produce actions, whereas the C-command may stay unchanged (Feldman 2015; see Chapter 3).

17.7. Missing pieces of the mosaic

This brief review has led to a number of counterintuitive conclusions that require further exploration and verification. In particular, the conclusion that "efference copy" may not be a copy of the efferent command makes this commonly used concept ill-defined. Efference signals may be corrupted or otherwise modified prior to their involvement in perceptual processes. Maybe *corollary signals* (cf. Sperry 1950) would be a better term to address involvement of neural control processes in perception. At least this term does not imply that these signals are a copy of any of those involved in the efferent process.

Another unexpected conclusion is that neural commands to antagonist muscle groups are unstable and prone to spontaneous drifts, as well as to changes under external stimuli, mechanical perturbations, or muscle vibration (see also Chapter 19). Moreover, they seem to be largely ignored in perceptual processes, at least in perception of force, although they obviously affect net force magnitude. It is possible that the hypothesized underestimation of RC_{ANT} is related to mechanisms associated with strong persistent inward currents generated in dendrites of spinal motoneurons in response to diffuse serotoninergic and noradrenergic descending signals (Heckman et al. 2004, 2008a). These currents have shown very high sensitivity to reciprocal inhibition between antagonist muscle groups. As a result, local reflex-mediated inputs can have a profound effect on control signals to individual muscles, in particular leading to a drop in RC_{ANT}. These effects could be amplified under muscle vibration, which leads to higher afferent input from the primary muscle spindles and correspondingly stronger effects of reciprocal inhibition.

Traditional views on the motor and sensory effects of muscle vibration have to be reconsidered. Indeed, a number of studies have shown that the effects of vibration are not limited to those expected from local responses of sensory endings (Lund 1980; Hayashi et al. 1981; Gurfinkel et al. 1998; Selionov et al. 2009). Vibration leads to shifts in signals reflecting neural commands (corollary signals!) and taking part in perceptual processes. Its effects seem to be more general, leading to perturbations of processes within the central nervous system and, depending on conditions, causing a variety of motor and sensory effects, largely independently of the site of vibration application. This conclusion is also supported by studies reviewed in Chapter 15.

The most glaring gaps in the current knowledge are in the field of possible neural structures involved in the emergence of corollary discharges and their interactions with afferent signals. As mentioned in the introductory part of this chapter, it seems unlikely that there are specific neural structures and circuits dedicated to these processes and interactions common across tasks, effectors, and species.

18
Equifinality and Motor Equivalence

The two groups of phenomena (and the two concepts) in the title of this chapter are related to each other and have sometimes been used interchangeably. The main difference is that *equifinality* most commonly refers to a feature of voluntary movements to show success in reaching the target when an unexpected transient change in the external force field takes place. *Motor equivalence* refers frequently to the ability of animals to reach the same desired outcome with various means (i.e., various involvement of the involved body elements), observed spontaneously, as a conscious change in the action (e.g., switching to another effector) or in the presence of smooth, longer-lasting changes in the external force field.

A classic example of motor equivalence is the ability of humans to show individual handwriting characteristics while writing with different involvement of joints and muscles (cf. writing with a pen on a piece of paper and with chalk on the blackboard) and with an implement held with the non-dominant hand, gripped by one's teeth, or attached to one's elbow or foot (Bernstein 1947; Raibert 1977). One can also consider such actions as eating with a fork, with a spoon, or with chopsticks held by the dominant or non-dominant hand. More subtle examples include varying contributions of elements, such as joints and muscles, that happen spontaneously and are not consciously perceived by the actor or noticed by an observer. This example makes motor equivalence closely related to the concepts of motor abundance and performance-stabilizing synergies (see Chapters 5 and 6).

Equifinality has been viewed as a natural consequence of the neural control of movement with changes in equilibrium states, as suggested by the equilibrium-point (EP) hypothesis (see Chapter 3). Indeed, equifinality may be expected within the EP-hypothesis, but only under certain conditions (Feldman and Latash 2005). Moreover, equifinality may be considered in different spaces, for example, in the space of task-specific performance variables, such as the trajectory of the hand during reaching, or in spaces of the involved elements, such as individual axes of joint rotation or muscle activation levels. Predictions of equifinality or its violations may differ across levels of analysis, types of changes in external forces, and a few other factors discussed later. In particular, under some conditions, equifinality may be observed at the level of task-specific variables but violated at the level of elements. In such cases, it is rather directly related to motor equivalence because various combinations of elemental variables (those produced by the elements) are used to reach about the same task-specific outcome.

Both concepts have to be considered in appropriate reference frames. For example, when a standing person is pushed not very strongly from behind, the body shows a sequence of posture-stabilizing reactions, and the projection of the center of mass returns to approximately the same location as prior to the push. If the push is strong, the person may have to take a protective step to a new coordinate in the external space, and the projection of the center of mass will move in that space (equifinality will be violated), but it may remain at about the same coordinate (i.e., showing equifinality) within the body-centered reference frame after the vertical posture is restored following the step.

The last example rather directly links equifinality to the concept of stability. As discussed later, stability of movements is also directly related to phenomena of motor equivalence. Relations among these three concepts have been discussed within the uncontrolled manifold (UCM) hypothesis (see Chapter 5), which has also provided a framework for the quantitative analysis of motor equivalence.

18.1. Examples of equifinality and motor equivalence

Many studies presented examples of equifinality across tasks, performance variables, and types of perturbation. In particular, when a person is asked to grip an object with two digits, the thumb and index finger, stopping one of the digits unexpectedly leads to a corrective movement of the opposing digit (Cole and Abbs 1987). As a result, the two digits grip the object successfully, showing equifinality with respect to the aperture between the digits. The task-related performance variable does not have to be mechanical. As shown in studies with perturbations applied to an articulator involved in the production of a sound, other articulators adjust their output at a very short delay, leading to uttering of a sound nondistinguishable from the one performed in trials without perturbations (Hughes and Abbs 1976; Abbs and Gracco 1984; Tremblay et al. 2003). So, equifinality can be observed in the field of acoustics.

Sometimes perturbations can be intrinsic, that is, generated within the central nervous system. For example, the same magnitude of an external force perturbation can be produced at different rates (Latash and Gottlieb 1990). Very fast perturbations, with the muscle length changes over 0.1 s or shorter, lead to phasic spinal reflexes, likely with a significant monosynaptic component, followed by the so-called preprogrammed reactions (a.k.a. long-loop reflexes, triggered reactions; see Chapter 14). Somewhat slower perturbations, over 0.2 to 0.3 s, do not induce phasic spinal reflex responses but lead to relatively short-lasting preprogrammed responses. Slower perturbations do not induce transient responses at all. In all cases, tonic reflexes contribute to responses of the muscle to the imposed length change, which leads to a shift from the initial equilibrium point (EP_0 in Fig. 18.1A) to the final equilibrium point (EP_1). The experiment showed that, in spite of the very different transient changes in muscle activations following perturbations applied at different rates, the final EP was about the same as long as the perturbation magnitude was consistent across the conditions (Fig. 18.1B). In other words, the transient muscle responses did not violate equifinality.

The classical example of motor equivalence is writing with different effectors, which has already been mentioned. Another example is reaching involving variable sets of effectors. Depending on the location of the target for reaching with respect to the body, different sets of effectors can be involved, including or not including the trunk and involving or not involving stepping toward the target. Experiments with unexpected perturbations applied to one of the effectors (e.g., limiting trunk motion)

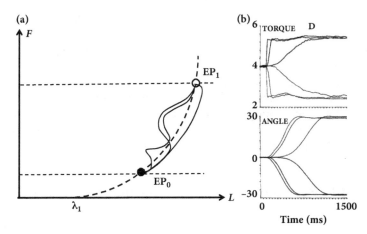

Figure 18.1 A quick torque perturbation results in equifinality in spite of the presence of transient reflexes and reflex-like responses during very quick perturbations and their absence when the perturbation time was ≥300 ms. This phenomenon is illustrated schematically in (a). (b) shows the experimental data (reproduced by permission from Latash and Gottlieb 1990).

have shown short-latency adjustments in the trajectories of other effectors, including involvement of effectors not participating under control conditions (Tomita et al. 2017). In particular, subjects made a step compensating for the lacking trunk contribution, which kept the hand trajectory in external space relatively invariant. As a result, the target was reached under those conditions with different sets of effectors—an example of motor equivalence.

Another example of motor equivalence is grasping objects with variable sets of digits. Depending on the object's size, weight, and possibly other properties (e.g., friction coefficient), humans use different numbers of fingers opposing the thumb when grasping the objects with the so-called prismatic grasp (Cesari and Newell 1999, 2002). The selection of digit combinations is not defined 100% by the object's properties, and sometimes subjects vary the number of involved fingers. Changing the number of fingers does not affect the grip force applied to the object; the force is shared differently across the fingers—an example of motor equivalence.

Motor equivalence can be used by humans to perform functional movements when motion of some of the effectors is limited, for example, as a result of cortical stroke. Following a stroke, the range of active motion of major arm joints, in particular distal joints, can be limited. Such patients perform reaching movements with a significant contribution of trunk motion, which is not seen in healthy controls during the performance of similar tasks (Cirstea and Levin 2000; Levin et al. 2002).

Under fatigue, various sets of motor units and various combinations of involvement of muscles and joints can be seen to achieve the same motor effect required by the task (Côté et al. 2002, 2008; Singh et al. 2010; Singh and Latash 2011; Ricotta et al. 2023a). In this example, spinal mechanisms likely play an important role in achieving motor equivalence. A more direct example of motor equivalence and equifinality achieved by spinal mechanisms was demonstrated in studies of the wiping reflex in the spinal frog (Fukson et al. 1980; Berkinblit et al. 1986a; Latash 1993). When an irritating stimulus was placed on the back of the spinal frog, it wiped it off with coordinated action by the ipsilateral hindlimb. Limiting the range of motion in one of the hindlimb joints did not prevent accurate wiping, which was achieved with different joint kinematics (motor equivalence). Accurate wiping (equifinality) was also achieved under conditions of additional loading of the hindlimb with a lead bracelet, which required very different joint torque profiles and, hence, different muscle activation patterns.

18.2. Equifinality and the equilibrium-point hypothesis

The EP-hypothesis got its name from one of its central notions, that of *equilibrium point*, a point where the muscle with its reflexes comes to an equilibrium when acting against an external load (see Chapter 3). Figure 18.2 illustrates the control of a movement produced by a muscle according to the EP-hypothesis. A shift in the control variable to the muscle (λ, threshold of the stretch reflex) leads to a shift in the EP from an initial combination of length and force values to some final combination, which depends on the external load characteristics. Different load characteristics are expected to lead to different movements produced by the same shift of λ. However, the final EP is expected to depend only on the final value of λ and the external load magnitude in the final state. Transient load changes during the movement within a reasonable range should not have effects on the final EP of the system—an example of equifinality.

This prediction was seen as a strong test of the EP-hypothesis. A number of studies confirmed this prediction by observing equifinality under a variety of transient perturbations in a variety of tasks (Schmidt and McGown 1980; Rothwell et al. 1982b; Jaric et al. 1999). However, at least two influential studies reported violations of equifinality under transient, velocity-dependent perturbations. One of the studies used

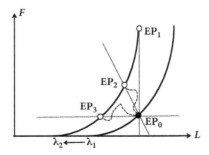

Figure 18.2 A change in the control variable (λ) to a muscle from λ_1 to λ_2 can lead to different peripheral consequences (new equilibrium points, EP_1, EP_2, and EP_3) depending on the load characteristic. If a transient perturbation is applied during the action (dashed lines), the trajectory can change but the EP can remain unchanged.

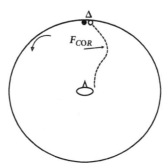

Figure 18.3 Reaching movements performed in the rotating centrifuge are characterized by trajectories deviating from the straight line due to the action of the Coriolis force (dashed trajectory) with a residual error (Δ) in the final position (the white dot).

the velocity-dependent Coriolis force as the source of transient perturbations during reaching movements performed in a rotating environment (Lackner and DiZio 1994; DiZio and Lackner 1995; Fig. 18.3). The subjects were positioned in the center of the centrifuge, so they did not feel the rotation, which accelerated very slowly, under the perception threshold. They performed reaching movements to a visual target in complete darkness, and the target was turned off immediately after the movement initiation. In the absence of rotation of the centrifuge, reaching movements showed straight trajectories landing accurately in the target. During rotation, the trajectories were curved due to the Coriolis force. They deviated from the straight line to the target and then reversed toward the target but stopped with a residual error, consistent across subjects. Note that the direction of the Coriolis force was always the same. So, the reversal of the trajectory toward the straight line reflected the tendency of the system toward the equilibrium state defined by the central nervous system of the actor. The question remained, however, why the target was not achieved accurately; as one would expect, because the Coriolis force was transient, it became zero as soon as the movement stopped.

Another study used a programmable motor to simulate a negative damping environment, that is, force proportional to movement velocity and acting in the same direction (Hinder and Milner 2003). This is the most unusual environment, which does not exist in inanimate nature. After the subjects learned to perform reasonably accurate movements in this environment (which was very hard since movements always ended up with major terminal oscillations), the negative damping was unexpectedly turned off. The movements in those trials were consistently inaccurate, undershooting the target. Since damping forces become zero when movement stops, this was seen as a violation of equifinality. An interpretation of those findings was suggested assuming force programming by the central nervous system, an assumption inconsistent with the design of the human body (see Chapter 4).

To address these violations of equifinality, one has to revisit the EP-hypothesis and its predictions with respect to equifinality. Equifinality is expected if the central neural control variables (λ for a muscle, reciprocal and coactivation commands [R- and C-commands] for effectors) remain at their unchanged values and the muscle force-coordinate characteristics remain unchanged under the action of unexpected transient external force changes. First, muscle force-producing capabilities can change following a brief period of activation—the so-called *catch property* of muscles (Burke et al. 1970, 1976). Second, as described further in more detail, central neural commands show a tendency to change unintentionally under a variety of conditions that do not involve very unusual manipulations such as rotation in a centrifuge or acting in a force field with negative damping. This latter factor remains the most plausible explanation for the aforementioned examples of equifinality violations.

Why unintentional changes in neural commands take place is still unknown. However, the existence of such changes is beyond doubt, as described in the next section (also see Chapters 13 and 16). It is possible that some destabilizing force fields are not tolerated by the central nervous system and it reacts to them by changing neural commands, even if such changes are detrimental for performance (reviewed in Feldman 2015). It is also possible that such drifts represent natural behaviors of the system for the neural control of movement that do not require the action of unusual force fields and tend to happen under many conditions unless corrected with the help of sensory signals.

18.3. Violations of equifinality in different spaces

Natural movements involve time functions of referent coordinates, RC(t), at different levels of the hierarchy, from the task level to the levels of elements such as extremities, digits, joints, muscles, and motor units (see Chapters 6 and 7). The abundance of solutions in the few-to-many transformations $\{RC_{TASK}(t)\}$ => $\{RC_{ELEMENT}(t)\}$ is the basis of performance-stabilizing synergies documented across tasks, effector sets, and populations (see Chapter 6). Note that unexpected transient forces can produce significantly different patterns of mapping in such transformations because their solution spaces, equivalent to the UCMs (Chapter 5), are inherently less stable than spaces where salient performance variables change. Hence, equifinality in the task space is naturally compatible with violations of equifinality at the level of elements.

This prediction was tested in experiments with smooth and relatively modest transient changes in external force (Zhou et al. 2014a). The seated subjects held a handle by the dominant hand in a comfortable position against a moderate bias force produced by a programmable robot and acting away from the body. The force increased modestly over 0.5 s and then returned to the initial magnitude. The subjects could not see their moving arm, and they were instructed not to react to possible hand movements produced by the robot. Following this perturbation, the hand coordinates in the horizontal plane were very close to the original ones—an example of equifinality (Fig. 18.4A). The vertical coordinate changed consistently—the hand stopped lower. Further, the framework of the UCM hypothesis (see Chapter 6) was used to estimate the inter-trial variance (V) in the joint configuration space within the UCM computed for the hand location or orientation and orthogonal to the UCM. In both analyses, V_{UCM} was consistently larger than V_{ORT}, suggesting that relative equifinality in the hand coordinate space was violated in the joint configuration space (Fig. 18.4B).

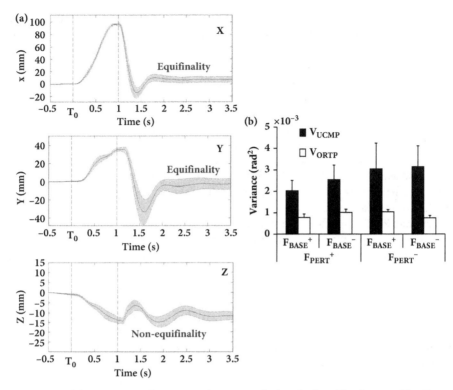

Figure 18.4 (a): A transient force perturbation applied to the hand leads to equifinality in the horizontal plane (X-Y) and a lower hand coordinate along the vertical axis (Z). (b): Analysis of variance across trials in the joint configuration space showed synergies stabilizing the endpoint coordinate. The larger V_{UCM} reflects violation of equifinality in the joint configuration space. The uncontrolled manifold was computed with respect to the handle coordinate along the force vector. Different pairs of bars represent series of trials with different directions of the bias force and perturbing force. Reproduced by permission from Zhou et al. 2014a.

When a dwell time between the force-increase and force-decrease phases was introduced, equifinality was also violated at the task level—the hand stopped short of the target (Zhou et al. 2014b, 2015b; ΔD in Fig. 18.5). These errors in the final hand position were large, about half the distance the hand had been moved from its initial location. The amount of this undershoot was relatively small for the dwell time of under 1 s, and it reached its maximal magnitude for the dwell time of about 2 to 3 s. These violations of equifinality suggested that RCs for the hand drifted during the dwell time toward the actual hand coordinate, and this assumption was confirmed in follow-up experiments that used additional small hand perturbations applied during the dwell period (Zhou et al. 2015a).

Taken together, these studies suggest that equifinality may be a relatively rare phenomenon observed under certain conditions but violated both in the task space and

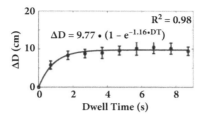

Figure 18.5 When a transient force perturbation applied to the hand was interrupted by a dwell time, equifinality was violated along the axis of force change in the horizontal plane (X). Longer dwell times were associated with larger violations of equifinality. The plot shows the dependence of the deviation of the final position from the initial position (ΔD) as a function of the dwell time. The exponential regression is shown. Reproduced by permission from Zhou et al. 2014b.

in spaces of involved elements. These violations do not contradict the EP-hypothesis but represent a natural consequence of the control with RCs organized into a hierarchy. Indeed, the unintentional drifts in performance discussed in Chapter 16 may be viewed as examples of violations of equifinality that happened spontaneously, in the absence of changes in the external force field, as well as in response to quick changes in the external force. All these drifts may be viewed as mechanical effects of unintentional changes in RCs happening as consequences of the tendency of natural systems to move toward states with lower potential energy.

18.4. Motor equivalence and the uncontrolled manifold hypothesis

Some of the examples considered earlier in this chapter link the notion of motor equivalence directly to the concept of performance-stabilizing synergies (see Chapter 6). Indeed, variable involvement of effectors to achieve the same result is equivalent to stability of task-specific performance variables under variable effector involvement or, in other words, under variable contributions of elemental variables to the task-specific salient performance variable. Figure 18.6 illustrates the concept of motor equivalence using the task of performing a certain magnitude of the sum of two elements variables, E_1, and E_2. The solution spaces (the UCM) for this task at two different magnitudes of ($E_1 + E_2$) are shown as slanted lines with negative slope. Imagine that an actor performs this task with some combination of E_1 and E_2 to produce the smaller sum (point a). Now the actor is asked to increase the sum quickly and then bring it back close to the initial magnitude. At the end of the trial, the actor will achieve a different point (b). If both points are projected onto the UCM and its orthogonal complement (ORT), the distance along the UCM is motor equivalent (ME) because it does not lead to changes in the performance variable. In contrast, the distance along the ORT is non–motor equivalent (nME). If the performance variable is stabilized in the space of E_1 and

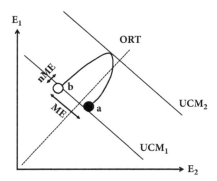

Figure 18.6 An illustration of the notion of motor equivalence in the task of performing a constant value of the sum of two elemental variables, E_1 and E_2. The action involved a quick increase in $(E_1 + E_2)$ and its decrease close to the initial level. Large deviations along the uncontrolled manifold (UCM, solid slanted line) for the task are motor equivalent (ME). The smaller deviations along the orthogonal to the UCM space (dashed line) are non–motor equivalent (nME).

E_2, one can expect ME > nME, because of the relatively low stability along the UCM as compared to the ORT.

The illustration in Fig. 18.6 suggests a nontrivial prediction: Quick actions or reactions should have a dominating ME component, which, by definition, does not change the salient performance variable. In other terms, it is wasteful because it spends metabolic energy to change elemental variables with no effect on the salient performance variable. This counterintuitive prediction has been confirmed in a number of studies exploring corrective actions to external perturbations across tasks and spaces of elemental variables, kinematic, kinetic, and electromyographic (Mattos et al. 2011, 2013, 2015a).

In one of the studies, the subjects performed pointing movements with a pointer to a spherical or cylindrical target (Mattos et al. 2011). They could not see their pointing arm in the initial position, and in some trials, the elbow joint was blocked with a rubber band with high or low stiffness (the left panel of Fig. 18.7). As soon as movement was initiated, the subjects felt the perturbation. Besides, the arm with the pointer got into their visual field. They were instructed to try to perform accurate pointing in every trial. So, they had to correct the perturbed movements. When the difference between the perturbed trials and average unperturbed trials was analyzed in the joint configuration space, it showed much larger components within the UCMs for the pointer orientation and tip coordinate as compared to the corresponding ORT spaces (the right panel of Fig. 18.7). In other words, most of the corrective actions were ME; that is, they did not correct the mentioned pointer characteristics. The ME > nME inequality was seen as early as the deviations between the perturbed and unperturbed trajectories became significant.

To explore the possible role of purely mechanical factors in this result, the same movements were analyzed in spaces of muscle activations, with a typical analysis

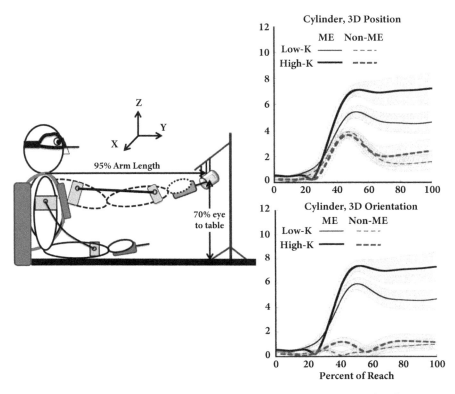

Figure 18.7 The subject performed pointing movements. In some trials, the elbow joint was blocked unexpectedly (left panel) by a spring-like load applied to the elbow joint. Corrections of the perturbed movements had large motor equivalent (ME, solid lines) deviations compared to the non–motor equivalent (nME, dashed lines) deviation (right panel). This was consistent across perturbations of different magnitude and pointing tasks to spherical and cylindrical targets. Reproduced by permission from Mattos et al. 2011.

involving two steps. First, individual muscle activations were grouped into a small number of muscle modes (M-modes), and then deviations in the M-mode space were compared between the UCM and ORT. The main result was confirmed, ME > nME (Mattos et al. 2013).

A series of studies explored similar phenomena in multi-finger kinetic tasks with positional perturbations applied to the effectors (fingertips). In those studies, the subjects pressed with the four fingers of a hand to produce a prescribed cyclical time pattern of the total force. The "inverse piano" device (Martin et al. 2011; see also Chapter 7) was used to lift one finger while the subjects were required to correct the total force time profile as well as they could. After a few seconds, the finger was lowered to its initial coordinate, while the instruction remained the same. The main results are illustrated in Fig. 18.8. Corrections of the total force were very quick and accurate, and one can hardly see performance deviations during the finger-lifting and

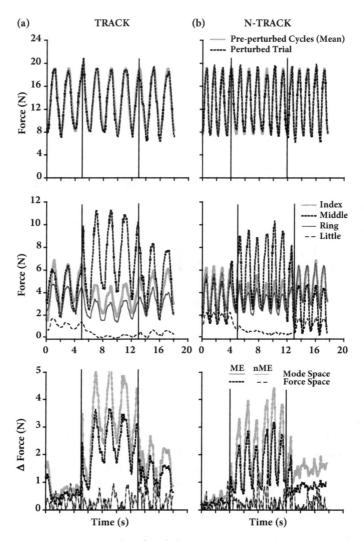

Figure 18.8 Large motor equivalent (ME) deviations are seen during corrections of an accurate sine-like force production task. The task was perturbed by the "inverse piano" (IP) lifting the fingers at time T_1 and lowering them at T_2. Note the major changes in the force-sharing pattern across the four fingers without a visible change in total force. Reproduced by permission from Mattos et al. 2015a.

finger-lowering phases. Lifting a finger, however, led to a major change in the sharing of the total force across the four fingers as compared to the pre-lift pattern. Lowering the finger back did not restore the original sharing pattern but generated a new one, different from both the original pattern and the one seen after the finger was lifted. As a result, in the space of finger forces, the same performance was obtained with significantly different patterns of finger forces, that is, with ME > nME.

One of the aforementioned examples of equifinality in the task space associated with violation of equifinality in the space of elemental variables (Zhou et al. 2014a) can be viewed as an example of motor equivalence, thus linking the two notions directly. To remind, when a person performed a positional task by the hand, a transient smooth change in the external force resulted in equifinality in the space of hand coordinates but not in the space of joint angles. In other words, about the same hand position was achieved with different joint configurations—an example of motor equivalence.

The UCM hypothesis makes predictions regarding relations between indices of motor equivalence and those of inter-trial variance. Lower stability along the UCM as compared to ORT is expected to lead to both inequalities, ME > nME and V_{UCM} > V_{ORT}. Moreover, if one assumes that ME and V_{UCM} reflect samples from a single normal distribution (the same assumption applies to nME and V_{ORT}), relations between these pairs of variables can be deduced analytically. Indeed, assume that there is a normal distribution of data points. If we assume that each measurement represents a sample from a normal distribution, the difference between two random samples will be another normal distribution with the mean $\mu_D = 0$ and standard deviation σ_D linearly related to the standard deviation of the original distribution, $\sigma_D = \sigma\sqrt{2}$. In the analysis of motor equivalence, absolute distances between data points are quantified. This makes the distribution of those distances non-negative (folded) with a new mean (μ_X) and new standard deviation (σ_X) (Leone et al. 1961):

$$\mu_X = \sigma\sqrt{2/\pi}$$
$$\sigma_X^2 = \sigma^2 - \mu_X^2$$
(18.1)

These equations can be applied separately to data in the UCM and ORT of the original space. They suggest that ME is expected to be proportional to σ_{UCM} ($\sqrt{V_{UCM}}$), whereas nME is expected to be proportional to σ_{ORT} ($\sqrt{V_{ORT}}$).

These predictions have been confirmed in a study of multimuscle synergies during cyclical whole-body sway by standing persons (Falaki et al. 2017a; Fig. 18.9B). It was violated, however, for the expected correlation between nME and V_{ORT} in a study of steady force production interrupted by a quick voluntary force pulse (Cuadra et al. 2018). In that study, the expected correlation between ME and V_{UCM} was observed. The lack of correlation between nME and V_{ORT} was interpreted as a consequence of the voluntary correlation of force magnitude by the subjects following the force pulse, which led to the data prior to the pulse and after the pulse sampled from different distributions along the ORT. Since no correction was required along the UCM, the data were likely sampled from a single distribution and the correlation held.

18.5. Motor equivalence as a promising clinical index

The observations of correlations between the indices of inter-trial variance and motor equivalence suggest that both can be used in studies of synergies stabilizing salient

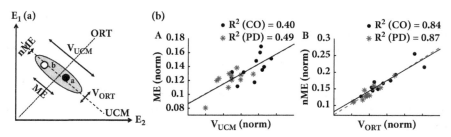

Figure 18.9 (a): The task of using two elemental variables, E1 and E2, to produce a certain value of their sum, C. UCM, uncontrolled manifold; ORT, orthogonal to the UCM space. Transient voluntary changes in C lead to deviations along the UCM and along ORT. Indices of motor equivalence (ME) are expected to correlate with variance along the UCM. Non-motor equivalent deviations (nME) are expected to correlate with variance along the ORT. (b): Such correlations were observed in the analysis of cyclical body sway tasks performed in the space of activations of muscle groups (M-modes) in both control subjects (CO) and patients with Parkinson disease (PD). Reproduced by permission from Falaki et al. 2017a.

performance variables, in particular in studies of populations with impaired motor coordination. Note that ME and nME can be quantified in individual trials, whereas V_{UCM} and V_{ORT} require analysis over clouds of data points. This gives analysis of motor equivalence a clear advantage in clinical studies because patients frequently show quick fatigue and cannot perform long series of trials. On the other hand, as mentioned at the end of the previous section, quantifying ME and nME between pairs of data points to estimate stability of the performance variable rests on the assumption that the pairs of data points belong to a single distribution, an assumption that can be easily violated even in young healthy persons.

Of course, quantifying ME and nME in a single trial may lead to spurious results. A study estimating the number of trials one has to collect to reach a criterion of reliability in multi-finger force production tasks (de Freitas et al. 2018; Pawlowski et al. 2021) has shown that analysis of variance requires about twice the number of trials as compared to analysis of motor equivalence ($\approx$20 vs. $\approx$10 trials). This is a significant reduction in the number of trials, which may be important in using analysis of motor equivalence in clinical studies. So far, only one study has applied both methods of analysis to explore performance-stabilizing synergies in the same clinical population (Falaki et al. 2017a). In that study, patients with early-stage Parkinson disease performed cyclical accurate whole-body sway, and analysis was performed across the same phases measured in different sway cycles. Both analyses confirmed lower indices of synergies stabilizing the center of pressure coordinate, $\Delta V = (V_{UCM} - V_{ORT})/V_{TOT}$ and $I_{ME} = ME/nME$, in the patients as compared to age-matched controls. More on clinical applications of the analyses of synergies can be found in Chapter 22.

18.6. Missing pieces of the mosaic

Necessary and sufficient conditions for equifinality have not been formulated yet. It is clear that this phenomenon cannot be expected across tasks and effectors simply based on the idea of control with spatial referent coordinates (or the EP-hypothesis). This theory has room for equifinality under some additional conditions, including "non-intervention" by the actor, which remains notoriously hard to ascertain in experiments. Violations of equifinality seem to be very common due to unintentional drifts in the neural commands leading to unintentional drifts in task-level performance variables (see Chapter 16). Alternative theories (Are there any? Really?) have to assume exceptionally fast computational processes within the central nervous system to account for equifinality under transient force perturbations. Equifinality is on par with the posture-movement paradox as litmus tests for theories on the neural control of movement.

Motor equivalence has been known for a long time, but only recently has it started to be incorporated into toolboxes available for analysis of synergies. Its advantages and underlying assumptions remain far from being clear. This phenomenon needs to be explored in more detail across tasks, effector sets, spaces of analysis, and populations. Recent studies quantified motor equivalence in spaces of control variables or, to be more exact, of mechanical variables (referent coordinate and apparent stiffness) reflecting the two basic commands, the R- and C-commands (Ambike et al. 2016b; Nardon et al. 2022). They have confirmed that this method works in such spaces using broadly varying tasks, from finger force production to whole-body sway. These results are promising, but they represent relatively disjointed first steps toward developing an analysis of motor equivalence.

It is very tempting to develop the notions of equifinality and motor equivalence to non-motor tasks. For example, consider solving a complex mathematical problem. If your attention is distracted in the middle of the process, it is feasible that, after the "perturbation" (distracting episode) is over, you will use a different route to solve the problem. Ultimately, the problem will be solved (equifinality) but using different sequences of mental steps. I am also confident that, while writing this paragraph, I am using only one of many possible word and phrase combinations conveying the same gist. The concept of perceptually equivalent motion in the multidimensional spaces of afferent and efferent variables has been already mentioned (Chapters 15 and 17). Indeed, perceiving the same magnitude of a variable can be associated with variable contributions of the elemental variables; for example, the same joint position can be perceived under different magnitudes of muscle coactivation and/or different forces applied to an external object that cannot move. Under all these conditions, all the relevant afferent and efferent variables change without affecting the veridical percept of the joint angle—an example of perceptual equivalence. Tools to analyze perceptual equivalence, gist equivalence, and, more generally, outcome equivalence are waiting to be developed.

19
Muscle Coactivation

The presence of muscles acting against each other, the so-called *agonist-antagonist pairs*, is a necessity given that muscles can only contract actively (i.e., act in one direction only). So, to be able to produce active movements in both directions along a coordinate, linear or angular, at least two muscles are needed. If an action requires the generation of net force or moment of force in a particular direction, one of the two muscles serving this kinematic coordinate has to be activated. Activating the opposing muscle, the antagonist, at the same time seems not only unnecessary but also wasteful: To achieve the same resultant force, stronger activation of the agonist is required, which leads to spending a larger amount of metabolic energy. Nevertheless, simultaneous activation of opposing muscles, their *coactivation*, is a very common phenomenon. The apparent non-optimality of muscle coactivation, at least with respect to most commonly used cost functions in the field of motor control (see Chapter 9), makes this ubiquitous phenomenon puzzling.

In the following text, we will always assume that there is an agonist (a muscle or muscle group generating force in the direction of a desired action) and an antagonist (a muscle or a muscle group acting in the opposite direction). Since the resultant force has to be in the desired direction, agonist activation is typically higher, and the magnitude of coactivation is defined primarily by the antagonist activation level, which is commonly normalized by the agonist level activation. This makes the neural control of antagonist muscles most salient with respect to issues related to the neural control of coactivation.

An increase in quantitative indices reflecting muscle coactivation is seen across populations with motor impairments, from the healthy elderly to persons with atypical development and neurological patients (Aruin and Almeida 1996; Lee and Ashton-Miller 2011; Richards and Malouin 2013; Hirai et al. 2015; Kitatani et al. 2016). Compared to adults, increased coactivation is also seen in typically developing children, particularly during multijoint dynamical contractions (reviewed in Woods et al. 2023), and in young, healthy persons performing motor tasks in challenging conditions (Berger et al. 1992; Shiratori and Latash 2000; Krishnamoorthy et al. 2004). Interpretations of these phenomena so far have not been very successful and have been limited to straightforward mechanical effects of coactivation such as changes in the apparent stiffness of the effectors. As shown later in the chapter, these interpretations do not look convincing, although the underlying idea that coactivation is linked to the issue of dynamical stability of movements remains valid. This idea, however, has to be formulated and analyzed at the level of neural control, not peripheral mechanics.

As is true for any phenomena in the field of motor control, their fruitful analysis is possible only after one accepts a feasible theory on the neural control of movements. Within this chapter, similarly to the other chapters in this section, we will accept the theory of muscle control with time changes in the stretch reflex threshold, that is, the equilibrium-point (EP) hypothesis (Chapter 3) and its generalization as the theory of movement control with time-varying spatial referent coordinates (RCs) for the effectors (Chapter 7). This analysis is not going to be smooth because, as we will see, one cannot equate typical indices of muscle coactivation used in experimental studies and one of the two basic commands, the coactivation command (C-command), in spite of the apparent similarity of the terms.

Problems associated with muscle coactivation emerge also in the field of kinesthetic perception. In particular, as described in Chapters 15 and 17, voluntary changes in muscle coactivation can lead to large errors in force perception, and these errors can even be qualitative; that is, reported force changes can be in the opposite direction compared to the actual force changes. Why coactivation is not readily incorporated into force perception remains an unsolved problem, but we will try to suggest possible reasons for these phenomena.

19.1. Surprising behavior of antagonist muscles

The most common interpretations of coactivation have been based on classical mechanics and linked this phenomenon to changes in the effector apparent stiffness, damping, and natural frequency. A relaxed muscle resists externally imposed stretch similarly to a compliant spring (Ralston et al. 1947; Feldman 1966; reviewed in Zatsiorsky and Prilutsky 2012; Latash and Zatsiorsky 2016), and the force-length dependence can be expressed in a local linear approximation: $\Delta F = -k\Delta L$, where F stands for force, L for length, and k is a coefficient termed *apparent stiffness* (Latash and Zatsiorsky 1993). Activated muscles show steeper dependences of force on length (i.e., higher magnitudes of k). This increase in k can be seen in the absence of reflexes (e.g., in deafferented muscles). Intact muscles with reflexes show even higher values of k, and the whole $F(L)$ characteristic becomes more monotonic and linear (cf. thick solid and dashed lines in Fig. 19.1; see also Nichols and Houk 1976).

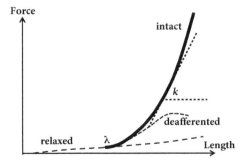

Figure 19.1 A schematic illustration of the force-length characteristics of a relaxed muscle (dashed line), muscle without reflexes (solid thin line), and muscle with reflexes (thick solid line). Local slope of the force-length characteristic (k) is the apparent stiffness of the intact muscle.

For a joint or another effector spanned by several muscles, agonists, and antagonists, apparent stiffness, in a linear approximation, is the sum of k values for the individual muscles. Hence, coactivating antagonist muscles increases k for the effector. This increase has direct implications for the ability of the effector to move at high speeds. Indeed, a linear mass-spring system is characterized by natural frequency: $\omega_0 = \sqrt{k/m}$, where m is mass (cf. Milner and Cloutier 1993). Fast movements are only possible in systems with sufficiently high natural frequency, ω_0, which requires high k. This predicts increased coactivation levels for very fast movements, confirmed in several experimental studies (Ghez and Gordon 1987; Corcos et al. 1989; Bennett et al. 1992). An increase in the inertial load (m) requires a proportional increase in k and, indeed, movements against increased inertial loads are characterized by higher indices of muscle coactivation (Gottlieb et al. 1989b).

Relations of increased apparent stiffness to stability are ambiguous. On the one hand, a system with higher k is expected to show a smaller deviation from a steady state caused by a small change in the external force. On the other hand, increasing k leads to a drop in the damping ratio: $\sigma = b/2\sqrt{mk}$, where b is damping, which can make the effector more underdamped and lead to longer-lasting oscillations following a brief external force perturbation—a sign of compromised stability. This problem is mitigated by an increase in the damping coefficient, b, which changes in parallel to an increase in k such that the damping ratio is kept nearly unchanged (Milner and Cloutier 1993; Milner 2002; Perreault et al. 2004; Lee and Ashton-Miller 2011; Heitman et al. 2012). Overall, this analysis has limited relation to behavior of actual biological effectors because of the inadequacy of second-order linear approximations (reviewed in Zatsiorsky and Prilutsky 2012; Feldman 2015). This is reflected, in particular, in the counterintuitive changes in postural stability with coactivation (see later in this section).

Arguably, the best-studied aspect of muscle coactivation is the so-called triphasic electromyographic (EMG) pattern observed over a variety of fast movements (reviewed in Gottlieb et al. 1989a). If a movement starts from a relaxed state, the first visible event is an EMG burst of the agonist muscle, accompanied by a relatively steady and modest coactivation of the antagonist (Fig. 19.2). Then, the antagonist muscle shows an EMG burst, while the agonist activation level drops. After the antagonist burst, there is a second burst of the agonist, and the effector stops in a new position where there is visible coactivation of the two muscles, which takes several seconds or even tens of seconds to subside. Interpretations of this pattern based on mechanics associated the first agonist burst with accelerating the effector, the antagonist burst with braking the movement, and the second agonist burst and terminal coactivation with stabilizing the final posture. The initial antagonist coactivation has been viewed as a means of increasing the natural frequency of the effector.

This interpretation assumes that the central nervous system can prescribe patterns of muscle activation—a highly questionable assumption discussed earlier (see Chapter 4). An alternative interpretation was offered within the EP-hypothesis, based on the velocity sensitivity of the primary endings in muscle spindles resulting in effects of muscle velocity on the reflex-mediated changes in muscle activation (Abdusamatov and Feldman 1986; reviewed in Feldman 2015). According to this interpretation, the triphasic pattern is a consequence of time shifts in two basic

19.1. SURPRISING BEHAVIOR OF ANTAGONIST MUSCLES   281

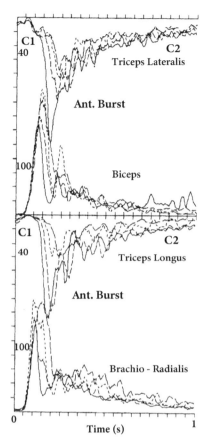

Figure 19.2 The typical triphasic pattern of muscle activation during fast elbow flexion movements over various distances has several phases with activation of the antagonist muscle (C1, early coactivation, antagonist burst; and C2, late coactivation). Modified by permission from Gottlieb et al. 1990a.

commands describing the control of a joint with one kinematic degree of freedom, the reciprocal and coactivation commands (R- and C-commands), and reflex-mediated effects of actual movement velocity on muscle activation. A change in the R-command was assumed to define the desired final position, and a change in the C-command played an important role in facilitating both effective movement acceleration and deceleration (described in detail in the next section).

Similar triphasic patterns of muscle activation have been described for quick isometric force production tasks (Ghez and Gordon 1987; Corcos et al. 1989). These studies documented large-amplitude antagonist bursts and large levels of coactivation at the final state. Note that the earlier interpretations of this pattern based on mechanics do not work. Indeed, since the effector acts in isometric conditions, no overt movement can take place, and it is not necessary to stabilize the joint trajectory. For the same reason, invoking velocity-dependent reflexes also does not help, because the velocity is close to zero. There is a possibility that these triphasic patterns reflect the fact that fast isometric force production is encountered very rarely during everyday life, while fast movements are performed commonly. As a result, the neural control patterns developed based on everyday actions are applied to produce similarly fast actions in relatively unusual, isometric conditions. This interpretation implies, however, that a non-monotonic change in the R-command is used leading to the triphasic EMG pattern across conditions (cf. Latash and Gottlieb 1991).

As of now, there are no reliable tools to measure in real time changes in the neural commands assumed within the idea of control with spatial RCs, which is equivalent to the control with $λ(t)$ for a muscle and with time-varying R- and C-commands for a joint (see Chapter 24). There were several attempts to reconstruct these hidden variables, and one series of studies reported the so-called N-shaped patterns of the equilibrium joint trajectories during fast movements, reflecting similar N-shaped patterns of the R-command (Latash and Gottlieb 1991, 1992; Latash 1992; Latash et al. 1999; Fig. 19.3). These studies have been criticized because of the likely underestimation of the damping coefficients in the mechanical model used to reconstruct equilibrium trajectories (Gribble et al. 1998). However, N-shaped patterns of the R-command have received support recently in another series of studies (Ramadan et al. 2022).

The N-shaped patterns of joint equilibrium trajectories were also reported in studies of movements of a joint in a two-joint kinematic chain involving the elbow and wrist (Fig. 19.4; Latash et al. 1999). Note that during fast movements of one of the joints, both joints show similarly timed triphasic EMG patterns (Koshland et al. 1991; Latash et al. 1995b; Shapiro et al. 1995). However, only one of the joints moves fast and allows invoking velocity-dependent reflex effects. In contrast, the other joint shows very small irregular deviations from the original angle, which can hardly be blamed for the consistent triphasic EMG pattern and suggests strongly that N-shaped control patterns are real. More importantly for our topic, N-shaped patterns of the R-command are readily compatible with the triphasic patterns across fast movements and isometric contractions.

As described in Chapter 13, stability of the vertical posture is a challenging and functionally very important task. When this task becomes even more challenging, as in cases of atypical development, various neurological disorders, and aging, one of the most commonly seen features is an increase in the coactivation levels of muscles acting at the main leg and trunk joints (Berger et al. 1992; Shiratori and Latash 2000; Krishnamoorthy et al. 2004). The most common interpretation of these observations has been based on consideration from mechanics: Higher coactivation was interpreted as an attempt to increase the apparent stiffness of the joints, k, and hence improve postural stability.

This interpretation is far from being obvious. First, increasing apparent stiffness about a joint in a multi-joint chain can improve stability of the whole chain under external force perturbations only if one of the ends of the chain is fixed to an external immovable object. Otherwise, higher stiffness will simply facilitate transfer of the perturbing forces to other segments and joints of the chain. During standing, the feet of the person are not glued to the ground. As a result, coactivating the muscles along the legs and trunk is unproductive. This is known very well to Tai Chi practitioners who

19.1. SURPRISING BEHAVIOR OF ANTAGONIST MUSCLES 283

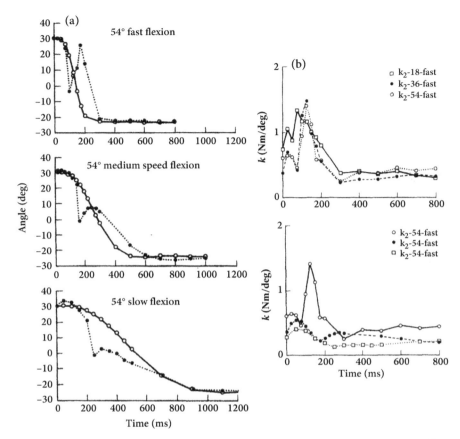

Figure 19.3 Changes in the equilibrium trajectory (a, dotted lines) and apparent stiffness (b) of the elbow joint reconstructed during elbow joint movements at various velocities. Note the N-shape equilibrium trajectory of the fastest movements and transient increase in the apparent stiffness. Modified by permission from Latash and Gottlieb 1992.

are aware that a slightly crouched, relaxed posture is more stable than an erect posture with high muscle coactivation.

The effects of voluntary muscle coactivation about the ankle joint on postural stability during standing have been explored in studies using several indices of stability (Yamagata et al. 2019a, 2021). An increase in muscle coactivation was shown to lead to faster spontaneous postural sway (both its rambling and trembling components; see Chapter 13), lower indices of multimuscle synergies stabilizing the trajectory of the center of pressure (estimated using the framework of the uncontrolled manifold hypothesis; see Chapter 5), and higher coordinate along the body where the ground reaction force vectors within a particular frequency range intersected (Boehm et al. 2019). All these findings point at detrimental effects of muscle coactivation on postural stability. So, why do people with impaired postural control coactivate their leg and trunk muscles? We will return to this question in a later section.

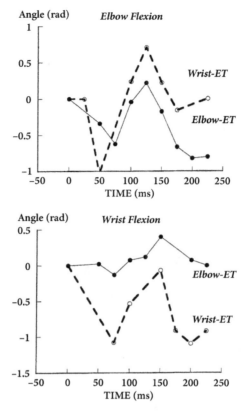

Figure 19.4 The non-monotonic (N-shaped) equilibrium trajectories (ETs) are seen in both wrist and elbow joint during voluntary movements of one of the joints. Top: During fast elbow movement. Bottom: During fast wrist movement. The non-instructed joint showed small, irreproducible deviations. Modified by permission from Latash et al 1999.

19.2. Features of the coactivation command

The concepts of reciprocal and coactivation commands were introduced within the framework of the EP-hypothesis to address the neural control of a joint with one kinematic degree of freedom (see Chapter 3). Later, these concepts have been generalized to the control of muscle groups contributing to the action of an effector along a spatial coordinate, linear or angular. The R-command defines the RC for the effector, that is, a coordinate where the effector is in equilibrium when the external load is zero. The C-command defines the spatial range where opposing muscles have non-zero activation levels. In terms of mechanics, it translates into apparent stiffness of the effector, that is, the slope of the force-coordinate characteristic. Recent studies have shown that patterns of recruitment of motor units (MUs) across an agonist-antagonist pair show grouping of the MUs into two robust groups suggesting direct reflections of the R- and C-commands (Madarshahian and Latash 2022a; Madarshahian et al. 2022;

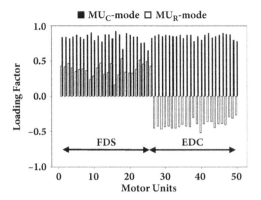

Figure 19.5 Two groups of motor units (MU-modes) identified with principal component analysis. One MU-mode (MU$_C$-mode) has the loading factors of the same sign for both flexor (agonist) and extensor (antagonist). The other MU-mode (MU$_R$-mode) has opposite signs of the loading factors for the two muscles. Modified by permission from Madarshahian and Latash 2022a.

see Chapter 10). Figure 19.5 illustrates the loading factors of individual MU firing frequencies for the two MU-modes. Note that one of the MU-modes is characterized by loading factors of the same sign for both flexor (flexor digitorum superficialis [FDS], agonist) and extensor (extensor digitorum communis [EDC], antagonist); that is, it reflects effects of the C-command. The other MU-mode shows opposite signs of the loading factors for the two muscles; that is, it reflects the R-command.

When a person coactivates muscles acting about a joint in an initial position, corresponding to a combination of commands $\{R_1; C_1\}$, and then makes a movement in isotonic conditions to another position, the level of muscle coactivation in the new position is about the same as in the initial one (Levin and Dimov 1997). So, a change in the R-command to perform the joint movement is associated with carrying the same C-command to a new spatial location (Fig. 19.6). These observations have been interpreted as reflecting a hierarchy of the two commands, with the R-command being hierarchically higher than the C-command. Studies of cyclical isometric contractions, however, have suggested that the C-command can be modulated with force more consistently than the R-command (see Chapter 11). Hence, the relation between the two commands, hierarchical or not, seems to be task specific.

During fast movements, the EMG patterns can be defined by both nonmonotonic changes in the R-command (cf. the aforementioned N-shaped equilibrium trajectories) and/or transient changes in the C-command. In particular, the well-documented transient increase in muscle coactivation during fast movements has been interpreted as a reflection of a transient change in the C-command leading to nontrivial consequences for the EMG phenomena. Figure 19.7 illustrates the control of a joint with two muscles, agonist and antagonist. To perform a fast movement, the R-command shifts to the new desired coordinate and initiates the generation of an agonist EMG burst. A simultaneous increase in the C-command (thick

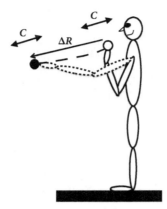

Figure 19.6 During a fast reaching movement, the reciprocal command (R-command) changes (ΔR) and defines the final coordinate, while the C-command is moved to the new spatial location with minimal changes.

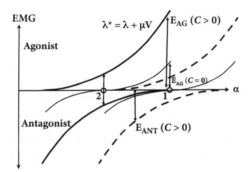

Figure 19.7 A transient increase in the C-command during a fast movement leads to both an increase in the first agonist EMG burst (E_{AG}) and an increase in the antagonist EMG burst (E_{ANT}). This is illustrated by the arrows showing the peak burst magnitudes without the change in the C-command (thin arrows) and with its increase (thick arrows).

solid lines) makes the agonist burst larger. (We assume that the actual movement lags the changes in the neural control variables for a number of reasons, including inertia.) Later in the movement, however, the agonist shortens while the antagonist stretches at high velocity. This leads to suppression of the agonist activation level and facilitation of a stronger antagonist EMG burst (dashed lines in Fig. 19.7). So, somewhat counterintuitively, an increase in the C-command contributes to both first agonist and delayed antagonist bursts.

This is not the only unusual feature of the C-command. Its effects on muscle activation depend strongly on the actual effector coordinate and external load. Figure 19.8A illustrates a situation when the C-command is positive but the effector is

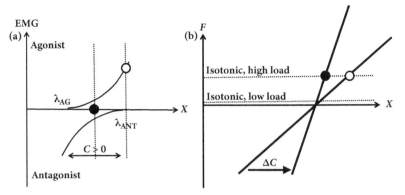

Figure 19.8 (a): Depending on the actual effector coordinate, the same value of the C-command can lead to different effects in the agonist and antagonist muscles. Only the agonist may show non-zero EMG (open circle) or both muscles can show comparable activation levels (black circle). (b): An increase in the C-command can lead to no effects on the effector position if it acts against very low external load. The same increase in the C-command leads to movement if the external load is non-zero.

at a coordinate where only one of the opposing muscles shows non-zero activation level. So, a relatively large C-command can be associated with no visible activation level in the antagonist muscles. Figure 19.8B shows that a change in the C-command produces no movement when the external load is zero, while it can produce displacement of the effector if the effector acts against a non-zero load. Moreover, if an effector acts in isometric conditions, changes in the C-command lead to effective changes in the force produced by the effector. This insight has been confirmed in the mentioned studies of finger force production tasks at the level of MUs across extrinsic flexor and extensor muscles (Madarshahian and Latash 2022a; Madarshahian et al. 2022).

19.3. Does negative coactivation exist?

According to the definition of the C-command, it can be negative (Fig. 19.9A). Indeed, if the RCs to the opposing muscles have no spatial zone where both agonist and antagonist muscles are active, formally this means $C < 0$. Note that, in such a situation, there is a spatial range where neither of the two muscles produces active force. Within that range, the behavior of the effector is defined only by passive force-coordinate properties of the tissues, including the relaxed muscles, tendons, and other tissues.

Consider an effector somewhere in the middle of that range. Small changes in the neural commands to the muscles will have no effect on the behavior of the effector. If the effector is outside the range of quiescence of one of the opposing muscles, shifting the neural control variable (λ) to only one of the muscles can affect its movement by

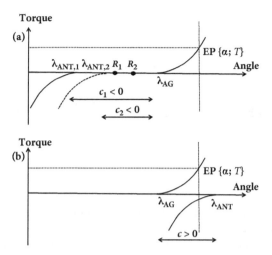

Figure 19.9 (a): An example of negative coactivation command (C-command). Changes in command (λ) to one of the muscles (λ_{ANT}) cannot change the state of the effector (e.g., its force in isometric conditions). (b): If the C-command is positive, changes in both λ_{AG} and λ_{ANT} can change the force in isometric conditions.

changing its active force. This makes the neural control of the effector degenerate. Normally, when $C > 0$, changes in two control variables, RCs for the agonist and antagonist, λ_{AG} and λ_{ANT}, can be used to move the effector or change its resultant force on the environment (Fig. 19.9B). Potentially, this makes room for covariation of those commands to stabilize desired performance (cf. Ambike et al. 2016a; see Chapter 7). When $C < 0$, only one of those commands has effects on the effector, if its coordinate corresponds to active force production by one of the opposing muscles. If the effector is in the quiescence zone, it is not actively controlled.

This analysis suggests that performance-stabilizing synergies at the level of neural control variables depend crucially on the availability of positive muscle coactivation. Formally, both R- and C-commands change when only one of the commands to the opposing muscles changes, λ_{AG} or λ_{ANT}. But, if the effector is outside the spatial range of the C-command or the C-command is zero, this abundance at the control level is only apparent, not real. To have a possibility to change the two basic commands independently of each other and, hence, to have abundance at the control level of these commands that can be used to stabilize a salient performance variable, the C-command has to be positive. For example, as illustrated in Fig. 19.10A, during isometric force production in the absence of coactivation ($C < 0$), force level can only be changed by changing λ_{AG}, and if this change is inaccurate, there is no chance of correcting the error since shifts in λ_{ANT} have no effects on the net force level.

This situation can also happen if the C-command is positive but the effector is outside the spatial range of muscle coactivation (Fig. 19.10B). If this happens, changes in only one of the two RC commands to the agonist and antagonist muscles can produce changes in the effector state. In such a situation, the neural control of the effector

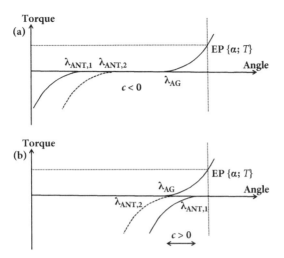

Figure 19.10 Examples of degenerate control during force production in isometric conditions. (a): When the coactivation command (C-command) is negative. (b): When the C-command is positive but a change in the command to the antagonist muscles (λ_{ANT}) cannot change the state of the effector.

becomes degenerate, and there is no room for covariation of λ_{AG} and λ_{ANT} or, equivalently, of the R- and C-commands.

We can conclude that muscle coactivation is a necessity for stable performance by any effector, a feature that may be more important from the evolutionary perspective than the net amount of metabolic energy spent on performing movements. This view on the functional role of muscle coactivation suggests that, in conditions of uncertainty, humans coactivate muscles (i.e., increase the C-command) to ensure that the effector never moves outside the coactivation zone. This could happen if an unexpected change in the conditions moves the effector outside the coactivation zone—a possibility if this zone is not large enough. If this happens, detrimental effects on performance stability can be expected because of entering the zone of degenerate control and the associated lack of performance-stabilizing synergies at the neural control level. A similar strategy of using increased magnitudes of the C-command can also be used by children in the process of motor development (reviewed in Woods et al. 2023), by healthy older adults (Lee and Ashton-Miller 2011; Nagai et al. 2011; Rozand et al. 2017), and by neurological patients with impaired postural control (Arias et al. 2012; Mari et al. 2014; Rinaldi et al. 2017).

19.4. Changes in the C-command and their (mis)perception

Changes in the C-command, voluntary or spontaneous, lead to a number of unexpected phenomena in the fields of both action production and perception (see Chapters 15 and 17). In particular, a voluntary increase in the C-command during constant force production tasks in isometric conditions leads to a substantial unintentional increase of force, which is not perceived by the actor (Cuadra et al. 2020). Moreover, subjects in such an experiment report that the force produced by the effector drops in spite of its increase, on average, by about 50%. To confuse the situation even more, when subjects are asked to match the force level with the contralateral effector, the actual force level is matched with an overshoot. Taken together, these phenomena suggest that effects of voluntary changes in the C-command are not adequately represented in kinesthetic perception, at least in its component involved in perception-to-report.

When subjects are asked to match force magnitude in finger pressing tasks with the contralateral homonymous effector, they do not show similar pairs of the R- and C-commands between the two effectors (Abolins et al. 2020; Cuadra et al. 2021a). Instead, they typically show smaller magnitudes of the C-command, as reflected in smaller apparent stiffness values, and larger magnitudes of the R-command, as reflected in the RC for the effector.

Similar conclusions on the inadequate reflections of changes in the C-command in kinesthetic perception have been reached in studies of unintentional force drifts toward lower magnitudes in the absence of visual feedback (see Chapters 16 and 17). Measuring mechanical reflections of the R- and C-command showed that the main cause of the force drop is the drop in the C-command, which is partly compensated by a change in the R-command (Reschechtko and Latash 2017; Fig. 19.11, left panel). A similar drop in the C-command was also observed when the subjects had continuous visual feedback on the force magnitude, which eliminated the force drift (see the right panel of Fig. 19.11). Under such conditions, adjustments of the R-command compensated perfectly the effects of the drifts in the C-command on force magnitude. Under all conditions, the subjects were unaware of force changes, even when these changes reached large magnitudes, up to one-third of the initial force level. The modest natural muscle coactivation levels in finger force production tasks

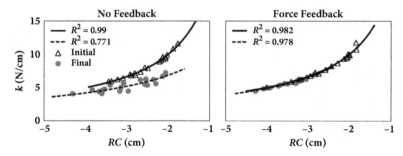

Figure 19.11 During continuous finger force production, there is a drop in the apparent stiffness of the effector (k) reflecting the C-command when the visual feedback on force magnitude was turned off (left panel) and when it was present (right panel). In the latter case, the decrease of k was nearly perfectly compensated for by changes in the referent coordinate (RC) of the effector reflecting the R-command. Reproduced by permission from Reschechtko and Latash 2017.

limit the range of possible C-command drifts and, respectively, the magnitude of the induced force drift.

The spontaneous drift in the C-command suggests its instability, which can be expected also to lead to faster and larger unintentional changes in this command in the presence of perturbing factors. Indeed, force drifts are accelerated if a finger involved in a constant force-producing task is lifted by the "inverse piano" device, causing an increase in its force (Wilhelm et al. 2013; Reschechtko et al. 2014). If visual feedback on the force magnitude is turned off, the initial force increase is followed by its relatively quick drop. The characteristic times of the force drop are on the order of 1 to 2 seconds, compared to 8- to 15-s exponents reported during spontaneous force drifts (see Chapter 16).

Instability of the C-command has also been invoked in studies of the effects of muscle vibration in force-matching tasks (Cuadra et al. 2021a; see Chapter 17 for more detail). The patterns of errors in force matching have suggested that vibration played the role of a nonspecific perturbing factor, and the less stable C-command showed a drift toward lower magnitudes, primarily defined by a drift in the RC to the antagonist muscles. This conclusion has been supported, in particular, by similar patterns of force-matching errors during the vibration applied to either flexor or extensor muscle groups.

19.5. Consequences of increased coactivation

Consequences of increased coactivation on motor performance can be classified into those leading to desired and undesired effects. The first group includes increasing apparent stiffness of effectors, thus facilitating their quicker transitions between steady states. Stabilization of the effector in the final state following a quick action may be viewed as another desired consequence of muscle coactivation, also mediated by an increase in the effector apparent stiffness.

A less obvious positive consequence of increased coactivation command is increased safety margin for ensuring performance stability by covarying changes in the two basic commands, the R- and C-commands. This may be particularly important for populations with impaired neural control of movements who commonly show increased muscle coactivation levels across tasks and effectors. Whether encouraging high coactivation levels in such populations is a good idea has to be considered in combination with the undesired effects of increased coactivation.

The list of undesired consequences of increased coactivation command includes the obvious increase in energy expenditure for the same net mechanical effects of muscle activation. Since energy is considered as a universal currency for the body (Yufik and Friston 2016), this may be an important factor to avoid excessive muscle coactivation levels. If coactivation levels reach large magnitudes, this may also lead to larger effects of fatigue (see Chapter 21). Larger coactivation command is also associated potentially with larger unintentional drifts in performance and larger errors in kinesthetic perception (see Chapters 15 to 17). These may be particularly important in tasks that involve object manipulation without continuous visual feedback. Last but maybe not least, there are documented detrimental effects of voluntary muscle coactivation on stability of the vertical posture (Yamagata et al. 2019a, 2021). Since increased coactivation is commonly seen in patients with balance impairment, these effects may be of particular clinical importance.

19.6. Missing pieces of the mosaic

There are several unexpected outcomes in the reviewed studies of muscle coactivation or, not completely synonymously, the *C*-command. One of them is the contrast between the idea of a control hierarchy, with the *R*-command being hierarchically higher than the *C*-command based on studies of voluntary movements (Levin and Dimov 1997), and the more consistent modulation of the *C*-command during force production tasks in isometric conditions. Note that purposeful force production in isometric conditions is not very common during everyday movements. Maybe the atypical control of this task in laboratory experiments is a consequence of the task being perceived as unusual. This explanation fits well the findings of large changes in muscle coactivation in populations with impaired motor control and in tasks perceived as challenging. The issue remains open, waiting for studies of the MU-modes and their modulation during more commonly performed movements.

The interpretation of the increased coactivation levels as reflections of ensuring action stability at the neural control level seen, in particular, under conditions perceived by the actor as challenging remains speculative. It is attractive because ensuring action stability is paramount during everyday functional movements and becomes even more important in persons with impaired motor control and in healthy persons moving in unusual, challenging conditions. So far, there have been no studies testing this hypothesis directly. On the other hand, alternative hypotheses based on changes in mechanical properties of peripheral structures seem inadequate.

Why is coactivation so poorly incorporated into force perception? This remains an open issue. First, it is necessary to find out if these phenomena can be seen in non-isometric conditions, that is, during voluntary movements. Kinesthetic perception of body kinematics in general, and joint configuration in particular, is usually robust: We can perform movements with acceptable accuracy without looking at the moving limbs at all times. If movements are unopposed, that is, we move without an external resistance, changing the *C*-command is not expected to have any effect on the effector steady state. The situation changes if movements are performed against a non-zero external load. Will changing muscle coactivation voluntarily lead to errors in contralateral homonymous effector position-matching tasks when both effectors move against non-zero external loads?

As is true for most issues discussed in this book, our current knowledge of neurophysiological mechanisms involved in the discussed phenomena is fragmented at best. We do not know what supraspinal structures generate descending signals reflecting the *C*-command. We also do not know how these signals interact with spinal circuitry and feedback loops from peripheral sensory endings to produce observed patterns of muscle coactivation. Mapping across the {R; C} command pairs at different levels of the assumed control hierarchy (see Fig. 7.3 in Chapter 7) remains another open issue. Do R-commands map on lower R-commands and do C-commands map on lower C-commands, or is there a more complex mapping of control RC vectors on lower control RC vectors? The latter option seems more likely and attractive, but so far, we have very little data addressing this question (cf. Ambike et al. 2015b; Reschechtko and Latash 2018).

PART V
IMPROVEMENTS AND IMPAIRMENTS

20
Improvements in Motor Performance

Improved motor performance occurs spontaneously during early development. Most commonly, it is described as an increase in the range of motor tasks the developing young persons can perform, in particular as related to functionally important activities such as reaching toward a target, standing, walking, and so forth. In grown-ups, improving motor performance is the main goal within such areas as learning motor skills required in one's occupation, athletic training, and motor rehabilitation of patients suffering from motor impairments. Sometimes the goal is not to improve motor performance but to avoid its deterioration expected from such factors as aging and progression of a disease. Across these areas, practice, which "makes better," is the most common tool.

In this chapter, we will focus on neural mechanisms of improved motor performance related to the control and coordination of movements, not on potentially very important other contributing factors such as increased muscle strength; optimization of energy resources and expenditure; improved states of joints, bones, and tendons; better functioning of the circulatory and respiratory systems; and thermoregulation. We will also not address topics that belong to the field of psychology, such as attention and motivation, which may require a different adequate language (see Chapter 1). It is not going to be a major surprise for the reader who has gotten this far that the discussion will be based on the theory of the neural control of movements with spatial referent coordinates (RCs; Chapters 3 and 7) and the principle of motor abundance (Chapter 5) linked to the concept of controlled stability of movements with the help of synergies (Chapter 6). Accepting this framework leads to reconsideration of some of the established views on skill acquisition and effects of practice. It also offers new directions of research with likely practical implications.

Most commonly, to achieve improved motor performance, practice is designed based on the accumulated experience summarized under the umbrella of "motor learning." Experience of the generations of practitioners is indeed a very powerful guiding factor, even if it is not based on a coherent theory on the neural control of movements. It has led to the emergence of influential schools in various areas including music performance, choreography, athletics, rehabilitations of neurological patients, and more. Even if many concepts used by practitioners are undefined and the prescribed practice regimes have little theoretical or neurophysiological validity, the general rules for motor learning show striking resilience.

In one of his very early studies, Bernstein (Bernstein and Popova 1930; see the English version in Kay et al. 2003) analyzed the mechanics of piano playing in top performers to check the predominant school of teaching music based on the so-called weight theory of playing, when students are encouraged to use the weight of the arms to produce sounds. He showed that gravity played a relatively minor role, only during soft sounds. The effects of this study on the practice of teaching music were, however, negligible. We see the main lesson from that study as follows: The goal is not to change established methods developed by the generations of practitioners but to understand why they are effective based on a valid theory of motor control and then, maybe, recommend theory-based modifications to these methods.

20.1. Bernstein's three stages

Arguably, one of the most influential schemes of motor skill development was introduced and developed by Bernstein (1947, 1996). This scheme considers building a motor skill as a staged process with three main stages. The first stage involves making the numerous elements participating in any action controllable by "freezing the redundant degrees of freedom" and alleviating the problem of motor redundancy. The second stage involves releasing ("freeing") the degrees of freedom and allowing them to be flexible in order to contribute to dynamical stability of action. The third stage involves using forces of interaction with the environment as contributors to the desired action. This three-stage scheme has been used as the framework in many studies of skill acquisition (e.g., Newell 1991; Vereijken et al. 1992; van Ginneken et al. 2018; Guimarães et al. 2020; Gray 2020), and it remains a commonly used framework for applied motor learning studies.

With all due respect to this scheme, we have to admit that the first two stages are formulated based on the not very well-defined notion of degrees of freedom. As described earlier (see Chapter 5), counting degrees of freedom in the most commonly used spaces, kinematic, kinetic, and electromyographic, may not be very informative. In particular, quite commonly, learning a motor skill is associated with an increase in the level of agonist-antagonist coactivation across muscle pairs during early practice, and the level of coactivation subsides during later stages when performance improves (reviewed in Ford et al. 2008; Brueckner et al. 2018). Interpreting this as freezing and freeing degrees of freedom can work at the level of joint kinematics. We should mention, parenthetically, that, even at the joint kinematic level, the number of degrees of freedom does not change; only their peak-to-peak excursions do. However, at the level of muscle activation and motor unit recruitment, the number of degrees of freedom obviously shows opposite trends: an increase during early stages of practice and a drop during later stages. To analyze what happens with practice with the variables manipulated by the central nervous system, such variables have to be defined explicitly. As has been emphasized many times (see Chapters 2, 4, and 5), the central nervous system cannot in principle prescribe patterns of peripheral variables recorded and analyzed in typical movement studies (kinetic, kinematic, and electromyographic). Changes in all these variables emerge in the process of dynamical interaction between the neural control process, interactions among elements involved in implementing the movement, and external force field (reviewed in Bernstein 1947; Latash 2007, 2019; Feldman 2015).

We have accepted the theory of hierarchical control with spatial RCs for the effectors, from the whole body to individual muscles and motor units (see Fig. 20.1 and Chapters 3, 7, and 11). RCs have a neurophysiological interpretation as subthreshold depolarization of a corresponding pool of neurons, down to individual

20.1. BERNSTEIN'S THREE STAGES 299

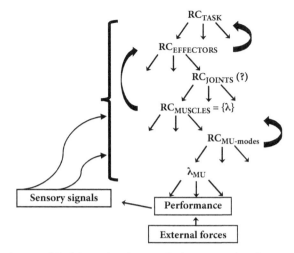

Figure 20.1 A scheme of the hierarchical control of action with referent coordinates (RCs) defined at different levels of analysis (effectors). MU, motor unit; λ, threshold of the stretch reflex.

alpha-motoneurons for the control of motor units and pools of alpha-motoneurons for the control of muscles. The question remains: At what level of the hierarchy should RCs be counted to reflect degrees of freedom? If one constructs a skill starting from scratch, the top, task-level, control is unavailable, and the skill has to be built using RCs (or, equivalently, pairs of the reciprocal and coactivation commands [R- and C-commands]; see Chapters 3 and 7) to the involved effectors at the highest available level. For example, to learn a skilled bimanual action, one may have to start with controlling each extremity individually, that is, prescribing time patterns of the R- and C-commands to the left and right arm independently. These may also be unavailable, for example, when just starting to learn how to play piano. Then, movement of each digit has to be controlled individually with a corresponding increase in the number of {R; C} pairs prescribed at the highest control level. So, we conclude that, during early stages of skill acquisition, the person has to define time profiles of a relatively large number of control variables at the task-specific level of control.

Further, these neural control variables have to project to RCs for the hierarchically lower sets of elements (as in Fig. 20.1). Such projections may be available for not very novel actions specified at the task level or may have to be developed in the process of practice (see later and reviews in Latash 2010b, 2019). A major purpose of the few-to-many mappings in Fig. 20.1 is stability of salient variables encoded by the corresponding hierarchically higher levels, that is, the development of corresponding synergies (Chapter 6). Changes in these mappings with skill development can be related, in particular, to the aforementioned trade-offs between synergies at different levels of the hierarchy (Gorniak et al. 2007, 2009; see Chapter 6). As discussed later (in section 20.3), these trade-offs can create an impression of freezing and freeing degrees of freedom at different stages of skill development.

20.2. Can the way our brain controls movements be changed?

So far, there has been no evidence that practice can lead to a change in the basic mechanism of movement control, that is, the control with spatial RCs (see Chapter 7). Observations of movements in deafferented animals and in persons with a rare disorder addressed as large-fiber peripheral neuropathy (Polit and Bizzi 1978; Bizzi et al. 1982; Rothwell et al. 1982a; Sainburg et al. 1993) suggest, however, that the principle of parametric control can be implemented in a different way if the stretch reflex loop is destroyed. After deafferentation, the only way to produce activation of alpha-motoneurons is to send a suprathreshold presynaptic input to the pool. This is illustrated in Fig. 20.2B. Note that if a suprathreshold input is maintained for a period of time, the target neuron will generate action potentials at the highest possible frequency defined, primarily, by the duration of the membrane refractory period. In the absence of reflex input reflecting the state of the muscle (cf. Fig. 3.6 in Chapter 3), the only way to make the neuron controllable and produce lower firing frequencies is to modify the time profile of the descending signal appropriately to ensure that the membrane threshold is reached at desired time intervals.

After some practice, animals after deafferentation learn this new method of controlling movements based on specifying muscle activation levels. This method has been addressed as the alpha-model (Polit and Bizzi 1978; Bizzi et al. 1982), in contrast

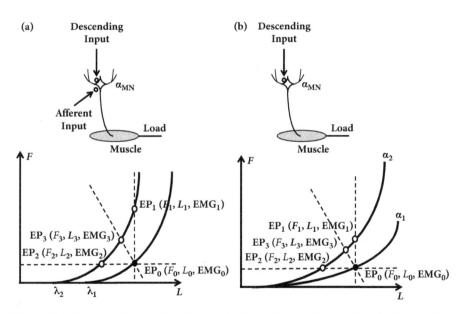

Figure 20.2 Control of a muscle with intact reflex pathways (a) and after deafferentation (b). Different sets of force-length characteristics are defined by the central nervous system (lower graphs). Note that, in the presence of reflexes, muscle activation (EMG) is different for conditions with different external load characteristics. Magnitudes of variables are reflected in the font size.

20.2. CAN THE WAY OUR BRAIN CONTROLS MOVEMENTS BE CHANGED?

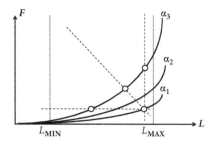

Figure 20.3 Within the α-model, a shift of the neural signal to an α-motoneuronal pool from α_1 to α_2 and to α_3 changes the shape of the muscle force-length characteristic and leads to different consequences (different equilibrium points shown with white circles) depending on the external load characteristic (dashed lines). Note that the muscle activation level is constant within the biomechanical range of muscle length values, from L_{MIN} to L_{MAX}.

to the lambda-model (Chapter 3; Fig. 20.2A). Note that specifying muscle activation level does not prescribe muscle force, which continues to show length and velocity dependence. The lack of reflexes makes the force-length muscle characteristic less steep and less linear (Nichols and Houk 1976). Nevertheless, the availability of such characteristics preserves some features of the equilibrium-point control (see Chapter 3 and Latash 1993). In particular, shifting the level of activity of a pool of alpha-motoneurons to another level (cf. α_1 and α_2 in Fig. 20.3) produces different mechanical consequences depending on the external load characteristics. In cases of transient changes in the external load, equifinality is expected because the final state of the system is defined only by the α-command and final external load values. A series of experiments on deafferented monkeys confirmed these main features of equilibrium-point control (Polit and Bizzi 1978, 1979).

Both alpha and lambda versions of equilibrium-point control belong to the class of parametric control. However, the nature of the parameters makes these two modes of control very different. By its nature, lambda is a spatial RC for the muscle, and changing lambda has the meaning of (indirect!) encoding of a change in the muscle spatial property (i.e., length). Depending on the external load properties, a change in lambda can lead to changes in both muscle length and force. But the principle of control is rooted in the notion of movement as a change in spatial coordinate. In contrast, alpha has no meaning related to ecologically valid variables. This may be one of the reasons that movements after the loss of reflexes are awkward, deliberate, and poorly coordinated. Unfortunately, a number of reviews did not distinguish between the two versions of equilibrium-point control and discussed the alpha-model as the main version of the equilibrium-point hypothesis (e.g., Shadmehr and Wise 2005).

Nevertheless, the alternative method of alpha-control can be learned as shown by observations of patients suffering from large-fiber peripheral neuropathy. Learning to perform functional movements takes time in these persons, and movements can only be performed under continuous visual control (Cole and Paillard 1995; Miall et al. 2018). There is major impairment in joint coordination (Sainburg et al. 1993), particularly pronounced during fast movements. Still, the central nervous system finds a way to switch to this alternative version of parametric control.

20.3. Changes in motor synergies with practice

Imagine that a person is asked to practice a motor task requiring accurate production of a performance variable with an abundant set of elements contributing to the variable (which is true for any movement; see Chapter 5). What can possibly happen with synergies stabilizing this variable in the process of practice? Figure 20.4 illustrates three possible scenarios using, as a cartoon illustration, the task to produce a value of total force while pressing with two effectors. Imagine that, prior to practice, there was already a total force stabilizing synergy as reflected in the inequality $V_{UCM} > V_{ORT}$, both variance indices computed over a sequence of trials (panel A in Fig. 20.4; see Chapter 6). Practice is expected to lead to more accurate performance, that is, to lower V_{ORT}. Changes in V_{UCM} are, however, not dictated by the explicit task since, by definition, this index reflects the variance component with no effect on the salient performance variable.

Panels B, C, and D of Fig. 20.4 illustrate three scenarios of change in V_{UCM}. First (panel B), V_{UCM} can stay unchanged or even increase, leading to an increase in the difference ($V_{UCM} - V_{ORT}$) and in the ratio (V_{UCM}/V_{ORT}), which can be interpreted as strengthening of the synergy, an increase in its index ΔV. Second (panel C), V_{UCM} can decrease proportionally to the decrease in V_{ORT}, which can be viewed as more accurate

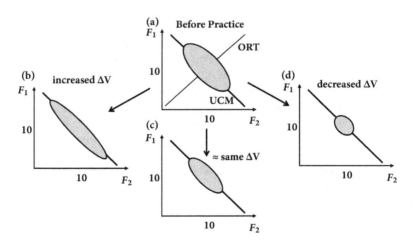

Figure 20.4 Possible changes in the structure of inter-trial variance in the space of two elemental variables, F_1 and F_2, in the task of producing the sum of the variables: PV = ($F_1 + F_2$). Data distribution before practice is shown in the top panel (a). (b), (c), (d): Practice is expected to lead to more accurate performance reflected in smaller variance (V_{ORT}) orthogonal to the uncontrolled manifold (UCM). Changes in variance along the UCM (V_{UCM}) with practice can be proportional, smaller or larger than the change in V_{ORT}.

performance without a change in the synergy. Third (panel D), V_{UCM} can decrease even more than V_{ORT}, which formally means that the synergy becomes weaker.

All three scenarios illustrated in Fig. 20.4 have received experimental support (reviewed in Latash 2010b; Wu and Latash 2014). When the task was relatively easy, practice led to either unchanged synergies (Domkin et al. 2005) or even a drop in the synergy index (Domkin et al. 2002). An increase in the synergy index was observed at early stages of practicing a novel task (Kang et al. 2004; Latash et al. 2003), which could lead to an increase in V_{UCM} associated with a drop in V_{ORT} (Wu et al. 2013). In the latter group of studies, since V_{UCM} was always significantly larger than V_{ORT}, total variance in the space of elemental variables increased, while variance of the salient performance variable dropped—a nontrivial outcome. Note that increasing variance at the level of elements has been associated with more accurate performance in a later study of synergies in the lower-body joint configuration space stabilizing the mediolateral foot trajectory during walking (Rosenblatt et al. 2014).

Taken together, these observations suggest two stages in the effects of practice. The first stage is associated with the creation and strengthening of synergies stabilizing the salient performance variable. This is likely associated with the emergence of mappings from RCs at the task level to abundant sets of RCs at the level of contributing effectors with the associated back-coupling loops (see Fig. 20.1, as well as Fig. 6.10 in Chapter 6). After the performance becomes acceptable in terms of its accuracy, further practice may be unable to reduce V_{ORT} even more. As a result, during the second stage, the actor optimizes not accuracy in the explicit task but other characteristics of performance (e.g., those related to energy expenditure, comfort, fatigue, etc.). This optimization process takes place within the uncontrolled manifold (UCM) and, naturally, leads to a reduction in the range of acceptable solutions, that is, a drop in V_{UCM}. Note that optimization and stability of performance are in competition (see Chapter 9), resulting in the counterintuitive observations of a reduction in the synergy index with continuing practice (Latash et al. 2003; Wu and Latash 2014).

There is one more aspect of possible effects of practice on synergies related to the idea of a trade-off between synergies at different hierarchical levels (Chapter 6). Imagine that a synergy stabilizes a particular performance variable, which results in its very low variability (Fig. 20.5). Another synergy stabilizes another performance variable. Now, consider a task of covarying the two performance variables to achieve a desired outcome. This may not be possible (see the right panel in Fig. 20.5) because the small variance in each of the variables does not allow large V_{UCM}, leading to stereotypical—and, therefore, not dynamically stable—performance. For example, imagine a multi-joint synergy stabilizing the direction of movement of a handheld object (e.g., a basketball) and another synergy stabilizing its speed. Accurate throw requires covariation of the direction and velocity at release. To ensure such a covariation, sufficient variability in both speed and direction is needed. As a result, synergies stabilizing each of the two performance variables may be attenuated to afford increased variance of those variables (as documented in Hasanbarani and Latash 2020).

Increasing variability at the level of elements to improve accuracy of performance is not unique to the mentioned study of basketball throw. It has been reported in a study of lower-body joint configuration synergies during walking (Rosenblatt et al.

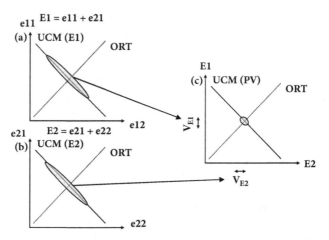

Figure 20.5 If two variables, E1 and E2, are strongly stabilized by synergies at the level of their elements (e11 and e12 for E1, and e21 and e22 for E2, a and b), their inter-trial variance is expected to be low. This creates a problem with organization of covariation of these variables to stabilize a hierarchically higher variable, (E1 + E2) (c).

2014) as well as in a study of practicing a multi-finger task in conditions that explicitly challenged stability of performance (Wu et al. 2013; Wu and Latash 2014).

20.4. Variability vs. stereotypy

A large number of studies have provided evidence in favor of performing trials in variable conditions (variable practice) as compared to performing large numbers of trials in a single condition (blocked practice) (Welsh and Elliott 2000; Keogh and Hume 2012; Herzog et al. 2022). Variable practice allows the actor to experience relatively large variations in the external forces and intrinsic body states as compared to practice in stereotypical conditions. Within the concept of performance-stabilizing synergies, the obvious benefit of variable practice is being exposed to a larger range of destabilizing factors, which contribute to the development and strengthening of corresponding synergies. The benefits of performance in destabilizing conditions have been documented in experimental studies in conditions when the task-related performance variable was purposefully destabilized using specially designed visual feedback (Wu et al. 2012, 2013), which made the task akin to walking along the ridge with steeper and steeper slopes. Under such conditions, a single 40-minute practice session led to a significant improvement in performance and an increase in the synergy index associated primarily with an increase in V_{UCM}. Similar effects were observed in a study of the effects of practice in persons with Down syndrome where a subgroup of subjects who were allowed to perform various tasks in the setup improved stability of their performance more than the other subgroup who practiced only the main task for the same total time (Latash et al. 2002a).

Note that stereotypical performance of a task may be viewed as more optimal with respect to some cost function because deviations from a single solution are minimal (see Chapter 9). Under predictable and reproducible laboratory conditions, stereotypy may be preferred. In more natural conditions, variable solutions are preferred because they allow stabilizing salient performance variables under the unavoidable variations in the external forces and intrinsic body states. This task- and condition-dependent trade-off between optimality (as reflected in stereotypy) and stability (as reflected in structured variability) may be the reason for different outcomes in studies of the effects of practice (reviewed in Latash 2023).

The importance of high variability in spaces of elemental variables confined primarily to the UCM for the salient variable has been demonstrated in studies of synergies in different populations, from professional golfers to neurological patients (see also Chapter 22). In particular, a study of golfers (Morrison et al. 2016) showed proportionally higher amounts of V_{UCM} in the total variance in the joint configuration space in higher-skilled golfers. Several studies of patients with Parkinson disease and multiple sclerosis have shown that the reduced synergy indices are primarily related to reduced amounts of V_{UCM}, not increased V_{ORT} (Jo et al. 2015; Falaki et al. 2016). Difficulties in adjusting V_{UCM} to the task are also reflected in the changes in the two variance components in preparation of a quick action, that is, during the anticipatory synergy adjustments (ASAs; see Chapter 8). The reduced ASAs in the same groups of patients were primarily due to the smaller change in V_{UCM} as compared to the control groups.

Benefits of variability have also been confirmed in studies performed outside the laboratory, in studies of professional butchers during their everyday labor movements (Madeleine 2003, 2008). In those studies, performed in a chicken-processing factory, the butchers had to perform very quick and accurate movements with a very sharp knife, which obviously imposed strong constraints on the allowed variability in the knife trajectory. Two findings are directly related to the importance of V_{UCM}. First, experienced butchers showed larger variance in the arm joint configuration space computed over repeated cuts compared to novices. Second, low variability in the same space was a predictive factor for the development of chronic pain. Overall, these studies illustrate the benefits of structured motor variability and emphasize the importance of its component along the UCM, which has no effects on salient task-specific variables.

20.5. Developmental changes

Mechanisms involved in early motor development are mainly unknown. There are some obvious contributors to changes in the way movements are controlled and performed. These involve continuing myelination of a number of relevant neural tracts involved in the sensory and motor functions and an increase in muscle mass and strength. However, neurophysiological mechanisms involved in the processes of development of hierarchical control with spatial RCs (Chapters 3 and 7) and the emergence of performance-stabilizing synergies (Chapter 6) are all but unknown.

Studies of walking-like leg movements across populations of different age, starting with newborns, have documented progressive development of muscle grouping in the lower part of the body (Dominici et al. 2011). Newborns can show cyclical leg movements when supported in the air over the treadmill. Their muscles form two major groups with parallel scaling of activation levels that resemble the first two of the four or five groups observed in the grown-ups during walking (cf. Ivanenko et al. 2013). The cited studies used the method of non-negative matrix factorization to identify the muscle groups and documented progressive emergence of the third and fourth groups in toddlers and preschoolers. Within the scheme of hierarchical control with RCs, these results illustrate the emergence of mapping from the task-level control expressed in RC(t) functions for the body in space to abundant sets of lower-level RC(t) functions to the involved effectors. Such mapping has to facilitate stable locomotor behavior given the variable and unpredictable reaction forces from the environment on the body, that is, develop synergies stabilizing salient performance variables.

Changes in synergies with development were studied in children who were able to perform accurate multi-finger force production tasks, starting from 4 years of age (Shaklai et al. 2017). This study used the framework of the UCM hypothesis and demonstrated a progressive, close-to-linear increase in the synergy index with age associated with a drop in the component of inter-trial variance affecting the salient performance variable (V_{ORT}) without a comparable change in the other variance component (V_{UCM}). An opposite finding was reported in a study of reaching in children aged 6 to 10 years old (Golenia et al. 2018): A drop in both variance components was found with a steeper drop in V_{UCM} (addressed as goal-equivalent variance). It is possible that developmental changes in performance-stabilizing synergies are task specific and show different changes in the two variance components in kinetic and kinematic tasks.

20.6. Motor rehabilitation: From magic to theory-based approaches

At this time, most approaches to motor rehabilitation are based much more on the accumulated experience than on theories on the control and coordination of movements. As a result, neurophysiological processes in the brain are directly compared to observed performance characteristics as if there were invisible wires from the brain to the effectors sending signals encoding desired movement mechanics and/or muscle activation patterns. The inadequacy of this approach has already been discussed (see Chapters 2 through 5).

This situation leads to predominance of ill-defined and frequently misleading concepts such as "upper motoneuron" and "muscle tone." The former term is frequently applied to cortical neurons with the axons forming the corticospinal tract. It implies that those neurons define activation patterns of "lower motoneurons" (alpha-motoneurons) in the spinal cord, which is, generally speaking, false. The corticospinal tract makes most projections not on alpha-motoneurons but on interneurons, thus mediating the reflex effects from sensory endings on alpha-motoneurons. These reflex effects play a major role in the activation of alpha-motoneuronal pools, as discussed in Chapter 4.

The concept of muscle tone was introduced by Bernstein as a relaxed muscle state reflecting its preparation to future action (Bernstein 1947; Meijer et al. 2001). The current practice of quantifying muscle tone involves asking the person to relax and moving examined joints passively over their respective ranges of motion. The examiners gauge the resistance of the moved joints based on their experience. So, the examined person is not asked to prepare for any future action, in obvious contrast to Bernstein's definition. Indeed, consider Fig. 20.6, which shows schematically the biomechanical range of muscle length (from L_{MIN} to L_{MAX}) for a muscle spanning a joint, which is being moved by the examiner. If the joint is at some starting position (L_0) corresponding to a relatively short muscle, relaxing at this position implies moving the value of λ to the muscle to the right of L_0. The exact value of λ is undefined. As a result, there is an infinite number of ways to comply with the imprecise instruction "to relax." Figure 20.6 illustrates three hypothetical healthy subjects who interpreted the instruction differently and established three different magnitudes of λ: λ_1, λ_2, and λ_3. When the examiner moved the joint passively to a new location corresponding to muscle length L_1, the three subjects showed very different magnitudes of active resisting muscle force, from close to zero (for λ_3) to a rather large magnitude (for λ_1). Does this mean that subject λ_1 has a neurological condition characterized by increased muscle tone, subject λ_2 has "normal" muscle tone, and subject λ_3 is hypotonic? Of course not. All three subjects are healthy and simply interpreted the vague instruction differently. Note that deep relaxation (illustrated with λ_3 in Fig. 20.6) is an art that requires practice, as is well known to professional athletes and masseurs.

A number of the approaches to motor rehabilitation have been based on a well-established relation between the amount of practice and functional outcome. Arguably, one of the best-known examples is the constraint-induced therapy (CIT; reviewed in Blanton et al. 2008; Kwakkel et al. 2015; Yang et al. 2023) applied in hemiparetic syndromes, commonly following stroke. The idea is relatively straightforward: to limit involvement of the less affected (ipsilesional) extremity and thus force the patient to use the more affected (contralesional) extremity during everyday

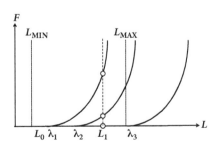

Figure 20.6 An illustration of the typical measurement procedure for "muscle tone." The force-length muscle characteristics are shown for three hypothetical subjects relaxed at the initial state (L_0). Moving the joint to a new position (L_1) leads to different resistive forces (corresponding to the equilibrium points shown with white circles) depending on the initial value of $\lambda > L_0$ (three values of λ are shown as λ_1, λ_2, and λ_3). The biomechanical range of muscle length is shown with vertical dashed lines marked L_{MIN} and L_{MAX}.

activities. When applied to the upper extremities, this approach involves putting a mitten on the less affected hand, forcing the other hand to do more than its fair share in hand actions. A number of studies provided evidence in favor of using CIT (see the referenced reviews), but other studies failed to document its advantage over using more traditional therapy (reviewed in Dromerick et al. 2006; Corbetta et al. 2015). Arguments against CIT involved, in particular, that bilateral movements of the two upper limbs are commonly used during training and in everyday actions, which become less common when the less impaired hand is constrained.

The CIT approach has been developed to lower extremities in the form of discomfort-induced therapy (Aruin and Kanekar 2013; Aruin and Rao 2018). After stroke, one of the challenges is to train the more impaired leg during everyday activities such as standing and walking. Patients commonly place more than 50% of the body weight on the less impaired leg and spare the more impaired one. This limits its training during natural everyday activities and potentially slows down the recovery. If one puts an insole into the shoe for the less impaired foot with sharp protrusions facing upward, placing much weight on that foot becomes uncomfortable, and the patient shifts the weight toward the other, more impaired, foot. Adjusting the insole can lead to controlled redistribution of the body weight between the two feet to optimize the involvement of the more impaired leg during the everyday activities.

Recently, an important development in the field of motor rehabilitation after stroke has been based on the theory of motor control with shifts in the spatial RCs (reviewed in Feldman 2015; see Chapters 3 and 7). This approach is discussed in more detail in Chapter 22. Briefly, it assumes that a major problem in the control of a muscle after stroke is the limited ability of the patient to shift the stretch reflex threshold (λ, analogous to RC at the single-muscle level) over its whole range seen in healthy persons. This problem is illustrated in Fig. 20.7, which illustrates the healthy range of shifts in λ (from λ_{MIN} to λ_{MAX}), which goes beyond the biomechanical range of muscle length

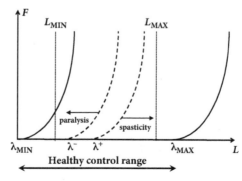

Figure 20.7 An illustration of typical consequences of cortical stroke with the theory of control with referent coordinates. A: The range of shifts of the stretch reflex threshold λ in a healthy person (from λ_{MIN} to λ_{MAX}) is broader than the biomechanical muscle range, L_{MIN} to L_{MAX}. B: The range of λ shifts is reduced (from λ^- to λ^+), leading to both paralysis within a range of muscle length changes and uncontrolled contractions (spasticity) in another range of length changes.

(from L_{MIN} to L_{MAX}). Using this range allows the person to relax the muscle even when it is long and to produce large force values even when it is short. After stroke, this range is reduced (from λ^- to λ^+), leading to an inability to relax the muscle when its length is beyond a certain value and an inability to activate it when it is too short. However, the muscle can be controlled voluntarily within the range $\{\lambda^-; \lambda^+\}$. Defining this range and training the patients to perform movements within the range can be beneficial for the recovery process, as shown in several recent studies (Turpin et al. 2017; Subramanian et al. 2018; Piscitelli et al. 2020; Frenkel-Toledo et al. 2021).

20.7. Missing pieces of the mosaic

The most obvious gaps in the current knowledge of mechanisms involved in movement improvements are related to the paucity of studies exploring the effects of practice using feasible theories of motor control and coordination (reviewed in Latash 2010b, 2019). Very few studies have explored approaches based on the theory of control with RCs for rehabilitation of neurological patients, and even fewer studies have explored the effects of training of healthy persons within the framework of this theory. It remains largely unknown what happens with practice with patterns of the *R*- and *C*-commands during everyday movements and whether these patterns can be improved in the process of motor rehabilitation. In contrast, changes in performance-stabilizing synergies with training have been studied in healthy persons but not yet translated into clinical research. It remains unknown whether the reported improvement in synergies during practice under controlled instability can be observed in patients with movement disorders and whether such improvements transfer to functional everyday movements.

The neurophysiological mechanisms involved in improvement of motor performance remain all but unknown. How can the range of voluntary λ changes be increased (cf., Piscitelli et al. 2020)? Can these processes involve substitution of effects of the injured descending pathways with other pathways, in particular those using uncrossed fibers from the contralateral hemisphere or from subcortical structures? How can an abundant mapping from task-specific RC functions to hierarchically lower RC functions be modified to bring about improved performance-stabilizing synergies? Can these processes involve changes in the gains of back-coupling loops (cf., Latash et al. 2005; Martin et al. 2009, 2019)? What pathways are used to modify those gains? These basic questions remain without a clear answer.

Given the inherent trade-off between stability and optimality of performance (see Latash 2023 and Chapter 9), what feature should be targeted by coaches and therapists in designing training schedules? Do such strategies depend on the desired outcome? For example, should rehabilitation approaches differ between mildly affected persons who want to recover the ability to perform everyday movements in the naturally variable environment and those who want to be able to play a musical instrument or to perform another highly accurate and sophisticated action in a much more predictable environment? The answer is probably yes. However, there are no theory-based studies exploring these questions. To conclude, the puzzle has many more pieces missing than are available.

21
Decline in Motor Performance

In this chapter, we will consider two common causes for impaired motor performance experienced by most people. One of them—muscle fatigue—leads to a transient decline in performance, commonly limited to specific groups of muscles and effectors, which recovers after a period of time. The other—aging—causes long-lasting and largely irreversible changes in the body, resulting in impaired performance across groups of tasks and many, if not all, effectors. We will assume that both causes happen in otherwise healthy persons, that is, in the absence of diagnosed musculoskeletal and neurological (including cognitive) disorders. As a result, we will not discuss chronic fatigue (reviewed in Farrar et al. 1995; Johnson et al. 1999; Nijs et al. 2011) and increased fatigability in neurological conditions, for example, in multiple sclerosis (MS) (Patejdl and Zettl 2022; Royer et al. 2022).

Definitions of fatigue include a drop in maximal voluntary force production by an effector (e.g., a muscle, a joint, a limb, or the whole body) in a standardized task after a period of activity by the effector, an inability to continue performing at a required level after a period of time, a shift in the spectrum of the electromyogram (EMG) toward lower frequencies, and an increase in the subjective estimation of task difficulty following an exercise (reviewed in Enoka and Stuart 1992; Vøllestad 1997; Enoka and Duchateau 2008). We are not going to contribute to discussions about different definitions of fatigue, and we assume that the reader knows what fatigue is from first-hand experience. Tools to quantify fatigue will be mentioned with respect to specific studies.

Aging is also a familiar condition, at least to those lucky to live long enough. Some impairment in the ability to perform at a high level, for example, in sports, starts relatively early in human life, between the ages of 30 and 40 years, and may be viewed as the first signs of aging-related changes in the body. We will, however, focus on possible mechanisms of impaired performance in persons who are considered "older adults" in most studies, that is, those aged 70 years or more. Many of these persons show changes across many functions of the body that may be viewed as borderline pathological. These changes span a wide range of functions, autonomic, sensory, and cognitive, that can contribute to the observed changes in the performance of motor tasks. Such persons are typically included in studies of "healthy aging" as long as they are free of diagnosed conditions that can, by themselves, affect performance in the studied groups of motor tasks.

We will not discuss here impaired motor performance by patients with diagnosed neurological conditions. Some of this material is going to be covered in Chapter 22. Note that some of the relatively common neurological conditions (e.g., Parkinson disease and stroke) are seen primarily in older persons, and some of the conditions are associated with increased fatigability in tasks where healthy persons do not report fatigue. So, material covered in this chapter may be viewed as forming the foundation for analysis of changes in motor control mechanisms in persons with movement disorders. As in earlier chapters, we will try to interpret the reported changes in aspects of performance and in certain neurophysiological structures and circuits within the general scheme of the neural control of movements and mechanisms ensuring dynamical stability of movements in the natural environment (see Chapters 6 and 7).

21.1. Fatigue: Peripheral and central effects

Fatigue is a multifaceted phenomenon that can potentially affect many body functions involved directly or indirectly in motor performance. Since we are interested in mechanisms of the neural control of movement, only relevant aspects of the spectrum of fatigue effects will be discussed. These aspects also span quite a range of structures that show fatigue-related changes, from properties of muscle fibers and patterns of recruitment of alpha-motoneurons to spinal reflexes and reflex-like reactions and coordination of multiple elements involved in typical movements (i.e., synergies; see Chapter 6). We will assume that fatigue is induced by a relatively short episode of exercise, on the order of a few minutes, involving a relatively small group of muscles, not by overall exhaustion caused by a long-lasting whole-body action (e.g., like after participating in an "iron man" competitive race).

Muscle fibers differ with respect to their fatigability. Typically, fibers within smaller motor units are relatively fatigue resistant, while those within large ones are relatively fatigable. This difference is related primarily to the different sources of energy used by the fibers. The more fatigable ones use adenosine triphosphate transformation and oxidative metabolism based on energy supplies stored in the fibers in the mitochondria. Less fatigable fibers use glycogen that can be delivered by the blood flow as it is being used to produce muscle contractions. Most muscles are composed of both fatigable and fatigue-resistant motor units. Due to the orderly recruitment of motor units from the small to the large ones (the Henneman rule; Henneman et al. 1965), sustained contractions at relatively low force levels involve primarily small, fatigue-resistant motor units and can be performed without visible fatigue effects over long time periods. At high levels of muscle activation, large, fatigable motor units have to be recruited, leading to fatigue.

Direct effects of fatigue include a drop in the peak contraction force of muscle fibers and a number of less obvious effects such as prolongation of the relaxation phase following a twitch contraction and slowing down the conduction velocity along the muscle fibers (Fitts et al. 1982; Bigland-Ritchie et al. 1983; Fuglevand et al. 1993). The former effect could be due to slowing down of Ca^{++} removal and changes in the time course of cross-bridge detachment. The latter effect may be related to an increase in the extracellular concentration of K^+ ions. Other peripheral consequences include accumulation in the muscle of the products of metabolism such as lactic acid, which can lead to reactions of free sensory endings, which can produce increased presynaptic inhibition of reflex projections.

The relative role of fatigue effects within the muscle itself is likely large. It has been estimated in studies with direct electrical stimulation of a muscle before and after a fatiguing exercise (Fig. 21.1). The logic of the study is rather straightforward. If

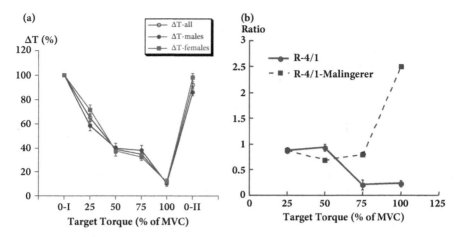

Figure 21.1 (a): Responses of a muscle (quadriceps femoris) to a direct electrical stimulus at different baseline levels of voluntary muscle activation. The data are shown for males, females, and both groups combined. ΔT, change in the knee extension torque in percent to the magnitude observed in a relaxed muscle. (b): The ratio between the response to the fourth stimulus (after 1 minute of exercise) and to the first stimulus (before fatigue). Note the drop in the response for the exercise at fatiguing levels of 75% and 100% of maximal. Note also an atypical behavior labeled "malingerer." Reproduced by permission from Latash et al. 1996.

muscle fibers are truly unable to produce large forces, they are expected to show a decline in the force induced by direct electrical stimulation of the muscle. If, however, a major role is played by fatigue-induced changes in neural processes involved in muscle activation, a drop in the muscle force with fatigue is expected to lead to a stronger response to direct electrical stimulation. Indeed, muscle response to direct stimulation in the absence of fatigue is the strongest for a relaxed muscle. It shows a drop with an increase in muscle voluntary activation (Latash et al. 1994, 1996), likely because a proportion of muscle fibers is involved in the ongoing voluntary contraction and cannot respond to an additional stimulus. So, if a person performs a maximal voluntary contraction (MVC) task and the MVC drops with exercise, a higher response to a standard electrical stimulus delivered to the muscle is expected if the drop is due to insufficient activation caused by the central nervous system, and a smaller response is expected if the muscle is truly exhausted and cannot produce larger force. The experiment on healthy participants showed a drop in the additional muscle contraction caused by direct muscle stimulation. Only one of the subjects (a "malingerer" in Fig. 21.1) showed a drop in the muscle force associated with an increase in the response to direct stimulation.

Such atypical behavior (i.e., an increase in the muscle response to direct stimulation during the fatiguing exercise) was also reported in a study of patients with MS (Latash et al. 1996) who were commonly complaining of fatigue during everyday activities out of proportion to the actual muscle involvement (Monks 1989; Royer et al. 2022). Since MS is a disease of the central nervous system, this may be viewed as an expected result, confirming that atypical fatigue in MS reflects its effects on the neural control of muscles, not on possible secondary changes in the muscle itself in the course of the disease.

The effects of fatigue on characteristics of motor behavior reflect changes both at the peripheral level and within the central nervous system. Some of the latter changes may be viewed as adaptive, that is, mitigating in part the direct fatigue consequences in the muscle. Figure 21.2 illustrates the possible effects of fatigue of stretch reflex characteristics. Note that these characteristics reflect both functioning of the reflex loop and response of the muscle to activation changes. Spinal reflexes are typically suppressed in fatigued muscles, possibly due to an increase in the presynaptic inhibition, in particular of the Ia afferent projections on alpha-motoneurons (Bigland-Ritchie et al. 1986a, 1986b; Woods et al. 1987). In addition, for a fixed amount of activation, a drop in the induced muscle force is expected. These two factors sum up to produce less steep changes in active muscle force with muscle length, illustrated in Fig. 21.2. To produce large MVC force with such a muscle, the person has to move the stretch reflex threshold as far as possible toward lower values of muscle length (to the left in Fig. 21.2). But this range is likely limited, although it is larger than that biomechanically accessible range of muscle length changes (cf. Feldman 2015; Turpin et al. 2017). As a result, a drop in the MVC force value is observed. The illustration in Fig. 21.2 is supported by the observations of changes in the ratio between muscle activation and force not only for MVC contractions but also for submaximal tasks (Garland et al. 1994; Christova and Kossev 1998). Typically, during isometric steady contractions, muscle force changes with muscle activation in a close-to-linear fashion. The slope of this relation, however, drops under fatigue.

Another typical consequence of fatigue is an increase in the variability of motor performance, for example, of force root mean square error during steady submaximal

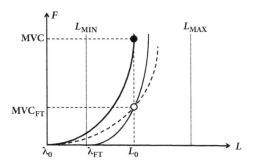

Figure 21.2 A schematic illustration of possible causes in the drop in voluntary force in isometric conditions (muscle length is L_0) with fatigue. The plot shows force-length stretch reflex characteristics. Solid thick line, before fatigue; thin solid lines, under fatigue assuming a change in λ; thin dashed lines, under fatigue assuming a change in the shape of the force-length characteristics. The dotted vertical lines show the biomechanical range of muscle length.

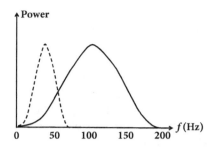

Figure 21.3 A schematic illustration of changes in the spectrum (power of the signal as a function of frequency, f) of surface electromyographic (EMG) signals under fatigue (dashed curve).

force production (Garland et al. 1994; Binder-Macleod 1995; Halperin et al. 2015). One of the causes of the increased motor variability may be the synchronization of motor unit discharges, which happens typically in fatigued muscles (Kadefors et al. 1968; Sato 1982). It is reflected, in particular, in a large shift of the power spectrum of a surface EMG signal toward lower frequencies (Fig. 21.3). This reaction may be seen as adaptive, allowing one to reach larger force magnitudes by synchronization of individual motor unit twitches. But it is also expected to lead to less smooth contractions, which may result in increased force variability.

Persistent inward currents (PICs) on the membrane of alpha-motoneurons have also been implicated in the effects of fatigue (Kirk et al. 2019; Kavanagh and Taylor 2022). In particular, the persistent nature of PICs makes them important in supporting steady muscle contractions. A drop in PICs can lead to a drop in not only the muscle contraction level but also the smoothness of contraction, resulting in higher variability indices.

The illustration in Fig. 21.2 suggests that a certain fixed amount of variability in the time changes in the control variable (λ) in isometric conditions is expected to lead to smaller variability in the force magnitude because of the smaller slope of the force-length characteristic. However, no action is controlled by a single muscle. A more realistic illustration views the neural control of an effector as a consequence of specification of two commands, the reciprocal command (R-command) and the coactivation command (C-command) (see Chapter 7 and Feldman 1986, 2015). For a fixed level of variability in the R-command, an increase in muscle coactivation (larger C-command) is expected to produce higher force variability in isometric conditions (Fig. 21.4). Increased levels of muscle coactivation have been reported in a number of studies of fatigue (Lévénez et al. 2008; Duchateau and Baudry 2014), which may be a reason for the documented increased motor variability.

Another set of phenomena observed under fatigue involve different changes in the typical responses to an unexpected mechanical perturbation applied to the effector. The earliest response in the muscle, associated with spinal reflex circuitry, is suppressed (Hagbarth et al. 1995; Nicol et al. 2003). Later responses, the preprogrammed reactions, show less suppression (the medium-latency response) or even facilitation (the long-latency response) (Windhorst et al. 1986; Duchateau et al. 2002). These

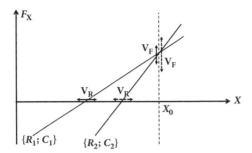

Figure 21.4 Effects of an increase in the coactivation command (C-command, C_2 compared to C_1) on force variability (V_F) during isometric force production at a coordinate X_0 due to variability in the R-command (V_R). Note the increase in force variability for the larger C-command.

observations support the hypothesis on different neural pathways involved in the three responses and also suggest strongly that the long-latency response is transcortical (e.g., Dietz et al. 1984). Indeed, to achieve the same baseline level of muscle activation, larger descending commands are needed (cf. larger values of λ in Fig. 21.2). Using larger descending input, in particular along the corticospinal tract (see also Raptis et al. 2010), is expected to lead to higher excitability of the neuronal population involved in the task. As a result, its response to a standard perturbation-induced afferent volley is increased, resulting in the increased magnitude of the long-latency response.

21.2. Changes in synergies under fatigue

The effects of fatigue of an effector on the variability of performance in a multi-effector task are relatively minor (Forestier and Nougier 1998; Gates and Dingwell 2008). For example, fatiguing muscles acting at one of the joints of a multi-joint extremity has only minor effects on the variability of performance in a multi-joint task (Côté et al. 2002, 2008). These observations suggest that fatigue may be associated with stronger inter-compensatory effects across the involved elements, that is, with stronger synergies stabilizing task-specific salient performance variables (see Chapter 6). Indeed, a number of studies exploring the effects of fatiguing an element on multielement synergies have documented increased indices of performance-stabilizing synergies under fatigue.

One group of studies explored the effects of fatiguing one finger of the hand on synergies stabilizing performance in multi-finger pressing and prehensile tasks (Singh et al. 2010, 2013). In those studies, most commonly, the index finger was fatigued to minimize the effects of the fatiguing exercise on other fingers because the index finger typically shows smaller indices of enslaving as compared to other fingers (Zatsiorsky et al. 2000; see Chapter 11). Synergy indices were quantified within the framework of the uncontrolled manifold (UCM) hypothesis (see Chapters 5 and 6) during cyclical and discrete four-finger accurate force production tasks. In both tasks, the synergy index increased after fatigue, reflecting a larger increase in the variance component along the UCM (V_{UCM}) compared to the increase in the orthogonal variance component (V_{ORT}). These observations were interpreted as an adaptive increase in force variability in the non-fatigued fingers.

Figure 21.5 illustrates this idea. Imagine that, before fatigue, there was a total force (F_{TOT}) stabilizing synergy produced by two fingers pressing in parallel. This is reflected in the ellipse of inter-trial data distribution elongated along the UCM (the dashed line with negative slope). If one of the fingers (F_1) is fatigued and produces larger variance in its force (V_{F2}) and variance of the other finger force (V_{F1}) is unchanged, the ellipse is expected to rotate, leading to a larger projection on the orthogonal to the UCM direction (ORT), that is, to a larger variance component affecting performance. One strategy to minimize the detrimental effects of an increase in V_{F1} is to increase V_{F2} while keeping the covariation between the two finger forces high. This strategy is illustrated in panel B of Fig. 21.5, which shows a larger ellipse elongated along the UCM and corresponding to an increase in the synergy index computed as the normalized difference between V_{UCM} and V_{ORT}. Later studies expanded these findings to more natural, prehensile tasks (Singh et al. 2012, 2013).

An increase in the synergy index under fatigue has also been confirmed in whole-body sway tasks analyzed in the space of muscle activations (Singh and Latash 2011; see Chapters 6 and 14). To remind, the first step of this analysis involves defining stable muscle groups (M-modes) with parallel scaling of activation levels. In the second step, the UCM framework is used to quantify variance in the M-mode space within the UCM and ORT for a salient performance variable, the center of pressure (COP) coordinate in this particular study. The tibialis anterior, a major ankle dorsiflexor, was

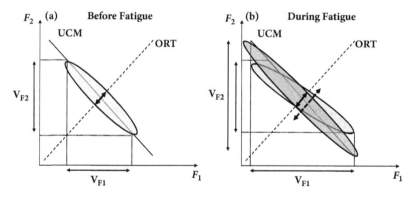

Figure 21.5 An illustration of possible direct effects of fatigue in force variability of an element (F_1) and adaptive changes in force variability of the other element (F_2) participating in the task of accurate ($F_1 + F_2$) production. UCM, uncontrolled manifold; ORT, space orthogonal to the UCM. (a): Inter-trial data distribution before fatigue. (b): Under fatigue, variance of F_1 is expected to increase. If variance of F_2 remains unchanged, the ellipse rotates, leading to larger V_{ORT} (dashed double arrow). Higher variance of F_2 makes room for rotation of the ellipse and aligning it along the UCM.

fatigued by an isometric exercise performed while sitting. Following the fatiguing exercise, there was an increase in the inter-cycle variance of the integrated activation indices of many muscles, including muscles of the trunk, which were not involved in the exercise. This was associated with stronger synergies stabilizing the COP anterior-posterior coordinate during the sway.

Taken together, these results may be illustrated with the following example. Imagine five persons who are asked to carry a very heavy object (e.g., a piano) upstairs. One of them gets tired and starts to stumble and apply erratic forces to the piano, thus complicating the task for everybody. There are two strategies to address this. First, the other participants can ask the tired one to step aside and take some rest. Meanwhile, they perform the task without the tired partner. The second strategy is more sophisticated. If the participants share a common controller and have adequate feedback (as is true for effectors within the body), they may try to predict the effects of erratic forces by the tired partner and also apply varying forces but out of phase. This would attenuate the expected effects of fatigue on the overall task performance. The human brain seems to prefer the latter, more challenging, adaptive strategy, at least across the studied tasks.

Intra-muscle synergies (see Chapter 11) are likely based on spinal circuitry and may be unable to benefit from adaptive features of the neural control, at least not at the time scale of the typical effects of fatigue studied in the laboratory. The direct effects of fatigue on spinal reflex loops are primarily inhibitory, likely via presynaptic inhibition (Bigland-Ritchie et al. 1986a, 1986b; Woods et al. 1987). In other words, at least some of the negative feedbacks that are likely to contribute to intra-muscle synergies are likely to have lower gains under fatigue. This leads to a prediction that fatigue may lead to reduced indices of intra-muscle synergies stabilizing muscle action. This prediction has been confirmed in a study of the effects of fatigue of the tibialis anterior on intra-muscle synergies stabilizing dorsiflexion moment of force during cyclical tasks (Ricotta et al. 2023a). The synergy index did not change following a non-fatiguing control exercise of a similar duration, but it dropped after the fatiguing exercise (Fig. 21.6). The contrast between the effects of fatigue on multi-muscle and intra-muscle synergies also supports the conjecture that these two major groups of synergies are based on different neural circuitry, supraspinal and spinal, respectively (see Chapter 10 and De et al. 2024).

21.3. Aging: Effects on muscles, neurons, and performance

Old age is something most people hope to experience (the alternative is less exciting), but few look forward to the experience. There is good reason: Aging is associated with a decline in the functioning of many important systems in the body, including the sensory and motor systems. As far as the production of movements is concerned, aging is associated with progressive loss of neurons in different structures within the central nervous system, from the cortex of the large hemispheres to the spinal cord (Campbell et al. 1973; Eisen et al. 1996; Roos et al. 1997). In particular, the progressive loss of alpha-motoneurons leads to denervation of muscle fibers from the corresponding motor units. These orphan fibers can be reinnervated by terminal axonal

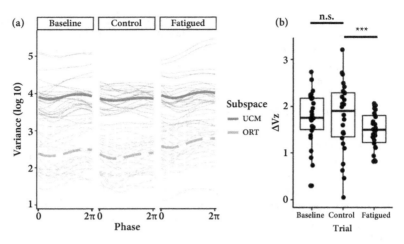

Figure 21.6 (a): Fatiguing exercise leads to an increase in the variance (V) component along the space orthogonal to the UCM (ORT). The profiles of V_{UCM} and V_{ORT} are shown within the cycle of force production for individual subjects (thin lines) and means (thick lines). (b): This leads to a drop in the force-stabilizing synergy index (ΔV_z) in the space of motor unit groups (MU-modes). No such drop is seen following a non-fatiguing control exercise. Reproduced by permission from Ricotta et al. 2023a.

branches of surviving alpha-motoneurons (illustrated schematically in Fig. 21.7). In such a case, they are incorporated into the host motor units, leading to a number of consequences. In particular, small motor units become larger, which may be causally related to a drop in the accuracy during force production tasks, in particular for low levels of force (Laidlaw et al. 2000). Less fortunate orphan muscle fibers do not get

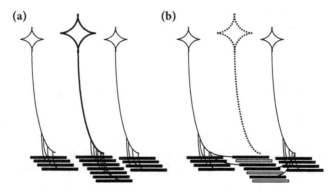

Figure 21.7 A schematic illustration of the processes of denervation and reinnervation during aging. Death of an α-motoneuron (dashed lines) leads to reinnervation of some of its muscle fibers (black bars), while the less fortunate ones (gray ones) turn into connective tissue.

reinnervated and, after some time, lose their contractile abilities and turn into fat and connective tissues. These processes contribute to sarcopenia, a drop in muscle mass with age.

Larger alpha-motoneurons are more likely to die early during aging (Owings and Grabiner 1998). Since they are part of the fastest motor units, muscle contraction speed drops. Taken together, these processes contribute to two well-known consequences of aging: slowness of movements and loss of accuracy. Other age-related changes involving supraspinal central nervous system structures, discussed later, also contribute to these features.

A number of studies have reported a distal-to-proximal gradient in the age-related drop in muscle force (Nakao et al. 1989; Shinohara et al. 2003a, 2003b). These processes may have implications for the neural control patterns developed over the lifetime, which ensure properly scaled involvement of muscles crossing proximal and distal joints in the serial kinematic chains involved in most everyday movements. In particular, intrinsic and extrinsic finger muscles have to be involved in parallel to ensure properly scaled moments of force in the finger joints during action by the fingertips (cf. Li et al. 2000). A few studies have suggested that more distal, intrinsic muscles lose force faster with advanced age (Shinohara et al. 2003a, 2003b; Cole 2006), which may require adjusting the learned mapping from the referent coordinate (RC) commands at the fingertip level to RCs related to involvement of specific muscle groups (see Chapter 7).

Some age-related features of changes in hand function look counterintuitive. In particular, indices of finger individuation (or enslaving; Zatsiorsky et al. 2000; see Chapter 11) show age-related changes across a range of tasks that seem to correspond to better individual finger control (Shinohara et al. 2003b). These findings, however, are not seen in other tasks (Oliveira et al. 2008; van Beek et al. 2019). As discussed in the next section, these changes may be related to loss of neurons in the brain involved in ensuring patterns of finger enslaving beneficial for stability of performance across a range of prehensile tasks.

Another major group of consequences of aging is seen in whole-body tasks (see Chapter 14) including those involving vertical posture. Some of the frequently observed phenomena are linked to compromised postural stability, including an increase in the postural sway (Maki et al. 1990; Melzer et al. 2004; Fujita et al. 2005) and delayed anticipatory postural adjustments (APAs; Inglin and Woollacott 1988; Woollacott et al. 1988; Woollacott and Shumway-Cook 1990; Rogers et al. 1992). The increase in sway with age is seen primarily in the rambling component (Sarabon et al. 2013), which means that it is unlikely to be related to the age-related increase in muscle coactivation. As discussed in Chapter 14, rambling is likely to reflect migration of the equilibrium point about which the vertical posture is stabilized. Loss of postural stability is reflected in difficulties with recovery of the vertical posture in cases of unexpected changes in the external forces and the increased probability of falls (Horak et al. 1989; Rosenblatt and Grabiner 2012).

A number of behavioral features of movements in older persons may be classified as excessive muscle activation. These include, in particular, increased levels of coactivation in agonist-antagonist muscle pairs (Tang and Woollacott 1998; Lee and Ashton-Miller 2011) and excessive grip force during manipulation tasks (Gilles and Wing

2003). These features are, however, not specific for aging and are seen across a variety of populations with impairments in aspects of motor control and coordination (see Chapters 12, 19, and 22). Whether they may be viewed as adaptive or maladaptive is a topic to be discussed.

21.4. Changes in synergies with age

A number of the mentioned features of movements in older persons may be seen as reflections of compromised stability of salient performance variables. Indeed, a number of studies have confirmed reduced indices of performance-stabilizing synergies in the elderly, in particular in multidigit tasks (Shim et al. 2004; Shinohara et al. 2004; Olafsdottir et al. 2007b; Kapur et al. 2010b). This is associated with reduced indices of optimality as assessed using the analytical inverse optimization technique (Park et al. 2011). So, although there seems to be a trade-off between optimality and stability (see Chapter 9), both can suffer in parallel with aging.

The weakening of multidigit synergies and the mentioned increase in finger individuation with age may be consequences of a single cause related to progressive death of brain neurons. Finger enslaving has been viewed as a reflection of purposeful minimization of the so-called secondary moment of force in pronation-supination (Zatsiorsky et al. 2000), a particular expression of synergic patterns of finger involvement developed over the lifetime. Note that synergies have two characteristics (Chapter 6): sharing of RCs at the task level to elements at lower levels and covaried adjustments of the element-level RCs reflecting stabilization of a salient performance variable related to the task-level RC.

The concept of "cortical piano" introduced by Schieber (2001) has direct relevance to both multi-finger synergies and enslaving. According to this concept, an unknown pre-M1 neuronal population projects onto M1 neuronal populations representing individual fingers with variable gains of the projections (illustrated schematically in Fig. 21.8). The combinations of those gains are developed over the lifetime to ensure stability of the rotational hand actions, which is crucial for a wide variety of functional actions. The importance of stabilization of the hand moment of force has been confirmed in studies of multi-finger accurate force production tasks that documented strong moment-stabilizing synergies, even when they were produced at the expense of force-stabilizing synergies (Latash et al. 2001; Scholz et al. 2002; see Chapter 12). As a result, the hypothetical pre-M1 neurons produce patterns of finger enslaving reflecting close-to-proportional involvement of individual fingers, that is, a single finger mode (see Danion et al. 2003).

Death of neurons within the hypothetical pre-M1 population is expected to have two effects. First, elemental variables forming multi-finger synergies (finger modes) stabilizing salient performance variables become not readily available. Second, enslaving becomes lower as actors have to switch to an alternative method of control, not with RC vectors reflecting finger modes and leading to simultaneous action by all the fingers but with individual, finger-specific RCs. In other words, the concept of cortical piano naturally brings about a trade-off between finger individuation and synergic control of the hand. A healthy amount of enslaving seen in young,

21.5. ADAPTIVE AND MALADAPTIVE CHANGES   321

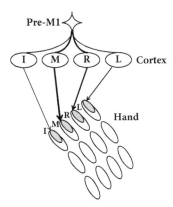

Figure 21.8 An illustration of the concept of "cortical piano." Unknown pre-M1 neurons project on cortical representations of fingers with variable gains, leading to the observed patterns of finger enslaving. A finger mode for the middle finger is illustrated. Strength of projections is reflected in thickness of the arrows. I, index; M, middle; R, ring; L, little.

neurologically healthy adults reflects properly functioning synergies forming the foundations for the everyday repertoire of hand actions.

Another feature of impaired synergic control is the reduction of anticipatory synergy adjustments (ASAs; Olafsdottir et al. 2007a, 2008; see Chapter 8), which may be seen as a specific example of impaired feedforward control. Other examples of reduced feedforward adjustments to actions include the mentioned reduction of APAs and reduced adjustments in the grip force during manipulation of hand-held objects. The combination of impaired stability, as reflected in lower synergy indices, and impaired agility, as reflected in delayed and reduced-in-magnitude ASAs, is common across a variety of neurological disorders, described in more detail in the next chapter.

21.5. Adaptive and maladaptive changes

The body rarely reacts to a local problem with a local adjustment (cf. Bernstein 1947). This is particularly true with respect to changes in the motor function following a decline in performance. In case of fatigue, the duration of the direct effects of a fatiguing exercise is relatively short, typically on the order of minutes or hours. Nevertheless, some changes in the neural control of a fatigued muscle may be viewed as adaptive (or maladaptive!). The most obvious example of adaptation to fatigue is rotation among motor units and among agonists when fatigued elements take a rest while other elements compensate for the lack of contributions from fatigued elements (Sjogaard et al. 1986, 1988). The difference between adaptive and maladaptive changes may not always be obvious. Whether a particular secondary change is useful (adaptive) depends on the group of tasks. For example, the aforementioned synchronization of motor unit firing observed commonly under fatigue and reflected in the shift of the spectrum of the EMG to lower frequencies may be viewed as adaptive if the task is to produce as large a force as possible. However, if the task is to produce a submaximal accurate force level, motor unit synchronization may be seen as maladaptive because motor unit synchronization leads to larger uncontrolled tremor. Along similar lines, prolongation of the relaxation time of twitch contractions of muscle fibers under fatigue (Bigland-Ritchie et al. 1983; van Groeningen et al. 1999) may be viewed as adaptive if the task is to keep a particular force level for a long time, but it may be seen as maladaptive if the task is to reduce muscle force quickly.

A number of secondary features typical of a variety of conditions with relatively mild motor impairments have already been mentioned. These include, in particular, excessive muscle coactivation (reviewed in Latash 2018b and Chapter 19). Is it adaptive or maladaptive? In cases of muscle weakness (common across many states including healthy aging), increasing agonist-antagonist coactivation looks maladaptive because it is associated with stronger resistance to the already weakened agonist. Moreover, for a given resultant force, larger coactivation is associated with larger energy expenditure, which is a questionable strategy across motor tasks. As discussed in more detail in Chapter 19, excessive coactivation may be seen as beneficial for the organization of performance-stabilizing synergies at the level of neural commands (RCs or R- and C-commands). So, if stability of performance is crucial, coactivation is an adaptive strategy. If, however, saving energy is more important, maybe coactivation should be considered maladaptive.

Similar reasoning can be used for other secondary changes in performance that accompany states with motor impairment. For example, using excessive grip force during prehensile tasks may be seen as maladaptive if the object is fragile (cf. Gorniak et al. 2011) or if the task is to be performed over a long time and can potentially lead to fatigue. It is adaptive if the main goal of the action is not to drop the hand-held object in spite of the decreased sensory function of the hand and excessive force variability.

Performing movements at lower speeds may also be seen as adaptive unless, of course, it is dictated by a decreased ability to produce muscle force. Slower movements afford more time to perform corrections and avoidance actions in unexpected situations. They are also associated with smaller magnitudes of impact forces in cases of collision with an unexpected obstacle. A typical example is slowness of movements in persons with Down syndrome who may be perfectly able to move faster as demonstrated in studies of the effects of practice over a few sessions on the peak speed of single-joint elbow movements (Almeida et al. 1994; Latash 2007). In that study, practice over 3 days led to doubling the peak speed in young adults with Down syndrome accompanied by the emergence of typical triphasic patterns of muscle activation, while in a matched group of typically developing persons, the speed increase was an order of magnitude smaller (Fig. 21.9). On the other hand, in a dangerous environment or during an athletic competition, slowness may be viewed as maladaptive.

21.6. Missing pieces of the mosaic

Both fatigue and aging are very complex phenomena, with effects not limited to the sensorimotor system. In particular, the effects of the cognitive components of fatigue and cognitive changes with age on the neural control of movements are all but

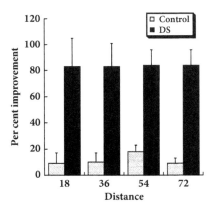

Figure 21.9 Effects of practice on the peak speed of single-joint elbow movements in persons with Down syndrome (black bars) and in typical control persons (light bars). Reproduced by permission from Latash 2007.

unknown. Speed and adequacy of perception and interpretation of changes in the environment are obviously crucial for many actions. Identifying and recognizing a slippery spot on the pavement is crucial for fall avoidance. Recognizing an approaching vehicle and predicting its future trajectory is crucial for adequate actions of one's own body and of the car one drives.

Is it possible to prevent the effects of fatigue and aging? For example, can one decrease fatigability? The answer seems to be positive if one considers the effects of training in athletes specializing in endurance sports (e.g., long-distance running, cycling, or swimming). However, the effects of such specialized training on performance of everyday tasks are unknown.

Can one prepare for age-related changes in the body to mitigate the effects of aging? In other words, is it possible to prepare for living with fewer neurons, fuzzier sensory signals, and weaker synergies? There are anecdotal stories about older adults who continue to be in perfect health and even participate in athletic competitions while being over 80 years old. These impressive videos, however, illustrate exceptions, not rules. On the other hand, one can always give counterexamples such as that of Sir Winston Churchill, who drank whiskey, smoked cigars, and never exercised throughout his long productive life. Exercising on a regular basis is indeed a good idea for one's well-being and avoiding certain conditions typical of advanced age such as hypertension. It has been shown to be associated with longer life expectancy, primarily because of the drop in cardiovascular mortality (e.g., Sarna et al. 1993). Improving the cardiovascular system is definitely very important. However, to our knowledge, there is no data describing the effects of exercise on the neural control of movements and its changes with age.

22
Motor Disorders in Neurological Patients

Studies of motor control and coordination in patients with various movement disorders are performed for a variety of reasons that range between "selfish" and "altruistic." Some studies aim at developing theories of motor control in healthy humans and use patients with dysfunction of specific parts of the nervous system to test predictions of existing theories leading to their confirmation, rejection, or refinement. Other studies have as their main goal maximal improvement of the state of specific patients and patient populations and development of new therapies. The latter approach looks more ethical because it puts the well-being of specific patients above theoretical advances. In the long run, however, the former approach may have a bigger impact on the field because understanding the main principles of functioning of the intact system is crucial for understanding the consequences of injuries to specific parts of the system and development of ways to recover lost functions.

Observations of movement disorders in neurological patients have been very important for the formulation and refinement of motor control theories and their physiological foundations. In particular, views on the role of different brain structures in a variety of functions oscillated between two extremes. Some researchers focused on the irreversible loss of specific functions following injuries to specific brain areas and developed a view that may be addressed as topographical: There are places in the brain specialized and responsible for supporting specific functions. Other researchers emphasized the ability of the brain to show plastic changes leading to partial or even complete recovery of an apparently lost function. In its extreme form, this led to a view that any area of the brain could potentially support any function. In other words, the brain was viewed as a *tabula rasa* shaped by the experience. As commonly happens in biology, both extreme views have been probably too extreme, and acceptable answers may be found somewhere in between.

Most of the current therapies and rehabilitation approaches to dysfunctions of the nervous system are based primarily on the accumulated experience, commonly of the trial-and-error type, and intuitive considerations by clinicians. There are relatively few approaches based on well-founded theories of motor control and coordination, and we will consider a couple of examples later in this chapter. As a result, a large number of clinical trials testing various methods of rehabilitation in neurological patients ended up with a null result: The dominant factor predicting outcome of rehabilitation remains the total number of hours the patient spends exercising under the supervision of a therapist, and it matters relatively little what the patient does during the exercise.

Within this chapter, we are going to adapt the theory of hierarchical control with spatial referent coordinates (RCs; see Chapters 3 and 7) and the concept of performance-stabilizing synergies (Chapter 6) to various pathological conditions. We will try to understand the meaning of some of the commonly used clinical concepts and scales within this framework and then consider possible direct consequences of injuries or neurodegenerative processes to specific parts of the nervous system, possible adaptive changes, and possible maladaptive changes. This topic is very broad, and we will review only a handful of clinical conditions ranging from peripheral, to injuries to the spinal cord, to neurodegenerative processes in subcortical loops related to the motor functions, to injuries to cortical areas. The choice of the specific conditions has been made to allow interpretation within the theoretical frameworks discussed earlier and drawing implications for those theories and concepts.

22.1. Large-fiber peripheral neuropathy

This rare systemic disorder is characterized by loss of conduction along large-diameter afferent fibers with little effect on efferent fibers. As a result, patients may have no kinesthetic perception in the whole body below the neck, including loss of joint position and motion sense as well as sense of contact forces with the environment, but can activate muscles in the affected areas to nearly normal levels. Direct consequences of the disorders also include lack of reflexes in the limbs and trunk, including the stretch reflex, originating in healthy persons from signals delivered by large-diameter afferent axons (Rothwell et al. 1982a; Cole and Paillard 1995). Sometimes these patients are addressed informally and imprecisely as "deafferented persons." Note, however, that conduction along some of the smaller afferent fibers is preserved, and these signals can be used for both sensory and motor processes. Persons with large-fiber peripheral neuropathy cannot stand independently and perform meaningful limb actions without continuous visual control. Even under visual control, movements of these persons are commonly slow and deliberate and show coordination deficits across a range of tasks (Heilman et al. 1987; Sainburg et al. 1993; Cuadra et al. 2019; Schaffer et al. 2021).

The lack of the stretch reflex disrupts one of the crucial components within the theory of movement control with changes in the threshold (λ) of this reflex in healthy muscles (see Chapter 3). The only available mechanism of control is using suprathreshold descending signals to the alpha-motoneuronal pools, which remain the only factor defining muscle activation levels. Note that the spring-like properties of deafferented muscles defined by their peripheral force-length and force-velocity characteristics (reviewed in Zatsiorsky and Prilutsky 2012) lead to an equilibrium-point type of control, although using a very different mechanism as compared to the original lambda-model. This type of control was introduced by the group of Bizzi (Polit and Bizzi 1978, 1979; Bizzi et al. 1982) under the name of alpha-model based on studies of movements in deafferented monkeys. Those studies demonstrated a few key features of the equilibrium-point control, such as equifinality of movements under transient force perturbations to the moving extremity and the gradual shift of the equilibrium point.

Figure 22.1 illustrates the main differences between the λ-control (panel A) and α-control (panel B). Both define force-length muscle characteristics. However, the λ-control defines the spatial RC for the muscle, that is, the coordinate where the muscle would come to rest in the absence of external load. This very important meaning of

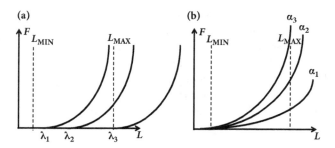

Figure 22.1 (a): An illustration of the control of an intact muscle with shifts of the stretch reflex threshold (λ). The variable λ has the meaning of spatial referent coordinate. (b): An illustration of the control of a deafferented muscle with changes in the level of activation of the alpha-motoneurons (α). The variable α does not have the meaning of spatial referent coordinate. Note that both types of control do not prescribe peripheral mechanical variables, which depend on the external load characteristic.

the neural signal, developed in the course of evolution, is lost in the α-control because, if α > 0, the muscle is active within its whole range independently of the external load. As a result, if the load were completely removed, the muscle would always come to its shortest possible length.

The effects of the lack of reflexes are not limited to the control of individual muscles but also affect muscle coordination. In particular, violations of interjoint coordination during arm movements lead to major changes in endpoint trajectories, which are particularly pronounced during fast movements. For example, when a person with large-fiber peripheral neuropathy tries to imitate the movement of cutting a loaf of bread, the dominant hand with the "knife" makes not a typical straight movement but a curved movement, which has been interpreted as a consequence of poor compensation for the velocity-dependent joint interaction moments of force (Sainburg et al. 1993, 1995). Note that intermuscle and interjoint reflexes mediated by large afferent fibers have been discussed as major contributors to joint interaction contributing to stability of the end-effector trajectories and forces (Nichols 2002, 2018). In other words, these reflexes contribute to synergies stabilizing end-effector action in the external world, and the lack of these reflexes forces the central nervous system to rely on visual information to provide corrections of ongoing actions. Learning how to use this pathological method of control is clearly very difficult. It takes weeks, months, and even years for persons suffering from large-fiber peripheral neuropathy to learn how to perform everyday movements without external help, even under continuous visual control.

There are more subtle changes in the control of movements in these patients. These involve, in particular, changed patterns on finger interdependence (enslaving; see Chapter 12) that have been discussed as reflecting possible adaptive changes within the central nervous system (Cuadra et al. 2019) leading to a change in the neural control of prehensile tasks. In particular, these persons seem to prefer overgripping the object to decrease kinematic consequences of possible errors in the applied moment of force. Another study documented deficits in bilateral arm coordination suggesting an important contribution of proprioception to typical patterns of limb coordination in bilateral tasks performed by healthy persons (Schaffer et al. 2021).

22.2. Spinal cord injury and spasticity

Injury to the spinal cord is, unfortunately, rather common, particularly among young males. The most common causes of spinal cord injury are results of reckless behavior and include driving accidents, acts of violence, and falls. Such injuries lead to a spectrum of pathological consequences related to disruption of the conduction of action potentials along descending and ascending pathways (white matter) and destruction of the neuronal spinal apparatus (gray matter). Depending on the level, severity, and location of the injury, it can lead to a spectrum of functional consequences, both sensorimotor and non-motor. Many non-motor consequences, such as dysfunction of the bladder, dysfunction of sexual function, and chronic pain, are seen by many patients as more important than disruptions of the motor function.

In the late 19th century, the great British neurologist Hughlings Jackson (1889) suggested a classification of typical consequences of spinal cord injury (and some other disorders) into negative signs and positive signs. The former group involved loss of features of movements that are seen in healthy persons including complete or partial loss of voluntary muscle activation (*weakness* or *paralysis*) and increased *fatigability*. The latter group involved muscle activations that are not typically seen in healthy persons such as increased *muscle tone* and muscle *spasms* that can occur spontaneously or be triggered by sensory stimuli or attempts to perform voluntary movements. Among other common signs of spasticity is clonus—alternating activation of agonist and antagonist muscles leading to cyclical movement of the effector at 6 to 8 Hz. Clonus likely represents auto-oscillation in the hyperexcitable stretch reflex loop. Taken together, the positive signs are commonly addressed as *spasticity*. Spasticity was later redefined by James Lance as a velocity-dependent increase in reflexes to muscle stretch. These changes are likely to get contributions from a decrease in both postsynaptic and presynaptic inhibition at the spinal level.

Over the years, the classification by Hughlings Jackson has been refined and tools developed to quantify spasticity. Some of the commonly used tools involve neurophysiological testing, for example, quantifying monosynaptic reflexes and suppression of the reflexes by muscle vibration. These tests provide information on changes in the excitability of alpha-motoneuronal pools and, separately, on likely changes in presynaptic inhibition. The most commonly used tools are, however, clinical scales, typically with five grades reflecting the generality and severity of spasms (the Spasm scale) and increased resistance of extremities to externally imposed motion (the Ashworth scale and its modified version). Although spasticity, particularly severe spasticity, is viewed as a very undesirable consequence of spinal cord injury, its presence is a positive prognostic factor for recovery of at least some motor abilities in persons with clinically complete paralysis. As shown in recent studies (Jo and Perez 2020; Jo et al. 2020, 2021; Sangari et al. 2021; Sangari and Perez 2022), patients with paralysis and spasticity are likely to have at least some descending pathways spared by the injury and are more responsive to neuromodulatory therapy (see later) than those who are paralyzed and show no spastic signs.

One of the most commonly used terms to define spasticity is increased muscle *tone* (or *tonus*). Although this term is used broadly, it is typically defined imprecisely as "increased resistance to passive motion." Note that tone was defined by Bernstein as muscle state reflecting its preparation to future action (Bernstein 1947). In clinical practice, however, tone is assessed when the patient is instructed to relax and do nothing (i.e., no future action is implied, in clear contradiction to Bernstein's definition). Consider the illustration in Fig. 22.2. It shows the biomechanical range of muscle length (from L_{MIN} to L_{MAX}). If a person is in a posture corresponding to a relatively short muscle length (L_0), the instruction to relax implies that the stretch reflex threshold (λ) should be to the right of L_0. This instruction, however, does not specify any specific value of λ. Consider three subjects who relax at L_0 with three different values of λ shown in Fig. 22.2 (λ_1, λ_2, and λ_3). Now, the examiner moves the joint to a new position corresponding to muscle length L_1. The first subject will show strong resistance (F_1) generated by the muscle, the second subject will show much lower resistance (F_2), and the third subject will show no muscle activation. Shall we conclude that the first subject has increased muscle tone, the second subject is normal, and the third subject suffers from hypotonicity? Of course not. All three subjects are healthy, and they simply interpreted the imprecise instruction "to relax" differently. Note that complete relaxation (as in the λ_3 example) is an art and may require extensive practice, as is well known to some athletes and masseurs. As of now, tone in clinical practice remains arguably the most commonly used ill-defined concept.

Based on Fig. 22.2, tone can be defined as depth of relaxation measured in spatial units—the difference between actual muscle length and its stretch reflex threshold, λ. This definition fits Bernstein's insight because the distance between λ and muscle length is an important factor in defining reaction time delay if suddenly an action

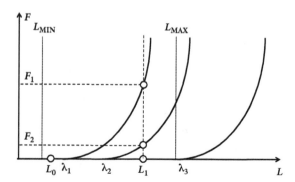

Figure 22.2 During clinical examination, the subject is required to relax at muscle length L_0. The stretch reflex threshold (λ) has to be $\lambda > L_0$. Three values of λ are shown corresponding to different locations of the force-length characteristics. Stretching the muscle to L_1 is expected to lead to different resistive muscle forces, from a very large (F_1) force to virtually no resistance.

22.2. SPINAL CORD INJURY AND SPASTICITY

is needed corresponding to activation of this muscle. In other words, this distance reflects a person's preparation to future action, which may or may not be needed.

The separation of clinical signs into positive and negative and the conclusion reached by Hughlings Jackson that treating one group of signs should not be expected to affect the other group dominated the field of clinical practice for nearly a century (e.g., Landau 1974; Sahrmann and Norton 1977; cf. Marsden et al. 2023). The invention of treatment with intrathecal delivery of a precursor of one of the major mediators of presynaptic inhibition (baclofen) allowed the reduction or even elimination of spastic signs (Penn and Kroin 1987; Latash et al. 1989, 1990; Penn et al. 1989). This was accompanied by unmasking of voluntary movements in some of the patients suggesting that treating positive signs was indeed able to improve negative signs of spinal cord injury (Corcos et al. 1986; Latash and Penn 1996). Figure 22.3 illustrates muscle activation patterns and mechanical characteristics of voluntary movements by a patient with spinal cord injury following a surgery to repair the intervertebral disc at L5 prior to and after the intrathecal delivery of baclofen. Note that the elimination of clonus led to much better-controlled movement and reduction of agonist-antagonist coactivation that was present before the drug delivery.

Spasticity accompanies disorders other than spinal cord injury. In particular, it is commonly seen in patients suffering from multiple sclerosis (MS) and in stroke survivors. Analysis of spastic signs after stroke affecting the cortex of the large hemispheres resulted in a hypothesis on the origin of spasticity directly linked to the aforementioned definition by Lance. This hypothesis assumes that spasticity is related to contraction of the range of voluntary changes in the threshold of the stretch reflex (λ). In a healthy person, the range of possible changes of λ is larger than the biomechanical range of muscle length changes (Fig. 22.4A). This allows the person to relax the muscle even if it is long and to produce large muscle forces even if the muscle is short. If the range of λ changes is reduced (to $\{\lambda^-; \lambda^+\}$; see Fig. 22.4B), the person becomes unable to produce voluntary movements in a range of muscle length values shorter than λ^- (shown as "paralysis") and to avoid muscle activation (shown as "spasticity") in a range of muscle length values longer than λ^+. This scheme based on the equilibrium-point hypothesis (see Chapter 3) links the positive and negative signs of spasticity and suggests that increasing the range of muscle voluntary control may be expected to lead to both better movements and lower spastic signs (cf. Fig. 22.3). We will consider the implications of this hypothesis for clinical practice later in this chapter.

Recent series of studies have provided an optimistic message suggesting that even persons with clinically complete paralysis and spasticity many years after the spinal cord injury can benefit from massive practice combined with a neuromodulation approach targeting synaptic connections from the surviving descending pathways to alpha-motoneurons (Jo and Perez 2020; Jo et al. 2020, 2021). These studies used an earlier-established method of synaptic facilitation with a peripheral stimulus that reaches the alpha-motoneurons a few milliseconds prior to a stimulus applied to the descending tracts (Bunday and Perez 2012; Feldman 2012). The peripheral stimulus is applied using electrical peripheral nerve stimulation, and the stimulus to descending pathways is applied using transcranial magnetic stimulation (TMS). The cited studies have shown that numerous sessions of such double-stimulation protocols can lead to

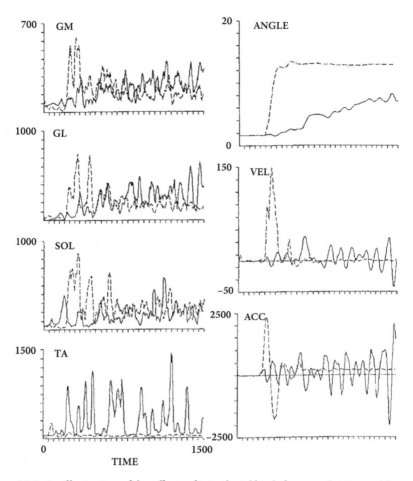

Figure 22.3 An illustration of the effects of intrathecal baclofen on voluntary ankle movement by a patient following a disc surgery at the L5 level. Note the elimination of clonus on baclofen (dashed traces) and much better movement kinematics compared to the condition off baclofen (solid traces). GM, gastrocnemius medialis; GL, gastrocnemius lateralis; SOL, soleus; TA, tibialis anterior; VEL, velocity; ACC, acceleration. Reproduced by permission from Latash and Penn 1996.

significantly increased responses of the alpha-motoneurons to TMS accompanied by visible improvements in functional activities such as ambulation.

As of now, no studies of changes in the synergic control of movements have been performed in patients after spinal cord injury. It seems likely that spasticity is associated with major problems with intramuscle synergies (see Chapter 11), which are supposed to rely on spinal circuitry (Madarshahian et al. 2022; Latash et al. 2023; De et al. 2024). The effects of spasticity on multieffector (multi-muscle) synergies, however, are hard to predict. On the one hand, in cases of spinal cord injury, the injury

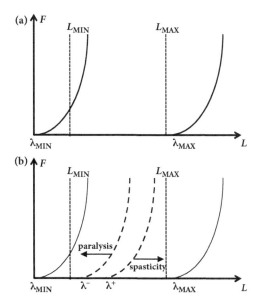

Figure 22.4 An interpretation of spasticity as reduced range of changes in stretch reflex threshold (λ). (a): A healthy person can change λ within a range larger than the biomechanical range of muscle length change (L_{MIN} to L_{MAX}). (b): A person with spasticity has a reduced range of λ change (from λ^- to λ^+), leading to both uncontrolled contractions ("spasticity") when the muscle is longer than λ^+ and weakness ("paralysis") when the muscle is shorter than λ^-.

does not have direct effects on supraspinal structures, which are likely involved in the synergic control of systems consisting of multiple effectors. On the other hand, salient sensory signals to the supraspinal structures may become unavailable. In addition, adaptive changes may lead to secondary modifications of the involved circuitry, leading to modified synergies involved in a variety of everyday movements.

22.3. Parkinson disease

Parkinson disease (PD) is a progressive neurodegenerative disorder associated with loss of dopamine-producing neurons in the substantia nigra, one of the nuclei of the basal ganglia. The circuitry involving the basal ganglia is complex, but a simple schematic in Fig. 22.5 offers a view on one of the cardinal features of PD, namely bradykinesia. There are two major loops via the basal ganglia and involving the thalamus and cortex, both modulated by signals from the substantia nigra (pars compacta, SN_{PC}). These are addressed as direct loop and indirect loop. Counting the inhibitory synapses within each of the loops suggests that the direct loop represents a positive feedback (even number of inhibitory synapses) and that the indirect loop represents a negative feedback (odd number of inhibitory synapses). Balance between the actions

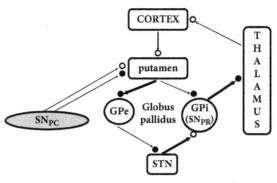

Figure 22.5 An illustration of the two main pathways via the basal ganglia. Note that loss of dopaminergic projections from substantia nigra in Parkinson disease leads to difficulty in movement initiation with contributions from both pathways. SN_{PC}, substantia nigra pars compacta; GP_i and GP_e, globus pallidus internal and external; STN, subthalamic nucleus; open circles, excitatory projections; filled circles, inhibitory projections; thick lines, stronger effects in PD; thin lines, weaker effects in PD. Another output structure, the substantia nigra pars reticulata (SN_{PR}), is shown in parentheses.

of the two loops is needed to ensure proper functioning of the circuitry and avoid both difficulty with movement initiation and poorly controlled spontaneous movements.

Poverty of voluntary movements in PD is viewed as a consequence of reduced inhibition of the output structure (shown only as the internal part of the globus pallidus, GPi, in Fig. 22.5; this schematic does not show separately the substantia nigra pars reticulata, which also serves as an output nucleus) via the direct loop and increased excitation via the indirect loop. The increased activity of GPi leads to excessive inhibition of its target—thalamic nuclei—and reduced excitation of the target cortical structures. This makes it difficult to initiate voluntary movements, resulting in such typical manifestations of bradykinesia as increased reaction time to initiate movements, overall movement slowness, shuffling gate, lack of mimics, and micrography.

Other cardinal signs of PD include rigidity, tremor, and postural instability. Rigidity is seen as increased resistance to externally imposed movement. It can lead to smooth externally imposed movements with consistently increased resistance (lead-pipe rigidity) or to jerky movements (cogwheel rigidity). Unlike spasticity, increased joint resistance in rigidity is not velocity dependent (Mullick et al. 2013). Tremor in PD is typically postural, can be seen in various effectors including articulators, and is partially alleviated by an attempt to perform a voluntary movement.

Postural instability and associated instability of gait are probably the most disabling features in PD that can lead to falls and associated morbidity. Patients with PD show increased postural sway during quiet stance at early stages of the disease, and the sway becomes dramatically reduced at later stages when the patient's ability to maintain vertical posture is strongly compromised. At later stages, the patient may stand with nearly no sway, like a log placed on one of its ends, and be unable to recover vertical posture even under a minimal external force perturbation. This example shows that sway may not be a proper marker of postural stability. Two contributors to postural instability have been consistently reported. The first is reduced and delayed anticipatory postural adjustments (APAs; Bazalgette et al. 1986; Latash et al. 1995a; see Chapter 14). The second is increased and poorly modulated preprogrammed reactions to perturbations.

Clinical studies frequently use the classification of PD stages suggested by Hoehn and Yahr (1967). According to this classification, stage I refers to cases when signs of PD are seen only on one side of the body. Note that PD commonly starts with strongly asymmetrical signs, so stage I most commonly refers to early-stage patients. Stage II refers to patients with bilateral clinical signs but without clinically identifiable signs of postural instability. Postural instability emerges at stage III, and stages IV and V refer to different degrees of functional disability of the patient.

Studies of synergies ensuring stability of movements have revealed major differences between PD patients and age-matched controls, which could be seen very early in the disease progression and even in apparently unaffected effectors (Park et al. 2012; Falaki et al. 2016). Figure 22.6 illustrates the time course of the synergy index

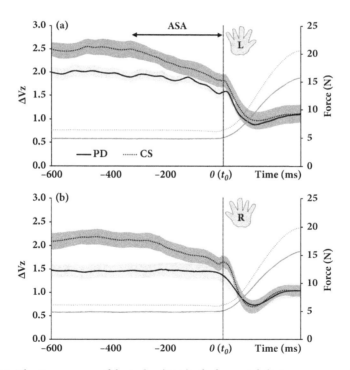

Figure 22.6 The time course of the index (ΔV_z) of a force-stabilizing synergy prior to the force pulse generation by the four fingers pressing in parallel. Note the smaller steady-state ΔV_z values in patients with early-stage Parkinson disease (PD) and smaller anticipatory synergy adjustments (ASA) as compared to the control group (CS). (a): Non-dominant (left) hand. (b): Dominant (right) hand. Reproduced by permission from Park et al. (2012).

(ΔV; see Chapter 6) computed with respect to total force produced by the four fingers of a hand pressing on individual force sensors. The subjects were asked to keep total force steady at a low level (5% of maximal voluntary contraction [MVC]) and then produce a very quick pulse into a target set at 25% of MVC force with visual feedback available at all times. The PD patients were at early stages of the disease (stages I and II); they were tested on their optimal medication.

It is obvious from this illustration that the two groups differ in at least two major characteristics of the $\Delta V(t)$ patterns. First, during steady-state force production, ΔV in the control group was larger than in the PD group. Note that both were higher than zero; that is, both groups showed force-stabilizing synergies by covariation of individual finger forces (cf. Chapter 12). Second, there were anticipatory synergy adjustments (ASAs; cf. Chapter 8) in the control group starting 200 to 300 ms prior to the force pulse. The ASAs were much smaller and shorter in duration in the PD group. In other words, the PD patients showed problems with stability during the steady-state phase and in agility during preparation of the planned quick action. Both problems were seen in both hands of PD patients at stage I, that is, when clinical examination was able to detect PD signs in one-half of the body only. Figure 22.6 also shows that ΔV in the steady-state phase was larger in the non-dominant (left) hand in both groups, in support of the dynamic dominance hypothesis (Sainburg 2005; cf. Chapter 10).

The two problems with the synergic control also have been confirmed for whole-body postural tasks (Fig. 22.7; Falaki et al. 2016). Those studies tested PD patients at stage II, that is, those who did not show signs of postural instability during clinical examination. The studies confirmed both lower indices of stability during steady-state postural tasks and smaller ASAs when the subjects prepared to drop a load from the extended arms (i.e., produced a self-triggered postural perturbation). Taken together, the studies suggest that synergy indices may be highly sensitive biomarkers of PD even at stages when routine clinical examination fails to detect signs of the disease in the examined tasks and effectors. Changes in the synergic control over both multi-finger and

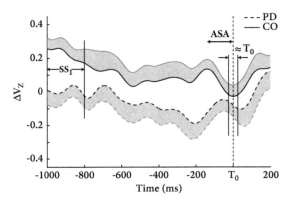

Figure 22.7 Lower steady-state index (ΔV_Z) of posture-stabilizing synergy and lower anticipatory synergy adjustments (ASAs) in patients with Parkinson disease (PD, dashed lines) without clinically detectable postural instability as compared to control subjects (CS, solid lines). Reproduced by permission from Falaki et al. 2016.

whole-body tasks have also been confirmed in *de novo* patients who have never been exposed to dopamine replacement therapy, showing that the tests are sensitive to the disease, not to long exposure to medications (De Freitas et al. 2020; Freitas et al. 2020).

The changes in ASAs suggest a particular problem with the feedforward control of movements. Other examples of disordered feedforward control include reduced and delayed postural adjustments to self-triggered perturbations (APAs; see Chapter 14; Bazalgette et al. 1986) and reduced grip adjustments in prehensile tasks with object manipulation (see Chapter 12; Gordon et al. 1997; Muratori et al. 2008). Note that ASA changes refer to adjustments of action stability, while other indices refer to changes in the magnitude of potentially important mechanical variables such as coordinates of the center of pressure during standing (APAs) and grip force applied to the hand-held object.

A hypothesis has been offered linking ASA deficits in PD to episodes of freezing, in particular freezing of gait (Latash and Huang 2015). This hypothesis has been indirectly supported by the observations of reduced ASAs prior to step initiation in PD (Falaki et al. 2023). It remains speculative because of the lack of longitudinal studies that could link reduced ASAs at early stages of PD to freezing-of-gait episodes that typically emerge at later stages.

Synergy indices in PD are sensitive to most commonly used treatments. Both indices (ΔV at steady state and ASAs) measured in multi-finger and whole-body postural tasks improve on dopamine replacement medications as compared to their magnitudes in the off-drug state (Park et al. 2014; Falaki et al. 2017b). In contrast, only ASA characteristics improve in PD patients treated with deep-brain stimulation, while the stability index (ΔV at steady state) remains unchanged (Falaki et al. 2018). The latter study also documented significant correlations between the indices of stability and agility measured in the multifinger tasks and whole-body postural tasks. Given that these tasks belong to two polar groups of fine and gross motor skills, these findings suggest that there may be general-purpose shared synergic circuitry involved in providing task-specific controlled stability across tasks and effectors.

Both aspects of disordered control of stability have been documented in neurologically healthy professional welders (Lewis et al. 2016). Welding performed on a regular basis over many years is known to lead to a specific brain disorder in a significant percentage of workers, which mimics in many aspects signs of PD and is commonly addressed as "manganism." This disorder is believed to be caused by exposure to metal fumes in the air, in particular manganese. The cited study also found a correlation between the ASA indices measured during multi-finger force production tasks and magnetic resonance imaging indices in the globus pallidus, one of the major nuclei of the basal ganglia. Given that neurological examination had failed to detect any disorder in the welders involved in the study, these findings confirmed that indices of synergic control are sensitive biomarkers of even subclinical changes in brain circuitry.

22.4. Other subcortical disorders

Two circuits involving the cortex have been discussed as major contributors to the control of movements, in particular synergic control (see Houk 2005; reviewed in

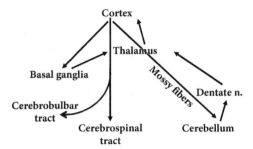

Figure 22.8 A simplified schematic of two major transcortical circuits via the basal ganglia and via the cerebellum.

Latash 2021a, 2021b). One of the circuits has been discussed briefly in the previous section; it involves the basal ganglia and thalamus. The other circuit involves the cerebellum, cerebellar nuclei, and thalamus. Both are illustrated schematically in Fig. 22.8. It has been suggested that the two circuits represent so-called distributed processing modules that can contribute to various movements (Houk 2005), a notion similar to the notion of brain operators introduced earlier by Bernstein and his colleagues (Bernstein 1935; Bassin et al. 1966; LP Latash et al. 1999, 2000). Disorders of either circuit are expected to lead to problems with synergic control. Such problems have indeed been documented in studies of patients with various brain disorders other than PD.

A study of patients suffering from multisystem atrophy (earlier diagnosed as olivopontocerebellar atrophy) showed that these patients demonstrated both problems with synergic control described for PD. In multi-finger tasks with accurate force production, these patients showed reduced indices of stability (ΔV) during steady-state portions of the task and reduced ASAs in preparation for the initiation of a self-paced force pulse (Park et al. 2013).

Another study (Jo et al. 2017) explored patients with MS, a demyelinating disorder that can affect various pathways within the central nervous system, including those involved in the two major transcortical loops mentioned. MS is an autoimmune disorder that is believed to be triggered by a viral infection. Its clinical consequences vary broadly depending on the affected tracts. The most common consequences include weakness, spasticity, increased fatigability, discoordination, and somatosensory problems. But problems with other sensory modalities, including vision, are also common. The cited study showed both aspects of disordered synergic control in MS (Fig. 22.9). Note that the reduced synergy index and reduced ASAs received contribution primarily from the component of intertrial variance that did not affect total force (V_{UCM} in Fig. 22.9; see Chapters 5 and 6). The patients showed only minor differences from the control group in the variance component that affected total force (V_{ORT}), but their steady-state magnitude of V_{UCM} was only half of that in the controls, and the slow drift in the V_{UCM} to lower values in the controls (reflected in their large ASAs) was all but absent in the patients. In other words, the patients had no problems with accurate task performance, but they performed the task in a more stereotypical way as compared to healthy persons. This conclusion was also reached in a study of posture-stabilizing

22.4. OTHER SUBCORTICAL DISORDERS 337

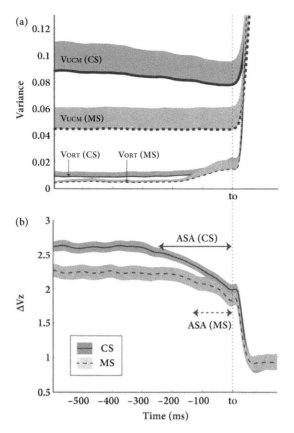

Figure 22.9 Patients with multiple sclerosis (MS) show reduced indices (ΔV_z) of force-stabilizing synergies (b) and smaller anticipatory synergy adjustments (ASAs) as compared to healthy controls (CS). (a): The differences are primarily due to the different magnitudes of inter-trial variance along the uncontrolled manifold (V_{UCM}) with minimal differences in variance orthogonal to the UCM (V_{ORT}). Reproduced by permission from Jo et al. 2017.

synergies in PD patients on and off their dopamine replacement medication (Falaki et al. 2017b). That study showed major effects of the medication on V_{UCM} but not on V_{ORT}, leading to a higher steady-state ΔV index and larger ASAs.

One of the lessons learned from all the cited studies is that ensuring proper stability of salient performance variables in functional tasks is based on proper functioning of the two aforementioned major subcortical circuits. Given the unavoidable unexpected perturbations during everyday movements, accurate stereotypical performance is not an acceptable solution. It fails to channel the effects of the perturbations into the uncontrolled manifold, which is necessary to protect the salient performance variables and ensure their stability. It also fails to manipulate stability in a task-specific way, as reflected, in particular, in the reduced ASAs. Increasing V_{UCM} should be one of the goals of therapy and rehabilitation programs, but unfortunately, so far there have been only a handful of studies exploring the feasibility of such approaches (see Chapter 20 and Wu et al. 2012, 2013).

22.5. Stroke

Stroke (a.k.a. cerebrovascular incident) is, unfortunately, a common cause of long-lasting disability and death. Here, we consider only mild to moderate cases of unilateral stroke affecting primarily or exclusively cortical structures. If parietal brain areas are affected, stroke leads to major sensorimotor disorders, which are particularly pronounced in the contralateral extremities. The ipsilateral (ipsilesional) extremities are sometimes addressed as unaffected, but this is incorrect because deficits in movements of the ipsilesional extremities have been documented (Colebatch and Gandevia 1989; Jones et al. 1989; Maenza et al. 2020). Long-term consequences of cortical stroke typically include paresis and spasticity in the contralesional extremities, discoordination across a variety of motor tasks, impaired kinesthetic perception in the affected areas, and, in some cases, hemineglect—inattention or lack of conscious perception of objects in the contralesional visual hemifield.

My former advisor, the brilliant scientist Gerry Gottlieb (see Corcos et al. 2022), suffered from a massive stroke affecting the parietal and partly frontal areas of one of the cortical hemispheres. When I visited him a few months later, we sat at the table and had lunch. Of course, he used only the ipsilesional arm. He ate food only from one-half of the plate, then moved the plate away, realized that there was food left, moved it back, and ate half of the remaining half. In three steps, the food was eaten, and he joked: "There is one good thing about stroke. You have each of your meals three times in a row."

Recovery from cortical stroke is rarely complete, and optimizing the outcome is a major goal of rehabilitation. For years, it has been accepted that massive practice of the affected extremities is the most effective way to promote neural plastic changes and motor recovery. A number of ingenious methods have been proposed to encourage patients to use their affected extremities more, beyond the time of direct contact with a therapist. One of the best-known ones is so-called constraint-induced therapy (CIT; reviewed in Taub and Morris 2001; Mark and Taub 2004). This method involves putting a mitten on the ipsilesional (less affected) hand, thus making it clumsy and forcing the patient to use the more affected hand during everyday activities. A number of studies reported the benefits of using CIT as compared to a traditional therapy method (reviewed in Fritz et al. 2012; Kwakkel et al. 2015). Other studies, however, failed to confirm the additional benefit (Dromerick et al. 2006; Corbetta et al. 2015). The idea of CIT also seems to go against the frequently used mirror-like movements of the two arms during practice (cf. Rose and Winstein 2004; Cauraugh and Summers 2005).

Along similar lines, so-called discomfort-induced therapy has been suggested to increase the amount of time patients spend loading their more impaired leg during walking (Aruin and Kanekar 2013; Oludare et al. 2017). This approach uses an insole with tack-like protrusions placed into the shoe for the less impaired foot. The patients feel pain or discomfort when they place too much weight on that foot and naturally shift the weight to the more impaired side.

As mentioned earlier in Chapter 20, motor impairments after stroke may be due to an inability to shift the RC for the involved muscles (their λs) over the whole range available for healthy persons. The contraction of the spatial range where voluntary control of movements is possible has led to the idea to design practice in such a way that the required movements were within the zone of available voluntary control. A few recent studies have shown that this is indeed possible and that practice within the range of available control can lead to improvements in motor function (Turpin et al. 2017; Piscitelli et al. 2020; Frenkel-Toledo et al. 2021).

Studies of action stability after stroke led to controversial results. A few studies failed to find significant differences in the indices of stability (ΔV) between the more impaired and less impaired extremities (Reisman and Scholz 2003; Jo et al. 2016b). These studies used arm reaching movements and multi-finger force production tasks performed by both arms, contralesional and ipsilesional. Figure 22.10 illustrates the ΔV time profiles for the task of constant force production followed by a quick force pulse in a group of stroke survivors, control subjects, and a matched group of PD patients. Note that the baseline ΔV values are very close in the stroke and control groups and are much higher than in the PD group. Note also that ASAs are reduced in both the PD and stroke groups as compared to the controls. A conclusion has been drawn that, after cortical stroke, stability of action may be unaffected while the ability to prepare for a fast action may suffer. On the other hand, a couple of other studies reported reduced indices of stability in a group of stroke survivors (Gera et al. 2016a, 2016b). It is possible that the differences were due to the different sites of stroke in different studies: For example, the latter studies could have more patients with the stroke affecting cortical areas involved in the transcortical loops via the basal ganglia and cerebellum.

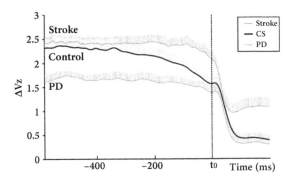

Figure 22.10 The time profiles of the index (ΔV_z) of force-stabilizing synergy during steady force production followed by a quick force pulse for patients after stroke, with patients with Parkinson disease (PD), and control subjects. Note the smaller anticipatory synergy adjustments in both patient groups and smaller steady-state ΔV_z in the PD group only. Modified by permission from Latash 2019.

22.6. Missing pieces of the mosaic

The current state of understanding neurological disorders is clearly unsatisfactory. This situation is a major impediment on the way to optimizing treatment and rehabilitation of patients. One of the major problems is that the language used in clinical studies and practice is based primarily on traditions rather than on understanding the nature of the disorders. The most pressing and challenging issue is the development of an adequate language based on an understanding of the mechanisms involved in disordered motor control. This is only possible if discussion of clinical issues is based on a clearly formulated scientific theory on the neural control of movements.

As of now, the discussions of motor disorders create an impression of a broken mosaic. There are fragments of established knowledge, but they are commonly disjointed. Arguably, the current understanding of the mechanisms involved in PD is one of the most advanced. But even this understanding ends at the level of interactions among brain structures and does not consider how descending signals from output structures in the brain are translated into movements. Such discussions and papers create an impression that the authors truly believe in the existence of invisible strings from the brain to muscles that prescribe their activation patterns or even mechanical output. But such strings do not exist, and the brain cannot prescribe what the muscles do (see Chapters 4, 6, and 7). Very few studies are based on theories of motor control based on the laws of nature. The examples of the analysis of spasticity based on the theory of control with spatial RCs and analysis of discoordination based on the theory of synergic control presented earlier in this chapter are exceptions rather than rules among clinical studies.

A related issue is the paucity of objective tools and biomarkers to quantify the effectiveness of treatment and rehabilitation. The currently used tools may be useful to estimate functional changes in the state of the patient but not in the state of neurophysiological mechanisms that mediate those functional changes. Technological advances are of course important for the development of new tools, but by themselves they may be of limited importance unless used to test specific hypotheses addressing the mechanisms of the disordered movements. In other words, it is time to switch from classification and topography of changes in structures of the body to laws of nature involved in transformations of those changes into movement patterns.

PART VI
METHODOLOGY

23
Types of Studies and Hypothesis Testing

The next three chapters are going to be even more subjective than the rest of the book because they address very basic issues that are rarely formulated or discussed explicitly. In particular, there is little agreement among researchers in the field of movement studies (as well as in many other fields of biology, psychology, neuroscience, etc.) on how to formulate and test hypotheses. This may be one of the reasons for the relatively modest progress in the field because researchers, in particular young ones, waste their time and resources trying to perform studies that are ill-conceived or ill-designed.

The first step in any scientific inquiry in natural science is to define the scope of phenomena the researcher wishes to address and the matched level of analysis. One may be interested in the Big Bang, the interaction of galaxies, the solar system, evolution of life on Earth, whales, *Homo sapiens*, the central nervous system, biological movements, perception in biological systems, cognition, muscle fibers, and so forth. Analysis can involve various classes of variables and parameters. For example, one may be interested in the neural control of vertical posture. Clearly, the functioning of the central nervous system in postural tasks is defined by numerous factors including human genetics, which in turn is defined by biological evolution, which reflects environmental factors on Earth, which is defined by the history of the solar system, all the way up to the Big Bang (and maybe beyond). Of course, studying the neural control of posture starting from the Big Bang is unlikely to be very productive. One has to draw borders and admit, at least to oneself: Everything outside the scope of the study is beyond our analysis and may be beyond our comprehension, which does not mean that those "outside the scope" factors are unimportant.

Within the main theory discussed in previous chapters, the neural control of movement is produced by a hierarchy of spatial referent coordinates (RCs) to involved effectors starting from RCs at the task level. The question "Where does RC_{TASK} come from?" is beyond the scope of this discourse. As discussed in Chapter 1, time patterns of RC_{TASK} may come from processes that are qualitatively different from those involved in the mentioned RC hierarchy.

The next crucial step is selecting an adequate language to address phenomena of interest. The notion of adequate language was introduced by the group of Israel Gelfand and then extended to the field of motor control (Vasiliev et al. 1969; Gelfand and Latash 1998). Examples of adequate language from the past include the introduction of the notion of *point* by Euclid as an object without width, length, or height, which started geometry as science, although intuitive geometry had been known, for example, to the Egyptians, who constructed perfect pyramids. Mechanics started with the introduction of the notion of *force*, although intuitive mechanics had been known, for example, to Archimedes. Note that both notions, those of point and force, are not truly defined: Points, as defined, do not exist, and force is defined in a circular fashion as something that changes the velocity of inertial objects (inertial objects are those that resist change in their movement by force!).

Do we have adequate terms for motor control? I believe that the answer is yes. Two candidates are described in detail in previous chapters. They are *referent coordinate* (or, equivalently, *threshold*) and *synergy*. Both may be applicable beyond motor control, in particular to perceptual processes and maybe cognitive processes (cf. Latash 2019; Kretchmar and Latash 2022). Further in this chapter, we start with classification of studies that are commonly performed in the field and then focus on studies that are hypothesis driven.

23.1. Types of studies

Motor control, just like many other fields of science exploring living systems, is still in the early stages of its development, which, in the subjective opinion of the author, makes it a very attractive field of scientific inquiry. Indeed, it may be compared to pre-Newtonian or even pre-Galilean physics: a wide-open field ready for general theories to be formulated. Theories, however, rarely emerge from nothing. They require reliable factual material, which in our times comes from descriptive, exploratory studies. The importance of these studies should not be underestimated. They may be compared to observations by Brage and Kepler on the regularities of motion of celestial bodies that formed the foundation for the gravity theory formulated later by Newton.

Exploratory, descriptive studies form the first group of studies that are currently dominating the field. This is expected because such studies are easy to perform and even easier to publish: Reviewers have little to disagree with! Of course, it is assumed that the studies are performed competently in all their aspects including design, subject selection, data collection, analysis (including statistical analysis), and presentation. Such studies may be very useful, in particular, in applied areas such as movement disorders and rehabilitation. Indeed, patients and clinicians cannot wait for a general theory of motor control to be formulated and confirmed, and studies comparing the effects of different medications and other methods of treatment and recovery are highly valuable. Unfortunately, many descriptive studies are not very imaginative. They apply developed methods to a new group of tasks or a new population or a new modification in external conditions, feedback to the subject, instruction, and so forth. As summarized many years ago by my late friend and colleague Slobodan Jaric: "And now let us paint the walls of the Laboratory blue and see how results change." The value of such studies is minimal. Indeed, modification of any aspect of any study is likely to lead to changes in some characteristics of subjects' behavior. What this means for the neural control of movement commonly remains unclear.

The second group of studies consists of those that are motivated by a specific theory, or at least hypothesis, in the field and strive to confirm the theory (hypothesis) by making direct predictions for specific experiments and showing that results are compatible with the predictions. These studies are commonly designed to corroborate a favorite hypothesis of the researchers. They rarely challenge the hypothesis but provide a broader experimental foundation compatible with it. Sometimes these studies produce unexpected findings, which may be valuable because they force researchers to reconsider some aspects of their favorite hypothesis. This does not happen very frequently, however, because the purpose of the studies is to support the hypothesis, not to question its validity.

The third group includes studies that try to develop a particular hypothesis or theory by extending it to yet unexplored areas. For example, a hypothesis can be developed to account for the neural control of vertical posture. Does it equally apply to the control of postural states of effectors such as the head, limbs, and eyes? The hypothesis may need modifications to account for the differences in typical tasks, salient sensory modalities, and mechanical properties of the effectors. Can the hypothesis account for postural disorders in a variety of patient populations? Can it address the famous posture-movement paradox (see Chapter 14)? These studies are usually very valuable because they contribute to refinement of the underlying hypotheses.

Finally, the fourth group of studies consists of those that strive to disprove an established hypothesis. If a study is designed based on deep understanding of the target hypothesis, it may be exceptionally valuable, impactful, and eye-opening. Some of such studies have been described in previous chapters, and we will focus on them later in this chapter. If a large number of such studies honestly try to challenge a hypothesis, test its nontrivial predictions, and fail to disprove it, the hypothesis may grow in its stature and become a theory. In the opinion of the author (see later), this has already happened with the equilibrium-point hypothesis (Chapter 3) and with the uncontrolled manifold (UCM) hypothesis (Chapter 5), leading to the theory of synergic control with spatial RCs (Chapters 6 and 7).

In the next sections, we assume that researchers are interested in exploring or testing a specific hypothesis. We ask a sequence of questions regarding hypotheses in the field. And we begin with the most crucial one: Is a hypothesis formulated in a way that warrants its experimental exploration?

23.2. Is a hypothesis worth testing?

There are certain basic requirements to a statement that have to be met to turn it into a scientific hypothesis. The first one relates to the formulation of the hypothesis: Is it formulated using exact, clearly defined terms? In one of our discussions, a great mathematician, Israel Gelfand, once said to me: "The worst method of discussing complex issues is using hints." Biology in general and motor control in particular deal with very complex issues related to the specificity of living systems and interactions within the bodies and between the bodies and the environment. Unfortunately, it is not uncommon to see or hear statements that create an impression that the author/speaker winks at you and says: "Oh, but we all understand what I mean, right?" It takes courage to say "no" and insist that all the words in the statement are clearly defined.

Some of such frequently used buzzwords have been discussed in previous chapters. For example, what is "internal model"? A model is a formal simplified representation of actual phenomena. It has an input and an output. When I asked one of the well-known champions of the idea of internal models in the brain participating in the control of movements: "What is the input and output of an internal model? In what units are they expressed?," the answer was "It does not matter. They can be any variables." So, internal model is what was commonly called centuries ago "God from the machine" (deus ex machina), an omnipotent being coming out of the blue during a theatrical performance and being able to do anything (Fig. 23.1A). In other words, the brain can accomplish any task (which is obviously wrong), and any experimental observation can be attributed to an internal model in the brain that generated the observed behavior. So, the hypothesis that the brain uses internal models to control movement fails the requirement to be formulated clearly and exactly in a way that allows its experimental verification.

Hypotheses in the field of clinical studies frequently use undefined or ill-defined but broadly accepted terms. Earlier, we discussed the example of "muscle tone." Another fuzzy commonly used term is "upper motoneuron" (Fig. 23.1B). It assumes, but this assumption is rarely spelled out, that the brain contains neurons dedicated to motor processes that send axons to "lower motoneurons" (alpha-motoneurons in the spinal cord) and define muscle activation patterns. First, even neurons forming the corticospinal tract project primarily on spinal interneurons, not on motoneurons. This is also true for brain neurons forming other descending tracts. Second, these projections are only one of the factors that define muscle activation patterns and movement mechanics.

A hypothesis should be specific; that is, it should address a set of well-defined objects and phenomena and not represent a general theory of everything. This makes the hypothesis testable and disprovable. The most general version of the so-called dynamical system theory (reviewed in Schöner and Kelso 1988; Kelso 1995) claims that human effectors and the associated neural structures are dynamical systems, which can show certain behaviors that cannot be accounted for by the motor programming approach (see Chapter 4). In particular, these behavioral features include spontaneous changes in the relative timing of effectors with a change in movement frequency and associated changes in stability of the relative phase of effector movement. Accounting for those phenomena was a major accomplishment of the dynamical systems approach. However, the formulation of the approach made it applicable to any object. Indeed, by definition, any material object is a dynamical system, even

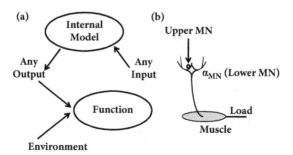

Figure 23.1 Common misnomers in motor control. (a): An internal model as a structure (neural network) that converts any input into any output. It resides somewhere in the brain, likely in the cerebellum. (b): Upper motoneuron is a neuron in the brain projecting on alpha-motoneurons in the spinal cord.

such a simple one as a brick, because its time evolution is defined by the laws of nature. So, the statement "the human brain is a dynamical system" is equivalent to "the human brain is made of matter," which is not a novel idea.

Finally, a hypothesis has to be novel, not a reformulation of earlier hypotheses produced by substitution of earlier terms and variables with another, equivalent set of terms and variables. An example of a hypothesis that fails this step is the idea of generalized motor programs (Schmidt 1975). It assumes that functions of undefined neural variables are developed as a result of practice and stored in the brain. They can be scaled in magnitude and used at different time scales to produce various actions. This definition makes generalized motor programs very similar to the concept of engrams introduced by Bernstein about 40 years earlier (see Chapter 2). Engrams were supposed to encode movement topology at the task level, not movement metrics. In contrast to engrams, however, generalized motor programs were linked to such peripheral variables as forces, torques, and muscle activations. This was an unfortunate step because the brain cannot encode these variables, which depend on the unpredictable changes in external force, as described in a number of earlier chapters.

23.3. Can a hypothesis account for the existing knowledge?

This question seems to be trivial, but a number of hypotheses prefer to ignore uncomfortable facts and consider only findings that are compatible with their predictions. This issue may be viewed as composed of three questions. First, can the hypothesis account for the well-established facts? Second, can it handle observations that competitive hypotheses struggle with? Third, can it question "well-established facts" and lead to their reconsideration?

Consider, as an example, the posture-movement paradox (von Holst and Mittelstaedt 1950/1973; reviewed in Feldman 2015; see Chapters 3 and 14). To remind, the paradox is that deviation of an effector (up to the whole body) from a postural state is quickly counteracted by involuntary posture-stabilizing mechanisms, but only when the deviation is produced by an external force. No comparable posture-stabilizing phenomena are seen when the person performs a very similar movement from the same posture to about the same new state voluntarily. What does happen with the apparently involuntary posture-stabilizing mechanisms during voluntary movements? This very basic observation has been known for many years, and most hypotheses in the field of motor control have tried to avoid it or brushed it away with obviously inadequate arguments. As a result, the posture-movement paradox has turned into a litmus test for motor control hypotheses.

Within the hypotheses of motor control with patterns of muscle activation and/or muscle forces, the posture-movement paradox has been discussed as a consequence of two potential mechanisms (Gottlieb and Agarwal 1988; Gottlieb 1996). First, the central nervous system was assumed to predict the posture-stabilizing reactions and add extra muscle activations (and forces) to move to the new desired position in spite of those reactions. This is theoretically possible, but, obviously, the person would be unable to relax in the final state because voluntary relaxation would stop counteracting the involuntary posture-stabilizing reactions and lead to movement to the initial position produced by these reactions. This prediction contradicts everyday observations that humans can relax after movements without triggering a reverse movement. Second, a hypothesis was suggested that posture-stabilizing reactions were turned off during voluntary movements and then reestablished at the new position. This hypothesis was a typical example of deus ex machina and it also failed experimental verification: A perturbation applied during a voluntary movement leads to strong involuntary trajectory-stabilizing reactions (Hayashi et al. 1990; Grey et al. 2001; Zhang et al. 2016).

The equilibrium-point hypothesis, in contrast, easily accounts for the posture-movement paradox (see Chapter 3). Within this hypothesis, movements are transitions between postural states, and posture-stabilizing mechanisms are of the same nature as movement-producing ones. The effects of posture-stabilizing mechanisms are seen when λ = const., while changing λ to produce voluntary movement converts these mechanisms into movement-generating ones (Fig. 23.2). So far, this hypothesis has been unique in its ability to handle this group of well-established experimental facts that other hypotheses struggle with.

As an example of challenging "well-established facts," consider the following. For years, it was accepted that fast voluntary movements start with a burst of activation in agonist muscles, which is produced by signals from the brain with no contribution from reflex feedback loops because of the conduction delays in those loops (reviewed in Gottlieb et al. 1989a, 1989b). The equilibrium-point hypothesis questioned two aspects of this statement and motivated experiments showing that it was essentially wrong. First, experiments showed that even the timing of the first agonist burst could be changed by small positional perturbations of the effector prior to the movement initiation (Adamovich et al. 1997). Second, when a standing person performs a fast, whole-body voluntary sway movement forward, the movement is initiated not by an activation burst in the agonist muscles but by a drop in the activation level of the antagonist muscles (Mullick et al. 2018; Nardini et al. 2019).

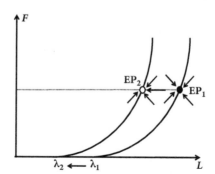

Figure 23.2 When the threshold of the stretch reflex λ = const., deviations from the equilibrium-point (posture) on the force-length (*F-L*) plane lead to involuntary posture-stabilizing reactions. When λ changes, the same reactions lead to the generation of movement to a new EP (from EP_1 to EP_2).

23.4. MAKING NEW TESTABLE PREDICTION 349

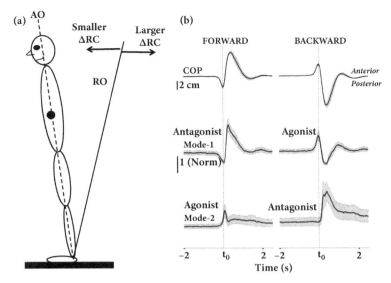

Figure 23.3 (a): During standing, body referent orientation (RO) is behind the body. Its shift backward leads to a burst of activation in dorsal muscles due to the increased difference (ΔRC) between RO and actual orientation (AO). Its shift forward leads to a drop in ΔRC and a drop in the activation of dorsal muscles with a delayed burst of ventral muscles. (b): Experiments confirm this prediction. Modified by permission from Nardini et al. 2019.

As illustrated in Fig. 23.3, this is a consequence of the fact that in the initial posture, the body referent orientation (RO) is behind the body actual orientation (AO). This allows one to generate active torque counteracting the moment of the gravity force. Shifting RO does not lead to activation of ventral muscles (the agonists for the body sway forward) until it moves beyond the AO, which takes some time. Until that time, the body sway is produced by the moment of the force of gravity, which becomes larger than the moment generated by the dorsal muscles. As a result, this voluntary movement starts with a drop in the activation of the antagonist muscle groups, and an agonist activation is seen at a time delay.

23.4. Can a hypothesis be used to make new testable predictions?

This feature is obviously very important because, in the absence of such new predictions, a hypothesis remains weak and indistinguishable from its competitors. Of course, the new predictions should be specific to the hypothesis; that is, they should not be easily made based on other hypotheses in the field. The predictions should be experimentally testable (i.e., falsifiable).

Let us consider the example of the dual-strategy hypothesis (Gottlieb et al. 1989a; see Chapter 4), which has assumed that the brain manipulates the height and width of the presynaptic "excitation pulses" to the alpha-motoneuronal pools for the agonist and antagonist muscles. By the time the dual-strategy hypothesis was formulated, a large amount of information had been accumulated on the muscle activation patterns during single-joint voluntary movement performed under a variety of instructions and conditions. The dual-strategy hypothesis not only was able to offer a compact description for those findings but also suggested experimentally testable predictions for movements performed under conditions and instructions that had not been used before. In particular, it suggested that certain predictable features of the muscle activation patterns would be observed when the subjects were asked to move "at the same speed" over different distances. Note that, if taken at face value, this instruction makes no sense because movement speed is a bell-shaped time function, which scales with movement distance and other factors (e.g., inertial load, target size, etc.). Naïve subjects, however, had no problems with this instruction and performed movements over various distances, with both peak and average speed scaling with movement distance (Gottlieb et al. 1990a). Features of the agonist muscle activation patterns confirmed the predictions of the dual-strategy hypothesis. These observations also suggested that there was an intrinsic variable corresponding to subjective understanding of unchanged "movement speed" even when actual speed varied across conditions. The dual-strategy hypothesis struggled, however, in trying to account for experimentally observed activation patterns of the antagonist muscles (which showed non-monotonic modulation of their burst amplitude; Gottlieb et al. 1989b, 1992; Fig. 23.4) and with movements performed under major changes in external conditions (e.g., in isometric conditions; Corcos et al. 1990). Accounting for those findings required the introduction not only of "excitation pulses" but also of "excitation steps" and even "excitation ramps." This was not completely unexpected, because the hypothesis did not consider changes in muscle activation patterns due to the action of reflex loops, which reflected the actual muscle kinematics. For example, the non-monotonic changes in the antagonist burst amplitude with movement distance was later attributed to the peripheral and reflex-mediated damping of the movement (Jaric et al. 1998): Movements over larger amplitudes were characterized by higher speeds, which produced larger damping effects.

The equilibrium-point hypothesis has generated many novel predictions, and a large number of those have been confirmed experimentally. First, consider the phenomena of equifinality under the action of transient changes in external forces (Chapter 18). They directly follow the main theoretical scheme of the equilibrium-point hypothesis under certain additional assumptions (Feldman and Latash 2005; Feldman 2015). In contrast, they present a problem for hypotheses based on central specification of muscle forces and joint torques because a transient force pulse in a certain direction is expected to violate the whole course of the movement and lead to a different final location. Accounting for these phenomena required champions of those models to accept additional assumptions such as very quick recomputation of the requisite forces and torques to ensure that a proper movement correction is introduced. Such assumptions are not easily compatible with experimentally observed changes in the reaction time under even relatively mild complicating factors. For example,

23.4. MAKING NEW TESTABLE PREDICTION 351

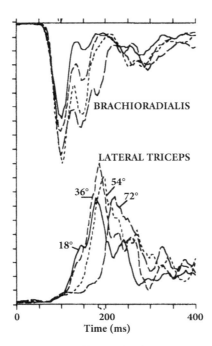

Figure 23.4 An increase in movement distance leads to a monotonic increase in the first agonist burst (brachioradialis), as predicted by the dual-strategy hypothesis, but to non-monotonic changes in the antagonist (triceps lateralis) burst. Modified by permission from Gottlieb et al. 1992.

simple reaction time to an imperative stimulus shows a major increase (by 70 to 100 ms) when two different stimuli can be presented, and the instruction requires different actions depending on the presented stimulus. This additional time reflects making a relatively simple choice, for example, between moving the instructed effector into flexion or extension. It is difficult to imagine that proper correction of forces and torques during a multi-joint fast movement in response to an unexpected transient perturbation can be initiated and implemented within a shorter time.

Other nontrivial predictions of the equilibrium-point hypothesis and its extension in the form of control with spatial RCs (Chapters 3 and 7) have been discussed in earlier chapters. The hypothesis has been able to make predictions with respect to a variety of motor and perceptual phenomena including unintentional force drifts (Ambike et al. 2016b; Abolins and Latash 2022b; Chapter 16), motor and sensory effects of changes in muscle coactivation (Latash 2018b; Cuadra et al. 2020; Chapter 19), force matching with the contralateral effector (Abolins et al. 2020; Chapter 15), kinesthetic errors and illusions (Feldman and Latash 1982a, 1982b; Cuadra et al. 2021a; Chapter 17), features of postural sway (Zatsiorsky and Duarte 1999, 2000; Chapter 14), features of movements and muscle activation in spasticity (Levin and Feldman 1994; Turpin et al. 2017; Chapter 23), synergies at the level of control variables (Ambike et al. 2016a; Nardon et al. 2022; Chapter 7), features of motor unit recruitment (Madarshahian and Latash 2022a; Chapter 11), responses to transcranial magnetic stimulation of the primary motor cortex (Raptis et al. 2010; Ilmane et al. 2013; Chapter 10), responses to galvanic stimulation of the vestibular system (Zhang et al. 2018), and many others.

An early precursor of the equilibrium-point hypothesis, the servo-hypothesis (Merton 1953), was novel, was formulated using exact terms and notions, and allowed making new testable predictions. This hypothesis assumed that the stretch reflex represented a perfect musclelength servo-control mechanism and the brain manipulated the target muscle length by signals to gamma-motoneurons innervating spindle sensory endings in the muscle (Fig. 23.5A). Changes in the external load were compensated by the stretch reflex and the muscle reached the set length value independently of load magnitude. This is illustrated in Fig. 23.5B with a set of force-length characteristics that differ by their threshold (set by the gamma-motoneurons) and are nearly vertical; that is, the gain of the stretch reflex was assumed to be very high.

Two predictions followed the scheme in Fig. 23.5A. First, movements were assumed to be initiated by a change in the signals to gamma-motoneurons, while alpha-motoneurons were expected to show changes in their activity after a substantial time delay due to the conduction delays along the relatively thin and slowly conducting axons of the gamma-motoneurons and thicker and faster-conducting axons of the spindle sensory endings. This prediction was falsified in studies showing the phenomenon of alpha-gamma coactivation (reviewed in Granit 1975; Hagbarth 1993), that is, simultaneous changes in the activity of alpha- and gamma-motoneurons. This finding required a modification of the hypothesis, which assumed that signals from the brain set target values of muscle length by signals to gamma-motoneurons and, simultaneously, accelerated movement initiation by signals to alpha-motoneurons

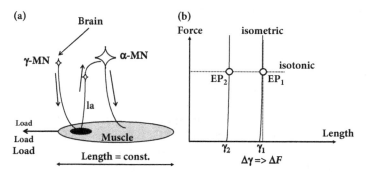

Figure 23.5 (a): A schematic illustration of the servo-hypothesis. Neural command to γ-motoneurons defines a value of muscle length, and the stretch reflex works as a perfect servo-mechanism moving the muscle to this length independently of possible changes in the external load. (b): The same hypothesis is illustrated as a set of nearly vertical force-length characteristics. Note that changing the γ-command leads to accurate movement in isotonic conditions. However, it cannot be used to produce accurate force changes in isometric conditions: A tiny change in γ (Δγ) leads to a disproportionately large change in force (ΔF).

(Matthews 1966). The second problem was more serious: Assessment of the stretch reflex gain showed its relatively modest gain values (Vallbo 1971, 1974), thus questioning the prediction of nearly vertical force-length characteristics. As a result, the servo-hypothesis was falsified. We have to note also that the hypothesis was designed to account for the control of movements, but it was unable to account for the control of force production tasks in isometric conditions. Indeed, as shown in Fig. 23.5B, in isometric conditions, nearly vertical muscle force-length characteristics led to very large force changes for very small command signals to the muscles; that is, the control of accurate force production was inherently unstable.

One should keep in mind that, if a particular novel prediction fails to be supported experimentally, this does not mean by itself that the hypothesis has been falsified. This may or may not be the case. Such observations may be highly valuable for developing and modifying hypotheses. In a sense, they may be parts of the next step, which is exploration of hypotheses and their development.

23.5. Exploration and development of a hypothesis

In branches of natural science, in particular those dealing with living systems, hypotheses are rarely complete. Any hypothesis may be viewed as having a core—a set of concepts and assumptions that distinguish it from other hypotheses—and implementation. Let us consider two examples that have been used earlier, the dual-strategy hypothesis and the equilibrium-point hypothesis.

The core of the dual-strategy hypothesis is assuming that the brain can specify presynaptic input into alpha-motoneuronal pools (Chapter 4). The original implementation of the hypothesis was using rectangular excitation pulses as inputs into the motoneuronal pools with the amplitude and duration of the pulses scaled to fit task requirements. Originally, the effects of changes in movement distance and inertial load on muscle activation patterns were considered only for movements performed as fast as possible (Gottlieb et al. 1989a, 1989b). Later, these rules were extended to movements performed at modest velocities representing development of the hypothesis (Gottlieb et al. 1990a). The concept of excitation pulses was unable to account for the muscle activation patterns observed during force production tasks in isometric conditions and forced the introduction of excitation steps and ramps (Corcos et al. 1990; cf. the pulse-step hypothesis, Ghez and Gordon 1987; an update in Scheidt and Ghez 2007). These steps can be seen as exploration and development of the hypothesis based on new experimental observations: They do not question the core of the hypothesis but offer its new implementation.

The core of the equilibrium-point hypothesis (Chapter 3) is the idea of parametric control of movements using the threshold (λ) of the stretch reflex as the parameter modified by the central nervous system to control a muscle. Movement patterns emerge in the interaction between the muscle and the external load. The original data used to support the hypothesis were obtained in human experiments when the subjects were instructed not to interfere voluntarily with external load changes (Feldman 1966). This assumption remains untestable, although indirect evidence suggests that humans are indeed able not to react to consequences of changes in the external forces

(Latash 1994). This ability, however, is not absolute: Some external forces, in particular destabilizing ones, may not be tolerated by the central nervous system, and it introduces corrections independently of the subject's intention. This is the likely reason for violations of equifinality (expected if the subject is not changing time patterns of λs to the involved muscles) under certain conditions such as performing movements in the rotating centrifuge (Lackner and DiZio 1994) or in the force field with negative damping, which was simulated with the help of a programmable torque motor (Hinder and Milner 2003). So, these violations of equifinality do not challenge the core of the hypothesis but suggest a new, unexpected finding: The subject cannot avoid changes in $\lambda(t)$ under certain conditions. Note that spontaneous changes in $\lambda(t)$ under much less unusual conditions—turning visual feedback off and/or using transient perturbations—have been observed in many studies and interpreted within the framework of the equilibrium-point hypothesis (Zhou et al. 2014b, 2015b; Ambike et al. 2016b; see Chapter 16).

Another example of exploration of the equilibrium-point hypothesis was the reconstruction of equilibrium trajectories and apparent stiffness patterns during a variety of single-joint and two-joint movements (Latash and Gottlieb 1991; Latash et al. 1999). These studies provided evidence for non-monotonic (N-shaped) equilibrium trajectories (Fig. 23.6A). Note that the mentioned studies were performed under certain assumptions regarding the mechanical properties of the moving effectors, including their damping properties. These assumptions were criticized later (Gribble et al. 1998), suggesting that the reflex-mediated damping was much stronger than that assumed in the studies reporting the N-shaped equilibrium trajectories. Since there is no direct method of measuring reflex-mediated damping (see Chapter 24), this issue remains unresolved. This method of reconstructing equilibrium trajectories was also applied to cyclical joint movements and produced nontrivial, meaningful results (Latash 1992). In particular, it showed nearly flat equilibrium trajectories during movements at frequencies close to the estimated joint natural frequency (Fig. 23.6B): One does not need

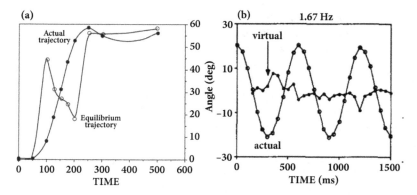

Figure 23.6 (a): The "N-shaped" equilibrium trajectory (labeled "virtual trajectory") reconstructed during fast elbow flexion movements. (b): During cyclical movement as a frequency close to the natural one, the equilibrium trajectory is "flat." Reproduced by permission from Latash and Gottlieb 1991 and Latash 1992.

to control a system, which moves at its natural frequency, only compensate for energy losses. Note that muscle activation patterns showed no special behavior during movements at the natural frequency, suggesting that they are emerging features of the movement, not true reflections of its neural control.

Studies of movements by a two-joint arm system (wrist and elbow), when only one joint was instructed to move, showed similarly timed N-shaped trajectories for both joints; one joint showed a large-amplitude fast movement, and the other joint showed small deviations under the action of joint interaction moments of force (Fig. 23.7). Damping effects in the apparently postural joint could not be strong because of the low velocity of its motion. So, it is hard to imagine that underestimating damping led to the observed large-amplitude N-shape. Evidence for the N-shaped equilibrium trajectories has also been reported in more recent studies (Ramadan et al. 2022). Note that the argument whether equilibrium trajectories are N-shaped or monotonic does not question the core of the hypothesis but explores it and contributes to its development.

The concept of synergic control becomes a testable hypothesis only after being formulated in a specific way and after sets of adequate tools are introduced to test its predictions (see Chapter 6). Two aspects of synergic control, following the classical works of Bernstein (1947), have been developed and tested. The first of them predicted grouping of elements participating in natural movements into a small number of groups, which have been explored using various analyses of covariation such as principal component analysis, independent component analysis, and non-negative matrix factorization (reviewed in Chapter 6). The second aspect related to ensuring dynamical stability of movements required the development of the UCM hypothesis and associated tools to become experimentally testable (Scholz and Schöner 1999; reviewed in Latash et al. 2007; Chapter 5). Note that the first studies used peripheral variables produced by apparent elements as elemental variables within the

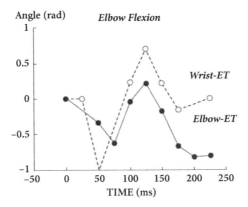

Figure 23.7 Non-monotonic "N-shaped" equilibrium trajectories (ETs) were reconstructed for both elbow and wrist joints during fast movements of the elbow joint while the wrist joint was supposed to stay motionless. Reproduced by permission from Latash et al. 1999.

UCM-based analyses, and only relatively recently has this method been expanded to analysis of variables at the control level, that is, RCs to individual motor units, muscles, and the whole body (Ambike et al. 2016b; Madarshahian et al. 2021; Nardon et al. 2022).

The idea of synergic control was able to account for a large body of data demonstrating dynamical stability of salient performance variables ensured by covariation of elemental variables, starting from the classical experiment of Bernstein on professional blacksmiths (Bernstein 1930; Chapter 2) and including studies of "error compensation" across tasks and effector sets (Abbs and Gracco 1984; Cole and Abbs 1987; Latash et al. 1998). This idea also offered an interpretation for the phenomena of motor equivalence (see Chapter 18). This theory has been able to formulate predictions and stimulate new experiments exploring stability of movements across tasks and populations. It led to the discovery of new phenomena such as anticipatory synergy adjustments (see Chapter 8) and trade-offs between stability and optimality of movements (Chapter 9). Its application to neurological patients and subclinical populations has shown that indices of synergic control are highly sensitive to changes within the central nervous system and can serve as biomarkers of certain neurological disorders (Chapter 22). Currently, the theory is at the step of exploration. A number of questions that have been addressed recently or are waiting to be addressed include the following:

Which neurological circuits are crucially involved in the synergic control of movement?
How can one improve aspects of the synergic control by exercise?
Are there effects of hand (foot) dominance on indices of synergic control?
How is movement stability shared between spinal and supraspinal circuits?
Is stability ensured by intra-muscle synergies different across muscles?

23.6. Missing pieces of the mosaic

The general rules and logics of offering new hypotheses and running experimental studies testing and exploring those hypotheses in the natural sciences have been established and followed by researchers in the physics of inanimate nature for years. Nevertheless, hypotheses and even theories are being offered in the field of life sciences, including motor control, using undefined terms and/or ignoring basic principles of the scientific method. This is not simply a missing piece of the puzzle but rather the most obvious factor contributing to the fact that the mosaic remains broken and incomplete.

Researchers frequently focus only on experimental findings compatible with their favorite hypothesis, while they brush uncomfortable facts under the carpet. Commonly, there are no honest attempts to disprove one's favorite hypothesis, only to support it. Arguments of opponents are not answered in an open debate but ignored or ridiculed forgetting one of Murphy's laws: "A witty statement proves nothing." There are a few encouraging examples where competing views are presented and analyzed based on discussions of experimental observations. Such examples are some studies with criticisms of the equilibrium-point hypothesis (Lackner and DiZio 1994; DiZio and Lackner 1995) and responses of champions of the hypothesis (Ostry and Feldman 2003; Feldman and Latash 2005). The same can be said about some of the discussions surrounding the UCM hypothesis (Müller and Sternad 2003; Sternad et al. 2010; Schöner and Scholz 2007; Scholz and Schöner 2014).

Of course, exploration of hypotheses, including the equilibrium-point and UCM hypotheses, is rarely complete. Researchers have to continue searching for the limit of applicability of the hypotheses, thus encouraging their exploration and development up to a point when these hypotheses are in principle unable to handle experimental observations and have to be replaced with alternative theories.

24
Measuring Hidden Variables

Measuring salient variables objectively and accurately has always been a prerequisite for any experimental study in the physics of inanimate objects. Much of the progress in physics has been associated with the invention and refinement of new measuring tools, from thermometers to microscopes and telescopes, and the construction of complex structures such as particle accelerators. About 100 years ago, it was recognized that the process of measurement always involves action on the object of study and, if the object is small, its state can be changed by the measurement process. This insight was formalized by Heisenberg as the uncertainty principle: One cannot determine precisely the position and momentum of a particle at the same time. More exactly, the product of the uncertainty in position and uncertainty of momentum cannot be smaller than the Planck constant divided by 4π. In more intuitive terms, if you want to see where a particle is, you have to direct a beam of light at it, and this beam will perturb the particle's momentum. So, you will not know precisely what the momentum was at the time you measured the position of the particle.

Problems similar to the uncertainty principle are inherent to measurement of salient variables in living systems. These problems can be summarized as follows: *Living systems react to processes of measurement by changing the variables that the researcher tries to measure.* While the uncertainty principle practically applies to measuring properties of only very small particles (because the Planck constant is very small), similar problems with living systems apply at a much larger scale. As we will discuss in the first section of this chapter, the biological version of the uncertainty principle makes it very difficult to estimate even very basic biomechanical properties of muscles and joints, such as their stiffness-like and damping-like characteristics (for detailed reviews see Latash and Zatsiorsky 1993, 2016; Zatsiorsky 2002). This problem makes the application of typical methods developed in engineering, such as system identification methods, inapplicable, or at the very least, they would require major modifications.

An even more serious problem is related to the lack of instruments for measuring salient neural control variables, such as λ for muscles and referent coordinates (RCs) for multi-muscle effectors: We have no reliable "lambda-meters." Given certain features of the system for movement production, in particular the multiple feedback loops from peripheral receptors with different characteristic time delays, it is even hard to imagine a mental experiment that would be adequate to measure changes in those truly hidden variables. There is no place in the brain where one could stick an electrode and measure $\lambda(t)$ in real time. Maybe this is the perfect time to mention the story about God's cousin that was told by Bernstein (I am very much grateful to Professor Vladimir Zatsiorsky, who told me this story).

> God's cousin was never very bright and somewhat upset by the fact that nearly all people knew about his cousin (the God) and respected him deeply, while he (the

cousin) remained unknown to most. So, he asked the God to help him by making him a famous scientist. To be more specific, he asked to be able to change his size, travel in space, and get information about the internal connections within any object. The wish was granted! First, the cousin decided to find out the structure of elemental particles. He became very tiny, crawled inside the electron, saw a few quarks, made notes, published a paper in *Science*, and got his first Nobel Prize. Second, he decided to find out whether there was life on other planets. He traveled in no time to all the planets, observed signs of life, made notes, published a paper in *Science*, and got the second Nobel Prize. Finally, he decided to find out how the brain controls movements. He requested information on all the connections of all the neurons, muscles, and sensory endings; got the blueprint; and, if the story is true, he is still sitting there looking at the blueprint.

This story is directly related to the problem of measuring hidden variables such as RCs. These variables likely reflect processes within vastly distributed networks, and there is no place to stick an electrode (or even an array of electrodes) to measure them. Much of this chapter is dedicated to sneaky ways of measuring such variables and associated problems.

24.1. Measuring "biomechanical" variables

Biomechanics is different from classical mechanics in a number of aspects, which makes application of many classical notions from mechanics complicated or even misleading. These include, in particular, the notions of stiffness and damping. The notion of stiffness is used in mechanics to describe objects that deform under the action of external force, accumulate potential energy, and release it when the external force is removed. The dependence of the magnitude of deformation on force change can be linear, at least locally: $\Delta F = -k \cdot \Delta X$, where ΔF is a change in force acting along a spatial coordinate X, ΔX is deformation, and k is a coefficient addressed as stiffness. Damping is defined with respect to force resisting motion as a function of velocity: $F = -b \cdot V$, where V is velocity, and b is a constant. To avoid confusion, we are not going to consider the notion of impedance, which is commonly invoked in human movement studies (e.g., Hogan 1985; Hogan and Sternad 2012; Sainburg 2014) but is defined differently across studies or not defined at all. Most commonly, it implies all three components of resistance to external force assuming a second-order linear system: inertial, damping, and stiffness, somehow lumped together.

Application of these notions to passive biological structures (e.g., bones and tendons) is relatively straightforward. The situation becomes complicated when one considers muscles with their connections to the central nervous system and complex structures, such as joints and limbs, involving muscles. Within a certain range of length and velocity changes, muscles, joints, and limbs can show behaviors similar to those of spring and damping elements. But their properties change significantly with muscle activation level, which is a function of length and velocity due to the action of reflex loops, in particular the stretch reflex. To avoid confusion, we will use the terms *apparent stiffness* and *apparent damping* to address stiffness-like and damping-like behaviors, which are produced by complex, multi-element systems.

Figure 24.1 shows schematically the dependence of muscle force on length (stiffness-like property; all measures are taken at steady states) for a passive muscle, active muscle with no reflex connections, and intact muscle. Note that the passive muscle behaves like a compliant spring; the active muscle behaves like a stiffer spring, which shows a non-monotonic force-length relationship; and the intact muscle shows a steeper force-length curve, which is monotonic and more linear as compared to the muscle without reflexes (cf. Nichols and Houk 1976). Sometimes the descending arm of the force-length characteristic has been addressed as "negative stiffness" (Dyhre-Poulsen et al. 1991). This is an impossible term reflecting the fact that the object of study is not a spring and cannot be approximated using the spring analogy, and the notion of stiffness is inapplicable.

Similar problems exist with the notion of damping. Muscles resist external forces in proportion to velocity, but the role of reflex loops in the magnitude of this resistance remains unknown. One can expect this contribution to be substantial due to the velocity sensitivity of primary spindle endings, but so far there have been no studies exploring this contribution quantitatively. An interesting and appealing notion of fractional power damping has been suggested implying that force acting against a velocity vector is proportional to velocity in the power of $1/n$, where n is an integer (Gielen et al. 1984; Novak et al. 2000). Note that this kind of damping produces values lower than "typical linear damping" when velocity is high and higher values when velocity is lower. So, effectively, it provides lower resistance during fast movements and helps dampen movements at their termination and avoid terminal oscillations. The importance of damping during fast movements is illustrated by the nonmonotonic change in the antagonist burst amplitude within the typical triphasic pattern (see

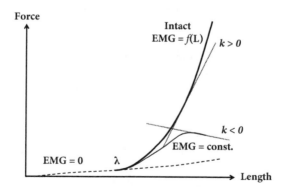

Figure 24.1 The illustration shows typical dependences of muscle force on muscle length in a relaxed muscle (the dashed line, EMG = 0), in a muscle at a constant level of excitation (the thin, solid line, EMG = const.), and the same muscle with intact reflexes (the thick, solid line, EMG = $f(L)$). Note that the slope of the curve can be positive ($k > 0$) or negative ($k < 0$). Negative k values mean that the system cannot be approximated as a spring.

Chapter 4 and Gottlieb et al. 1989b) with an increase in movement amplitude. For movements over larger amplitude, velocity in the middle of the movement becomes proportionally larger and damping forces increase, thus helping the antagonist muscles to brake the movement (Jaric et al. 1998).

Consider Fig. 24.2, which shows schematically a few mechanisms leading to muscle behavior with changes in its length. Note that a few of those involuntary mechanisms act at various time delays on the order of 50 to 100 ms, keeping in mind the electromechanical delay (cf. Corcos et al. 1992). Some of the reactions are transient (phasic reflexes and reflex-like reactions), while others are maintained over time (tonic reflexes). Note also that muscles are relatively sluggish and take considerable time to reach peak force in response to a change in activation level. Imagine now that a muscle activated to a particular steady level is quickly stretched by an external force. It will generate force related to both length and velocity at close to zero time delay due to properties of its tissues. Note, however, that both force-length and force-velocity relations are activation dependent. While the muscle is generating force due to those peripheral properties, signals from spindle endings sensitive to length and velocity will travel to the spinal cord and produce a sequence of changes in the muscle activation level due to their monosynaptic and polysynaptic projections on alpha-motoneurons. These changes will lead to changes in the force-length and force-velocity properties of muscle tissues, which will be superimposed on the ongoing peripheral muscle reaction to movement. As a result, the muscle response will represent a superposition of heavily filtered (due to sluggish muscle force generation) sequences of responses with various time delays. Neither the time delays nor the gains in those loops are known a priori.

A number of studies used system identification techniques borrowed from engineering, applied very brief and small random perturbations to the muscle length, measured its force changes, and used second-order linear models to fit the behavior.

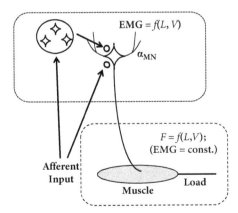

Figure 24.2 A schematic illustration of several mechanisms leading to muscle behavior with a change in its length. F, force; L, length; V, velocity; EMG, muscle activation. There are no reliable estimates of the reflex-mediated EMG-velocity and force-velocity dependence.

Note that correlating force with length and its derivatives measured at the same time instants effectively turns the muscle deafferented because reflex loops are given no time to exert their effects. Effects of these loops are, however, not negligible (e.g., see Fig. 24.1). As a result, these methods likely produced significantly underestimated values of apparent stiffness and apparent damping, as compared to those relevant to natural movements. On the other hand, comparing magnitudes of length and force measured at different times is also not of much use because the muscle activation levels at different times differ and, as a result, the magnitudes of apparent stiffness and apparent damping differ too: The muscle reacts to the process of measurement by changing its characteristics that the experimenter tries to quantify!

One of the lessons learned so far is that to measure muscle characteristics relevant to natural movements, one has to apply smooth, sneaky perturbations, which do not give rise to transient reflexes. It would be even better to apply small, slow changes in the external force field that do not lead to unintentional, reflex-like movement corrections but produce measurable changes in movement mechanics.

24.2. Measuring lambda

The original experiments that formed the basis of the equilibrium-point hypothesis (Feldman 1966; see Chapter 3) involved partial unloading of muscles under the instruction to the subjects not to intervene voluntarily with movements induced by the changes in the external load. Such unloading produced movements of the effector from the initial to some final equilibrium state (equilibrium point [EP]; Fig. 24.3). Using unloading of different magnitudes produced a sequence of EPs, which differed in muscle force, length, and level of activation. These EPs were interpolated to produce an intercept with the coordinate axis when muscle activation was zero—the threshold of the stretch reflex (λ). Note that the slope of the force-coordinate curve increased with deviations of muscle length from λ, reflecting the increase in the

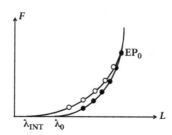

Figure 24.3 An illustration of experiments with muscle unloading assuming "non-interference" by the subject. Sequential equilibrium points are shown with black circles. Note that if the subject reacts (shifts λ) in a consistent way (open circles), estimates of the apparent stiffness (slope of the force-length curve) and λ (λ_{INT}) will differ depending on the way the subject adjusts λ to the unloading.

muscle activation level, which produced higher magnitudes of the apparent stiffness (Feldman 1966, 2015).

In human experiments, limiting manipulations to a single muscle is impossible due to the anatomical constraints. So, unloading and movement always affected other muscles crossing the joint, in particular the antagonists. Muscles crossing other joints were also likely affected, in particular due to the presence of biarticular and multi-articular muscles as well as inter-joint muscle reflexes (reviewed in Nichols 2002, 2018; Miyake and Okabe 2022).

Figure 24.4 shows an illustration of two muscles crossing a joint with one kinematic degree of freedom (e.g., the elbow joint), a flexor, and an extensor. Sequential unloading applied to one muscle at a time would allow reconstructing the muscle's force-length (torque-angle) characteristic. This is, however, impossible because the antagonist muscle always reacts to changes in the external load, even if its activation is zero. If, however, in the initial state both muscles show non-zero activation levels (i.e., the coactivation command is not zero; see Chapter 3), changes in the external load would affect the activation levels of both muscles and lead to new equilibrium states of the joint following the algebraic sum of the two muscle characteristics (the thick line in Fig. 24.4). Removing the external load completely would reveal a joint coordinate corresponding to the reciprocal command (R-command in Fig. 24.4)—the RC for the joint. The coactivation command (C-command) is reflected in the slope of the joint characteristic (i.e., its apparent stiffness, k), but to measure it directly, one would have to change the load within a wide range and identify points where each of the two muscles stops generating active force (i.e., its electromyographic [EMG] signal disappears). The spatial range between λ_{FL} and λ_{EXT} (λs for the flexor and extensor) is the C-command.

The assumption that the subjects were able not to intervene with the unloading procedures remained a weakness of the approach. However, similar results were obtained in studies of animal preparations with the neural axis cut at various levels (Feldman and Orlovsky 1972) when the possibility of intentional interference was minimized. In those experiments, measurements could be limited to individual muscles: The

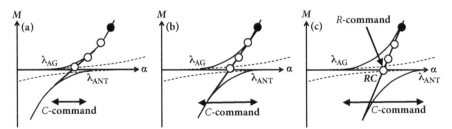

Figure 24.4 Effects of changes in coactivation of the antagonist muscle (changes in its stretch reflex threshold, λ_{ANT}) on the reconstructed torque-angle characteristic. The initial equilibrium point is shown with the black circle. Equilibrium points following unloading are shown with white circles. Compare the situation without coactivation (a), with low coactivation (b), and with high coactivation (c). Panel (c) shows the referent coordinate of the joint (RC).

distal tendon of the muscle was separated from the point of its attachment and connected to the measurement device, which allowed changing the length of the muscle without affecting the length of other muscles of the limb. Besides, a study explored the behavior of humans under the instruction "do not intervene" and "intervene as quickly as possible" and showed more consistent behavior under the former instruction (Latash 1994). These observations suggested that non-intervention was indeed a consistent strategy among the subjects, more consistent than intervention.

Note that the described method of muscle unloading requires several measurements to identify a single value of λ. This method had to be developed to allow its practical application to study time functions $\lambda(t)$ and possible changes in λ across trials. To be more exact, in human studies, the method had to be developed to allow estimating the R- and C-commands. Two developments of the method have been proposed to estimate these hidden control variables, both limited in the range of tasks they could be applied in.

One of the methods could be applied only to steady-state tasks when the subjects could be assumed to keep values of the R- and C-commands unchanged during the trial duration. The method involved the application of a smooth, small, short-lasting positional perturbation to the involved effector and measuring the time profiles of its force and coordinate (or torque and angle). Using a smooth perturbation allowed avoiding phasic reflexes and reflex-like reactions (cf. Latash and Gottlieb 1990). A small magnitude of the perturbation was used to allow linear approximation of the effector's behavior. The perturbation was short-lasting to facilitate "non-intervention" by the subjects. The method was applied to a range of effectors, from a single finger to the whole body (Ambike et al. 2016b; Nardon et al. 2022). The idea of the method is illustrated in Fig. 24.5A. It is assumed that the commands to the agonist and antagonist muscles lead to some values of the R- and C-commands, which define a linear dependence of the force and coordinate variables. If the effector is moved slowly, its coordinate and force are expected to follow this characteristic because the acceleration and velocity may be assumed to be very low, resulting in close-to-zero inertial and damping forces. Indeed, in most trials, the effector showed a close-to-linear trajectory in the force-coordinate space (Fig. 24.5B).

Practically, the perturbation lasted for about 500 ms and was small. For example, in finger studies, it lifted the finger by 1 cm only. Only the middle portion of the effector trajectory was used to avoid edge effects that could be brought on by reflex delays and non-negligible acceleration early in the movement and the subject's unintentional interference later in the movement. The 250 to 300 ms in the middle of the movement commonly produced strongly linear changes in the force and coordinate across trials and subjects.

One of the first direct applications of the method was the demonstration and quantification of synergies in the space of the mechanical variables reflecting the R- and C-commands (RC and k; see Fig. 24.4 and Ambike et al. 2016b). The measurements of RC and k across repetitive trials at the same task have demonstrated large variability in both variables with a strong hyperbolic covariation reflecting stabilization of the task variable (force or moment of force) at the initial state (Fig. 24.6 and Chapter 7). This is not a trivial observation. For a given level of inter-trial force variability (shown as dashed lines in Fig. 24.6), accurate performance does not require large inter-trial

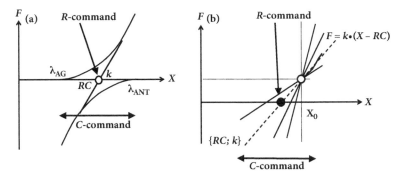

Figure 24.5 An illustration of the method to quantify mechanical reflections of the R- and C-commands (RC and k, respectively). (a): A combination of commands to the agonist and antagonist (λ_{AG} and λ_{ANT}) can be represented as values of the R-command and C-command, which define referent coordinate (RC) and apparent stiffness (k) of the effector, respectively. (b): Producing a value of force in isometric conditions is compatible with numerous {RC; k} pairs. A few of the corresponding force-coordinate characteristics are shown as straight lines. Changing the coordinate of the effector smoothly in a trial is expected to lead to a close-to-linear changes in force with coordinate following the actual characteristic (thick line). The slope of the regression is k, and its intercept is RC. RC, referent coordinate; k, apparent stiffness.

variability in both RC and k along the solution space (the uncontrolled manifold for the task). It is compatible with variance in either RC or k only. The limitation of the method only to steady-state tasks remains its significant drawback.

Another method was developed to estimate the time profiles of the R- and C-commands during single-joint movements performed at different speeds (Latash

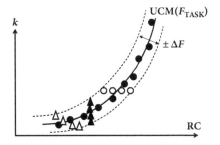

Figure 24.6 When the method described in Fig. 24.5 is used multiple times across trials when the subject produces the same force at the same coordinate, pairs of data points {RC; k} are expected to be constrained to a hyperbolic line (filled circles). This is not a trivial result: Similar accuracy of performance can be obtained with different data distributions shown as open circles (only RC varies), filled triangles (only k varies), and open triangles (both RC and k vary but their covariation is opposite to that predicted by the UCM hypothesis).

and Gottlieb 1991, 1992; see also Chapter 7). This method used analysis across multiple trials at the same task assuming that the subject was reproducing the same (or nearly the same) time functions, $R(t)$ and $C(t)$, across trials, which differed in possible time changes of the external load. The load changes were unpredictable, slow (with the characteristic times close to the movement time), and relatively small, and the subjects were instructed and trained "not to intervene" with movement changes that could happen within individual trials. The idea of the method is rather straightforward (Fig. 24.7). If all the trials are aligned, and the joint angle and moment of force are estimated at the same movement phase (time t_i), these data points are expected to be confined to the moment-angle characteristic defined by the instantaneous values of $R(t_i)$ and $C(t_i)$. Figure 24.7A illustrates three torque and angle trajectories performed when the external torque was unchanged and when it decreased or increased slowly over a time interval comparable to the movement time. The pairs of computed joint torque and measured joint angle variables are shown in Fig. 24.7B with the regression line. In a linear approximation, the intercept of the characteristic will produce an estimate of $R(t_i)$ and its slope will reflect $C(t_i)$. This analysis can be performed for different t_i values, thus yielding an estimate of the time evolution of $R(t)$ and $C(t)$.

The method hinges, however, on adequate estimation of joint torque. One has to accept a mechanical model of the moving joint, which is far from trivial. While external torque can be measured and inertial torque can be estimated relatively accurately, the damping torque component remains estimated poorly at best. During fast movements, the damping component cannot be ignored, in particular during movement phases at high speeds. This point was emphasized in a paper (Gribble et al. 1998) criticizing the original studies, in particular the reported N-shaped $R(t)$ functions (Fig. 23.6A in Chapter 23). In the original studies, the damping was estimated using published damping coefficient values, but those had been obtained in

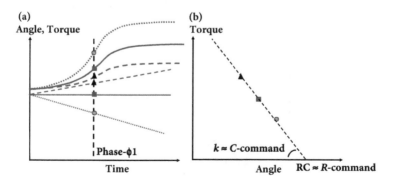

Figure 24.7 (a): A schematic illustration of three joint trajectories (red, thick lines) and corresponding external torques (blue, thin lines) assuming that the subject did not react to the change in the external torque. At a selected phase (φ1) joint torque was computed and angle measured in each trial. (b): These three points are shown on the torque-angle plane. They can be interpolated by a straight line, resulting in estimates of its intercept (referent coordinate, RC) and slope (apparent stiffness, k), which are mechanical reflections of the R- and C-commands.

biomechanical studies that effectively ignored the reflex-based contributions to damping mentioned earlier in this chapter (cf. Latash and Zatsiorsky 2016). Whether underestimating damping could lead to the N-shaped control functions remains debatable. Non-monotonic N-shaped $R(t)$ functions were also reported in studies of joints that moved over relatively small amplitudes and at relatively low speeds. This was done during movement of one joint within a two-joint wrist-elbow system; the other, non-instructed joint moved a little, and N-shaped $R(t)$ functions were reconstructed for both joints (Latash et al. 1999; see Fig. 12.3 in Chapter 12). Similar results were obtained in a study attempting to reconstruct joint-level equilibrium trajectories during a whole-body movement (Domen et al. 1999). Such functions have also been reported in more recent studies (Ramadan et al. 2022).

Overall, the lack of a reliable method to estimate $\lambda(t)$, $R(t)$, and $C(t)$ in real time across various movements and effectors remains a major problem slowing down the accumulation of the experimental foundation necessary for progress in the development of the theory of motor control with spatial RCs (described in Chapter 7). It also slows down applications of this theory to the control of movement in special populations and to changes in the neural control with such factors as fatigue, exercise, development, aging, and rehabilitation. A step in this direction has been made recently in studies of spasticity based on the equilibrium-point hypothesis.

24.3. Measuring lambda in clinical studies

In Chapter 22, we discussed a hypothesis on the origin of spasticity in patients after cortical stroke (see also Levin and Feldman 1994; Mullick et al. 2018). According to this hypothesis, the cause of both components of muscle spasticity—loss of voluntary control and presence of involuntary contractions—is the inability of these patients to shift the control variable to the muscle (λ) within its normal range, which is larger than the biomechanical range of changes in muscle length. To test this hypothesis and turn it into a practically useful tool, a method of measuring λ was needed. This method was introduced based on one of the basic features of spasticity, as formulated in its definition by James Lance (1980): the velocity sensitivity of the stretch reflex loop.

The idea of the method is illustrated schematically in Fig. 24.8. If a muscle is stretched quickly, its effective stretch reflex threshold (λ^*) changes because of the sensitivity of primary spindle endings to muscle velocity. In other words, a stretched muscle is expected to show the first signs of activation at smaller length values as compared to the same muscle with the same descending command tested in static conditions. A linear relation has been suggested for this dependence: $\lambda^* = \lambda - \mu V$, where V is velocity and μ is a coefficient reflecting velocity sensitivity of the stretch reflex (Abdusamatov and Feldman 1986; reviewed in Feldman 2015). As described earlier (Chapter 3), the dependence of λ^* on movement velocity is an important factor leading to the emergence of triphasic patterns of muscle activation during quick movements.

Imagine now a person who is asked to relax a muscle and then an external force moves the joint crossed by the muscle and stretches it. The muscle will show the first

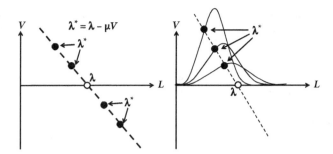

Figure 24.8 Left: When a muscle is moved at a non-zero velocity, its effective stretch reflex threshold is not λ but $\lambda^* = \lambda - \mu V$, where V is velocity (λ^* values for different velocities are shown with filled circles). Right: Stretching a muscle multiple times at various velocities produces different values of λ^* estimated as the coordinate when the first visible muscle activation emerges. Interpolating these values allows defining the intercept when $V = 0$ and $\lambda^* = \lambda$.

signs of electrical activation at some combination of length (L) and velocity (V); this is the value of λ^* in that particular trial. If such trials are repeated starting from the same initial state but using different external forces leading to different velocity time profiles, muscle activation is expected to start at different muscle length values. In other words, one gets a group of data points following a relation between λ^* and V. Assuming the linear dependence between these two variables, $\lambda^* = \lambda - \mu V$, one can perform linear regression of those data points on the $\{L; V\}$ plane and obtain the intercept corresponding to zero velocity, that is, when $\lambda^* = \lambda$. A number of studies have been performed using this method and reporting a smaller range of voluntary changes in λ in patients with spasticity as compared to healthy controls (Turpin et al. 2017; Piscitelli et al. 2020; Frenkel-Toledo et al. 2021).

While the method is based on a solid theoretical foundation, its practical application obviously hinges on a few assumptions. The first and most obvious one is that the relation between λ^* and μ is linear. The number of data points obtained in typical studies corresponds to the used number of different velocity profiles and is not very large. It may be on the order of 6 to 10 points in patients with spasticity, but it may be only 2 or 3 in a healthy person (see Turpin et al. 2017). Using linear regressions for such sets of data points is not very convincing. On the other hand, assumptions of linearity are commonly made in studies of human behavior, and one should not be excessively critical of this step.

The step of identifying the first signs of electrical activity in a stretched muscle is also far from being trivial. In such experiments, muscles rarely show abrupt activation bursts comparable to those during very fast voluntary movements. More commonly, the buildup of muscle activation is slow, and its baseline value is inherently noisy. Identifying a moment of time when the first motor unit in the muscle shows its first action potential (true stretch reflex threshold) obviously depends on the method of measuring the muscle activation signal (EMG), the location of the "first motor unit" with respect to the recording electrodes, and the noise level.

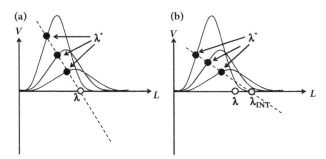

Figure 24.9 Assume two subjects tested using the method illustrated in Fig. 24.8. One of them (a) follows the instruction "do not interfere" perfectly. The other subject (b), an identical twin of the first one, shifts λ in proportion to the perceived velocity of muscle stretch. Note that the subjects may produce different data sets (values of λ*), leading to a conclusion that their initial λ values are different.

Still another assumption is more serious. It assumes that, when a person is asked to relax, he or she specifies the same value of λ larger than the muscle length across trials. However, as discussed earlier with respect to the concept of muscle tone (see Fig. 22.2 in Chapter 22), the instruction "to relax" is imprecise and allows the same person to use a wide range of λ values across trials. For example, following a strong perturbation in a specific trial, the subject may unwillingly specify a smaller value of λ in the next trial. This would be a natural reaction, conditioned by the evolutionary process, to facilitate a quick reaction to another strong perturbation if and when it comes. This could, by itself, lead to a steeper λ*(V) relation, as illustrated in Fig. 24.9 assuming two subjects, a healthy person who is not surprised by strong perturbations and does not adjust λ, and a patient who does.

In spite of the mentioned problems, the method remains unique in the field in using a theory-based method of estimating a hidden neural control variable in a clinical population. It has led to a number of results, including correlations of the estimated range of active control of λ and clinical indices of spasticity such as the modified Ashworth scale. It has also been predictive of the ranges of joint angles where patients were able to perform voluntary movements without triggering spastic contractions.

24.4. Missing pieces of the mosaic

The impressive progress in the development of methods to study activation patterns in individual neurons, individual motor units, and large neuronal populations (reviewed in Merletti et al. 2008; Patil and Thakor 2016; Luo et al. 2022; Farina and Enoka 2023) has not been accompanied by a comparable progress in methods able to measure "hidden variables" relevant to the process of the control of movements. Arguably, the main problem is that researchers are hesitant to accept a physiological theory on the control of movements (or come up with their own one!), and without accepting a theory, methods will remain limited by their exploratory nature. They may be adequate to accumulate lots of information, but this seems to be a good time to recall the story about "God's cousin" mentioned earlier in this chapter. Collecting more data and hoping that big data methods will provide answers by themselves is futile: Answers become useful only when questions are formulated exactly.

So far, the available methods of estimating hidden variables remain indirect and based on a number of assumptions that may be incorrect or incomplete. Merging mechanics-based methods with electrophysiology seems a promising direction on the route to developing better tools to quantify such variables because they typically have two facets, mechanical and neurophysiological. For example, the basic variable λ has been discussed as a measure of subthreshold depolarization of alpha-motoneurons (a neurophysiological variable) and as a threshold of the stretch reflex measured in units of muscle length (a mechanical variable). R- and C-commands are defined in spatial units (see Chapters 3 and 7). On the other hand, they are reflected in patterns of activation within agonist-antagonist muscle pairs and have been estimated using EMG-based indices (Slijper and Latash 2000; Piscitelli et al. 2017). Recently, relations between these two basic commands and patterns of recruitment of motor units have been demonstrated (Madarshahian and Latash 2022b). These explorations, however, remain sporadic and disjointed, not combined to produce a method of measuring these basic commands as time functions in the course of natural movements.

The situation is even less promising in the field of applied movement studies. While the language of control with RCs and the related concept of performance-stabilizing synergies have been used in clinical settings (see Chapter 22), estimating "hidden variables" in impaired populations is even more challenging because of the likely impaired ability of these persons to follow instructions consistently across multiple trials and conditions. To be clinically useful, methods have to consider such factors as increased motor variability and quick fatigue, both physical and mental, in patient populations.

In spite of the generally bleak current situation regarding adequate methods to measure salient variables, it may be seen as highly promising for young researchers. It is waiting for inventors of new "lambda-scopes" and "$\{R; C\}$-meters," which will lead to revolutionary discoveries.

25
Writing Papers

Every laboratory is different, as is every researcher. Some of my colleagues view writing papers as a chore that distracts from the flow of thinking and experimental studies. Others view writing papers as a necessary, important, and useful step. This author belongs to the latter group: Writing and publishing a paper allows one to summarize a particular step in the research program, run it through the peer review process, and use it as a stepping stone for designing and running the next studies. I am not a fan of any pseudo-scientific metrics, including the number of publications, h-index, impact factor, and other indices, which are unfortunately used in the promotion and tenure process, awarding of grants, and so forth. All of my friends-scientists agree that the only measure of one's stature in science and quality of research is the attitude of colleagues, not any pseudo-objective index, including the number of published papers.

Writing papers should never be turned into a competition of who publishes more. A single good paper may be more important for stimulating ideas and general development of the field than a dozen papers describing routine experimental studies with results compatible with a number of theories or hypotheses, not challenging them.

There are too many types and styles of papers to offer a single set of advice on writing them. A review, a theoretical paper, a model, an experimental study, and an applied study require different formats, not to mention that journals specializing in various types of studies have very different requirements. In this chapter, I assume that the paper reports an experimental study driven by a certain set of theoretical views and tests explicitly formulated specific hypotheses. It is being prepared for a traditional journal without a preset limitation on the number of words or pages, the number of illustrations, and so forth. I also assume that the purpose of the described study is to test specific hypotheses challenging predictions of a particular theory in the field. In other words, the paper is about ideas and concepts, not about presenting data for data's sake.

Recently, discussions have started related to various artificial intelligence (AI) tools, such as ChatGPT, and their ability to "write papers." As of now, papers produced by AI are verbose and full of general, nonsubstantial statements, ridiculous exaggerations, and sometimes factual errors. I have had the misfortune of reviewing some of such manuscripts written by authors who opted to use these tools to avoid mistakes in their nonnative English. This is understandable, but still, a well-written nonsubstantial text remains a nonsubstantial text, not a publishable contribution to the field. However, the field of AI is developing quickly, and by the time this book is published, AI likely will be able to produce much better, more coherent texts. On the other hand, this author does not believe that AI will ever generate new meaningful ideas and theories. In a sense, "artificial intelligence" seems to be an oxymoron: It is either artificial or intelligence. Computers are only quantitatively different from the good old abacus as far as intelligence is concerned: Both cannot do much without humans. What looks like "intelligence" was programmed by truly intelligent biological systems (programmers), but even the smartest ones are not able to predict the development of specific fields of research and write software routines producing new nontrivial ideas and theories.

Being a rather conservative person, this author is not a big fan of novelties in the field, even such obviously useful ones as reference management tools. For the author, going through the reference list and introducing small stylistic changes is another opportunity to check whether the references have been selected properly. Besides, these tools typically import references with the associated journal-specific formatting, for example, using capital first letters for each word in the title or only for the first one. These have to be corrected by hand anyway. Of course, there is a price to pay, for example, if a specific journal requires citations to be organized by their appearance in the main text, not alphabetically. There are other pieces of advice in this chapter that may not be shared by most colleagues. As in other chapters, please feel free to ignore them.

25.1. Inviting coauthors and selecting a journal

Producing a paper based on an experimental study involves efforts by many persons. The study has to be conceived and designed, hardware and software have to be put together, sensors have to be checked and calibrated, pilot studies have to be run to ensure that everything works smoothly, experimental procedures have to be performed and the data collected, data processing software has to be written, data have to be processed and summarized, figures and tables have to be created and optimized, results have to interpreted with respect to the specific hypotheses of the study and to theories and models in the field, and the text has to be written, rewritten, rewritten, and rewritten again until it is ready for submission. In very rare cases, all the steps involved in producing a paper are done by a single person. In most cases, a decision has to be made about who should be on the list of authors. Making this decision is not easy and is sometimes painful. Actually, an even earlier question is: Who will make this decision? Most frequently, this is done by the leader of the group, the senior, most experienced person, based on traditions.

The traditions of individual groups and laboratories are different. In some groups, only those who participated in the design of a study and interpretation of its findings are considered qualified to be on the list of authors. Other contributors are mentioned in acknowledgments, even if their time commitment was very significant. This sometimes relates to engineers, programmers, and students involved in crucial technical aspects of the study but not in its most important intellectual aspects: conception, design, and interpretation. In some groups, authorship is offered to persons funding aspects of projects or administrators in charge of larger units, such as departments, centers, and institutes, even if they were only marginally aware of the project. In my subjective opinion, this borders on dishonesty and compromises the research process.

In the Motor Control Laboratory at Penn State, we have never been rich enough to afford dedicated engineers and programmers: These aspects of studies have been commonly performed by students and postdoctoral researchers who also run experiments. Frequently, lab members team up and help each other perform experiments. Fewer team members participate in the data processing stage, and even fewer in the figure and table generating stage. Ultimately, only the leading person who knows how the experiment was designed and performed and how the data were collected, processed, and summarized teams up with the advisor(s) to work on interpretation of the results. Our tradition is to offer authorship to everybody who contributed significantly to any of the main stages of the project. The decision hinges on the word *significantly*, which is, of course, subjective. However, we have managed to avoid arguments on authorship of nearly all of the 400+ papers produced by the lab.

Selecting a journal is the next nontrivial step. Frequently, choosing the wrong journal results in a rejection of the submission even if it is scientifically solid and contains new ideas. Here, we assume that the paper is written for a respected journal in the field, not necessarily the most fashionable one (as reflected, for example, in an enormous impact factor). Of course, the main factor at this stage is the match between the journal and the contents of the paper. Other factors include personality of the editor and editorial board members (who are likely to participate in the reviewing process), fairness of the review process given previous experience with the journal, and time delays in the review and publication. We rarely select journals that do not publish papers we enjoy reading.

Over the years, a large number of open-access journals have emerged that charge substantial, sometimes even exorbitant, fees for publication. Some of these journals use rigorous reviewing processes and criteria; others publish more or less anything as long as the authors pay the fee. Personally, I have a very strong conviction that the "pay to publish" scheme is detrimental to science, discriminatory to those with limited funds, and ridiculous in business terms: It charges those who spend the time and effort to produce the product (authors) to deliver the product by the middlemen who pocket all the profit (publishers) to consumers who get it for free (readers). This scheme survives only thanks to the "publish or perish" policy dominating most universities; it encourages researchers to literally buy publications. I have for years refused to edit, write, and review for such journals even when the publishers offer not to charge the fee for an invited paper. To me, this is bribery.

25.2. Introduction

There are a few general rules that apply to sections of the paper. Avoid boring, trivial, and repetitive statements (e.g., "The ability to keep vertical posture is important," "Our findings do not contradict Newton's second law," "More research is needed in this direction," etc.). Do not assume too much knowledge on the part of the readers. Try to be exact and define all the important concepts, in particular those used in the field in various meanings (e.g., "synergy," "postural sway," "muscle tone," etc.). Use special indices and abbreviations sparingly; try to use intuitive ones (e.g., V for velocity, F for force, etc.; avoid something like "$RFT_{i,x}$"). Use a conversational style as if telling a well-informed friend about your study. Try to avoid lab-specific jargon (e.g., "leg stiffness," "good variability," etc.). Illustrate difficult concepts with schematic figures. An example is presented in Fig. 25.1 for a hypothetical study exploring the control

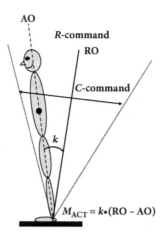

Figure 25.1 An illustration of a number of concepts that might be used in a study of the control of vertical posture. AO, actual orientation of the body; RO, referent orientation; k, apparent stiffness; M_{ACT}, active moment of force.

of vertical posture within the theory of neural control with spatial referent coordinates. This figure illustrates the concepts of reciprocal and coactivation command (R-command and C-command), referent orientation (RO), and apparent stiffness (k).

The opening part of any paper has several main goals. First, you would like to entice future readers (including reviewers), to whet their appetite, by showing that the topic is important, is nontrivial, and offers more than just getting more data. Second, you have to show that you have done your homework: You know the field, have read recent papers on the topic, and are aware of the main theoretical views and controversies. Refer as much as you can to studies from other labs exploring related topics and hypotheses. Finally, you want to explain how your study contributes to the field.

Personally, I start with a brief plan for the Introduction: one line per future paragraph. A scheme for a hypothetical paper is presented in Fig. 25.2. For example, the first paragraph may be used to introduce the general topic of the study and explain

Plan (Introduction)
Vertical posture within the theory of control with referent coordinates.
Two basic commands, R and C, and their effects on mechanics.
Quiet standing and background muscle activation.
Traditional views on muscle activation patterns during fast movements.
Muscle activation patterns within the RC theory.
Hypothesis-1:
Voluntary sway backward will show the classical tri-phasic pattern.
Voluntary sway forward will start with suppression of activation of dorsal muscles.
Hypothesis-2:
Voluntary sway backward will be faster (active sway vs. gravity-induced effects)
Experimental design to test the hypotheses.

Figure 25.2 A plan for the Introduction written for a study of muscle activation patterns during fast voluntary body sway.

why it is important. The second paragraph introduces the main concepts and definitions. Next you introduce the immediate basis for the study (i.e., recent publications and findings that motivated your study). After the three opening paragraphs, the reader is ready to be exposed to your specific hypotheses. It may take a few paragraphs to explain where the hypotheses came from and how you plan to address them experimentally. A small trick: Try to introduce a reasonably sounding hypothesis or two that will be falsified by your results. Since you already know the results, this should not be impossible. If you succeed, reviewers would be surprised that they were fooled into believing a hypothesis that happened to be wrong. It would be very difficult for them to state that the results are trivial and predictable from earlier published studies.

25.3. Methods

There is a simple rule to follow when writing the Methods section: *The information should be sufficient for another researcher to repeat your experiment and get similar results.* I have never seen a criticism from a reviewer: "Methods are described in too much detail." Nearly always, reviewers ask for more information on the selection of subjects, hardware and software, experimental procedures, data processing, statistical methods, and so forth. Let us assume that this section is structured in a way that is rather common; that is, it has the subsections Subjects, Apparatus, Procedures, Data Analysis, and Statistics.

In the first subsection, information on subjects should be included regarding their total number (sometimes reviewers ask for a formal power analysis, which was run to define this number), gender distribution, and population (e.g., young healthy adults, healthy elderly, professional athletes, children within a certain age range, etc.). Further, this section is expected to present anthropometric characteristics of the subjects, such as their mass and height. If hand or foot movements were studied, handedness and footedness should be mentioned, including the methods used to define it. If vision is involved, mention whether the subjects had normal or corrected-to-normal vision. Always end with a statement about the consent procedure and whether it complied with the Helsinki declaration.

The description of the apparatus has to be very detailed. Describe all the pieces of equipment and include their manufacturers and item numbers. If special construction was used (e.g., see Fig. 25.3), describe it in sufficient detail to allow another person to copy it. If special procedures were used, for example, skin preparation in studies using electromyography, these also have to be described in detail. End this subsection with information on raw data collection such as filtering, amplification, frequency of analog-to-digital conversion, and its resolution (in bits).

The description of the procedures used in the study is expected to be even more detailed. First, describe the position of the subject with respect to the apparatus. Such details as position of parts of the body not explicitly involved in the tasks should also be mentioned. Describe whether there was a specific instruction with respect to the gaze direction and whether there was a visual target. Further, if the task involved movements under visual feedback, describe how visual feedback was organized. If a computer monitor was used for feedback, explain how it was located with respect to

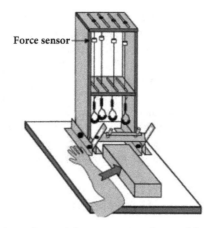

Figure 25.3 An illustration of a special construction device, "the suspension system," which allows recording forces applied by fingers at different phalanges. Its description should mention the dimensions, materials, type of force sensors, connectors, wire cables, hand fixation device, and so forth. Reproduced by permission from Shinohara et al. 2003a.

the subject, how large the monitor was, and even the colors of the cursors, lines, and objects used to provide targets and feedback.

Then, describe the structure of the experiment: How many blocks of trials were performed by the subject? How many trials were there per block? How long were the time intervals between trials and between blocks? Were different conditions fully randomized or block-randomized? What was the total duration of the experiment? Could fatigue or boredom be an issue? And so forth. This subsection can also be used to introduce (sparingly!) abbreviations, special terms, and symbols useful in further sections. If you introduce an abbreviation or a symbol, use it consistently in the rest of the paper, including in figures and tables.

It is useful to structure the section on data analysis into smaller subsections with subtitles related to the computation of specific groups of the main outcome variables. This section reads better if its structure is linked to the specific hypotheses in the Introduction. Explain what was done with the raw data, from their digitization to the computation of outcome variables. This information should include such commonly used steps as filtering, rectification, differentiation, integration, alignment, normalization, and averaging. Further, move to study-specific analysis. Use typical raw data, for example, a typical time series of salient variables recorded in a trial (or averaged across a series of trials) performed by a representative subject (e.g., Fig. 25.4), to illustrate what was measured and when to compute outcome variables linked to the specific hypothesis. This is also a good time to add special abbreviations and symbols for the main outcome variables. Stick to these abbreviations and symbols in the rest of the paper.

The last subsection of Methods is statistics. Explain how the data are going to be presented in the Results section (as well as in figures and tables), for example, means

25.3. METHODS

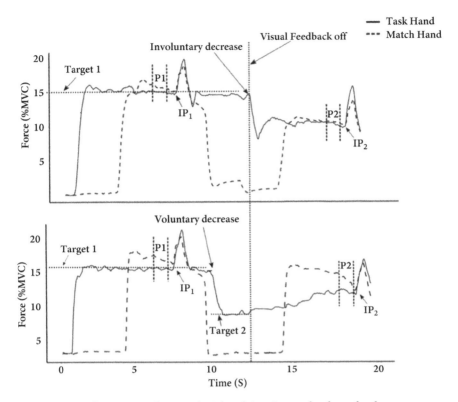

Figure 25.4 An illustration of a typical trial and time intervals where the data were quantified. In this study, one hand ("task hand") produced a constant force level followed by voluntary or involuntary (induced by lifting or lowering the fingers) force changes. The other hand ("match hand") tried to match the force produced by the task hand. IP stands for "inverse piano" episodes with finger lifting-lowering to estimate the referent coordinate and apparent stiffness values (see Chapter 7). Reproduced by permission from Abolins et al. 2020.

and standard deviations or medians and quartiles. If methods of parametric statistics are used to test the main hypotheses, explain how the data were tested for normality and sphericity of the distributions and what was done if those tests showed violations of normality or sphericity assumptions. For example, were the data log-transformed to make the distributions normal? Were degrees of freedom adjusted if the sphericity assumption was violated? Further, describe the hypothesis-testing tools. Frequently, these involve analysis of variance (ANOVA) of some kind (e.g., with or without repeated measures). Describe all the factors and their levels. Explicitly link these analyses to the main hypotheses of the study. Explain what tools were used to explore the significant main effects of ANOVA for factors with more than two levels and significant interactions. If multiple similar analyses were performed, mention if the significance level was adjusted to account for the multiple comparisons.

25.4. Results

My advice is to write this section in a conversational style as if telling your colleague about the results of the study. Unfortunately, too commonly this section is turned into an exercise in statistics without telling the reader what the described statistical effects mean with respect to the main hypotheses. In fact, what matters is what your data say about the hypotheses, not how large F values and how small p values are. Of course, statistics are important, but in most studies, their importance is in demonstrating to the reader that your results are valid within the acceptable criteria. I like two ways of presenting results. The first starts each meaningful subsection (paragraph) with the main result described in simple words as related to your hypotheses. Then, the result is illustrated with a figure and/or table, statistical support is presented, and all other effects (secondary analyses) are described in detail. The second method is to incorporate statistical support, usually in parentheses, into the story about the main results. Sometimes this becomes impractical, in particular if the design of the study is multifactorial and significant statistical effects are many and include interactions. To follow are examples of writing following the "exercise in statistics" style and two "better examples."

A bad example: "Three-way RM ANOVA on ΔV_z has shown significant main effects of Factor-1 ($F_{[n,m]}$ = X.XX; p < 0.00X) and Factor-2 ($F_{[n,m]}$ = X.XX; p < 0.00X), as well as a two-way interaction ($F_{[n,m]}$ = X.XX; p < 0.00X). Pairwise contrasts have confirmed that Level-1 of Factor-1 > Level-2 and Level-3. In addition, the interaction reflected.... These results are presented in Table 1."

A better example: "To test the effects of the support area on postural stability (Hypothesis-1 in the Introduction), we explored the effects of Stability on the index of synergy, ΔV_z, computed in the space of muscle modes for the COP trajectory. Overall, the subjects showed COP-stabilizing synergies (ΔV_z = XX.X±Y.YY > 0) for Level-1, but not for Level-2 (ΔV_z = XX.X±Y.YY) and Level-3 (ΔV_z = XX.X±Y.YY). These results are illustrated in Figure 3, which shows averaged across subjects ΔV_z values with standard deviation bars. Note the higher values for the larger support area as compared to the other two conditions. RM ANOVA confirmed these results (main effect of Factor-1, $F_{[n,m]}$ = X.XX; p < 0.00X). In addition, it showed a dependence of ΔV_z on Factor-2 (visual feedback), which was present only for Level-1 of Factor-1 (F1 × F2 interaction, $F_{[n,m]}$ = X.XX; p < 0.00X). Pairwise contrasts confirmed..."

Another better example: "To test Hypothesis-1 on effects of the support area on postural stability, we explored the effects of the support area (Factor-1) on the index of synergy, ΔV_z. Overall, the subjects showed COP-stabilizing synergies (ΔV_z = XX.X±Y.YY > 0) only for Level-1 (main effect of Factor-1, $F_{[n,m]}$ = X.XX; p < 0.00X) but not for Level-2 (ΔV_z = XX.X±Y.YY) and Level-3 (ΔV_z = XX.X±Y.YY). The synergy index was larger in the presence of visual feedback (Level-1 of Factor-2), but only for the data when Factor-1 was at Level-1 (F1 × F2 interaction, $F_{[n,m]}$ = X.XX; p < 0.00X). These findings are illustrated in Figure 3, which shows averaged across subjects ΔV_z values with standard deviation bars. Note the higher values for the larger support area, which are particularly pronounced in the presence of visual feedback."

25.5. Tables and figures

One of the basic rules for preparation of figures and tables is that they have to be understandable without reading the text. In our times of astronomical numbers of publications in nearly any field, rarely readers have time to go through the whole paper. Most frequently, they limit themselves to the Abstract and, if interested, look at the illustrations. So, even if this seems excessive, all abbreviations and symbols have to be defined in the captions for each figure and each table. Let us start with tables, which are typically less subjective in their structure. Tables are usually used for large data arrays. If you want to show only four values, even with standard deviations, creating a table would be too much; simply present the numbers in the text. Each table should have an informative title, not simply "data" or "average values." Each table should have a caption on the bottom explaining what is presented and all the symbols and abbreviations. Figure 25.5 presents examples of a poorly organized table and a better-organized table.

It is much harder to give advice on designing figures. Besides, I do not feel myself an expert in this field. Many of my colleagues make much better-organized and more attractive illustrations for their papers. So, I am going to describe only the most basic rules.

Do not overload figures with information. Their messages should be clear and easy to see. Use the same terms and abbreviations as in the main text and do not forget to define them again in the caption. When presenting graphs, do not forget to label each axis with the corresponding variable and units in parentheses (when applicable). Use double-Y-axis design if you want to show two variables with very different magnitudes

(a) Table 1. Force data

Finger	I	M	R	L
Mean	18.6	18.1	16.2	13.6
SD	2.1	2.0	1.8	1.7

(b) Table 1. Individual finger forces in Task-1

Finger/Subject	I	M	R	L
S1	15.3 (4)	14.2 (3.5)	12.1 (2.2)	10.0 (2.1)
S2	20.4 (4.2)	21.1 (4.1)	17.2 (3.3)	14.2 (2.3)
S3	17 (3.4)	16.5 (3.6)	14.5 (3.0)	12.3 (2.1)
S4	16 (4.2)	15.5 (4.1)	15.1 (3.3)	14.0 (2.3)
S5	21.1 (4.6)	21.0 (4.5)	19.2 (4.1)	14.6 (2.8)
S6	16.5 (4.1)	16.0 (3.9)	15.4 (3.3)	12.0 (2.3)
Mean (SD)	18.6 (2.1)	18.1 (2.0)	16.2 (1.8)	13.6 (1.7)

Individual values for subjects across trials with standard deviations (SD) in parentheses are shows as well as the mean and SD values across the subjects. I – index, M – middle, R – ring, and L – little.

Figure 25.5 (a): A poorly organized table. (b): A better-organized table.

on the same plot. Select scales wisely to avoid large empty areas. Use titles over panels. Use both colors for the electronic version and stripes/patterns for the printed version (I assume that you are stringent with funds and do not want to pay for printed color illustrations).

Show at least one figure with examples of the time series of the important variables. These can be from a "representative subject" or averaged across subjects data with standard deviation (or standard error) shades. Depending on the goals of the study and its design, one or the other may be preferred as more informative and illustrative. This illustration can be used to remind the reader what was quantified and within what time intervals. Then, illustrate all the important results, particularly those that test the specific hypotheses of the study. These may be time series, bar graphs, regressions, and so forth. Do not present the same result multiple times (e.g., in a figure, in a table, and in the main text). When describing a figure in the main text, mention the important values, not only "$p < 0.01$." These values may be hard to read in the figure after it appears in print. Figure 25.6 presents examples of a "bad figure" (top) and the same data presented in a better format (bottom).

Figure captions should be informative and sufficient to understand all the details. Bring the reader's attention to features of the presented data that you consider particularly important. However, avoid repeating the main text in the caption verbatim. A bad figure caption for the data presented in Fig. 25.6 would be: "Variance within the UCM and within ORT for the conditions without visual feedback (A) and with visual feedback (B)." A better caption would be: "Indices of the intertrial variance computed within the uncontrolled manifold (UCM) and orthogonal to the UCM space (ORT) for the conditions without (panel A) and with visual feedback (panel B). Means across

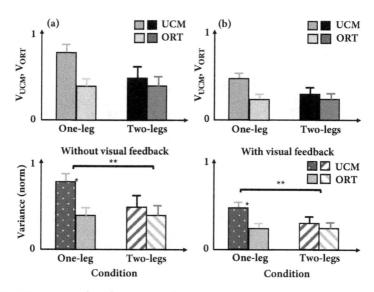

Figure 25.6 Two examples of presenting the same data. Top: A poorly designed figure. Bottom: A better-designed figure.

subjects and standard error bars are presented. Note the higher variance within the UCM during the one-leg condition only. * $p < 0.05$; ** $p < 0.01$."

25.6. Discussion

This is arguably the most subjective part of the paper. You should use this section to explain your views on the topic, ideas, concepts, hypotheses, and models in the field. You can use this section to go beyond the data and speculate how the body produced the explored class of behaviors. However, try to keep such speculations under control and relate them to studies published by others.

One of the most common mistakes in papers that I have reviewed has been turning the Discussion into an expanded version of Results. Unfortunately, very frequently, the authors structure their Discussion explicitly by their own findings, which have already been presented. They repeat the findings and then compare them to earlier studies. Statements like "Our findings are compatible with those by Johnson and Johnson 2055" or "Our results differ from those by Doe and Doe 2044, likely because of the different characteristics of the subject groups" are marginally informative and add nothing to new lessons the reader expects to learn from your study. The purpose of the Discussion is not to repeat the findings but to interpret them in the frameworks offered by competing hypotheses in the field. So, instead of using subtitles such as "Increased coactivation during standing on narrow support" and "Stronger effects of the support surface under visual feedback," use something like "Effects of muscle coactivation of the control of vertical posture" and "The role of visual feedback in challenging postural tasks." Of course, you can refer to specific findings of the study and even to specific figures in the Results to support your conclusions and conjectures.

Personally, I prefer to start the Discussion with an opening part reminding the reader of the specific hypotheses from the Introduction (the reader does not have to memorize them by heart) and saying explicitly whether the results confirm or falsify the hypotheses (the latter is always preferred!). This part should use only minimal referencing to earlier publications as needed to be polite and respectful of your colleagues who might have contributed to the same topic of research.

The rest of the Discussion, its main body, should be structured by ideas, hypotheses, and theories, not by your findings (Fig. 25.7). It is better to use subtitles telling the reader upfront what specific hypothesis or theory you are going to discuss. Some journals encourage ending the Discussion with a summarizing statement emphasizing one more time the main findings and the main lessons these findings teach about the neural control of movement. Other journals do not have this requirement and even request that the authors remove such statements. Use this final section or a similar one, for example, "Methodological aspects and future plans," to admit drawbacks of your study (I have not yet seen a study without drawbacks); explain why you view the drawbacks as relatively minor, not disqualifying your results and conclusions; and explain how you plan to develop this line of research, maybe including ways to address the drawbacks.

Discussion – A

Increase in postural sway with muscle coactivation.
Effects of vision on sway characteristics.
Age effects on muscle coactivation during standing.
Age effects on sway characteristics.
Conclusions.

Discussion – B

Introduction part: Answers to the specific hypotheses.
Control of vertical posture with spatial referent coordinates.
The role of muscle coactivation in postural stability.
Postural sway and postural stability.
Why do older persons coactivate more?
Methodological issues and limitations.

Figure 25.7 Two versions of structuring the Discussion. A: A bad example—structuring by findings. B: A better example—structuring by ideas and hypotheses.

25.7. Responding to reviews

Remember that reviewers are busy people who carve out a piece of their time to read and review your paper without any reward, except being the first to learn about your study. It should not be surprising that sometimes they miss an important bit of information or misinterpret something you wrote. Generally, they try to be helpful and write productive comments. Of course, there are exceptions when reviewers use their anonymity and position of power to write unfriendly comments that carry no useful message—but these are rare exceptions. Most likely, even very critical reviews contain useful critiques that should allow you to improve the paper and resubmit it to that or another journal. However, if you feel strongly that a specific review is unproductive and personal, write a polite letter to the editor of the journal, present examples from the review, and suggest that this reviewer not be used again. The chances that the editorial decision will be changed are close to zero, but at least there is a chance that, in the future, your submission to this journal will not be reviewed by the same person.

The first reaction to any strongly critical review is that of frustration: You spent so much time performing the study and writing the paper, and now an unknown person tells you that it is no good. Let the review sit on your desktop and your feelings cool down. Then, copy the review to a Word document and consider it comment by comment. After each meaningful comment, write "RESPONSE" and leave the space blank. Cool down a bit more. Then go through the comments, one at a time, and write responses that, in your opinion, would be adequate. Do not write "Done" or "OK," even if the comment refers to a typo. To be polite, respond: "The error has been corrected." It is very useful to have the submitted manuscript open on the computer desktop because commonly reviewers' comments come with references to specific sections, lines, figures, and so forth in the text. Do not start revising the manuscript at this stage. After you are done responding to each and every comment, send the updated document to your coauthors (if you have them) and ask for suggestions,

Responding to a critical comment:

The authors have not defined explicitly one of the central concepts of their study, that of "synergy".

(a)
This concept was defined on page 2 of the Introduction in the original submission.

(b)
We apologize for not making our understanding and definition of "synergy" explicit. In the revision, we added a paragraph describing the current definitions of this concept in the literature and justify the definition we use in this particular study.

Figure 25.8 Two examples of responding to a critical comment. (a): A bad example. (b): A better example.

editorial changes, additions, and so forth. If you do not know how to handle a specific comment, state this explicitly and ask for help from your coauthors. After this first stage is over and you have a "Response to Reviews" file ready, modifying the manuscript should not be too complicated.

A typical criticism and adequate and inadequate responses to it are illustrated in Fig. 25.8. Note that telling the reviewers that they have misunderstood you or missed an important explanation is not productive: If the reviewers misunderstood you or missed an explanation, this is your mistake. Likely other readers who have even less time to read your paper will have the same problem. Admit a mistake and modify the text to make the contentious point as clear as you possibly can. If reviewers tell you that you have missed important publications, add references to those in the paper even if you feel that their relevance is minimal. So, the first general rule should be: "The reviewer is always right." And if you feel that the reviewer is wrong, reread the first general rule.

25.8. Rejection: What to do next?

It is no fun to have one's paper rejected with no possibility of resubmission. First, you feel mistreated. Second, you want your paper to be out quickly, and a rejection adds a time delay. As mentioned earlier, arguing with the editor that the reviews were subjective and, hence, the editorial decision should be changed is not going to work. The editor selected the reviewers and feels that the reviewers did the editor a favor by accepting the invitation to review your paper and providing reviews within a reasonable time. Frankly, more commonly than not, when a paper gets rejected, it deserves it. Of course, there are more selective and less selective journals. The former may reject a paper unless both reviews are strongly positive and asking for relatively straightforward changes only. The latter may be willing to allow resubmission even with two critical reviews that recommend "major revision."

I am not even considering here fashionable journals that reject 90% of submissions without reviews based on the personal opinions of the editors, who may view the topic of your study as not broad enough, not novel enough, or not appealing enough. I stopped wasting time submitting to those journals over 30 years ago after two of them rejected a paper entitled "Relativistic effects in voluntary movements" without reviews with a standard editorial letter that the topic was not of interest to their broad audience. Parenthetically, the paper was published in a "normal journal."

If you feel that your paper deserves to be published, even a very negative set of reviews should be seen as useful. Since you are planning to submit to another journal, you will very likely deal with a new set of reviewers. So, comments that you view as subjective can be ignored. However, productive criticisms should be taken into consideration when preparing a new submission. Add new references, add explanations and maybe illustrations of complex concepts that confused the former reviewers, modify figures, fix statistical presentation, and so forth. Basically, your new submission should be very close to satisfying all the criticisms of the first set of reviewers.

Think of the next target journal. Maybe you should submit to a less competitive or less specialized (or more specialized!) journal. On the other hand, if the fatal flaws mentioned in the first set of reviews are valid, it makes sense to start a new study and not to waste time trying to publish the flawed one.

Good luck!

References

Abbs JH, Gracco VL (1984) Control of complex motor gestures: Orofacial muscle responses to load perturbations of the lip during speech. *Journal of Neurophysiology* 51: 705–723.

Abdusamatov RM, Feldman AG (1986) Description of the electromyograms with the aid of a mathematical model for single joint movements. *Biophysics* 31: 549–552.

Abolins V, Cuadra C, Ricotta J, Latash ML (2020) What do people match when they try to match force? Analysis at the level of hypothetical control variables. *Experimental Brain Research* 238: 1885–1901.

Abolins V, Latash ML (2021) The nature of finger enslaving: New results and their implications. *Motor Control* 25: 680–703.

Abolins V, Latash ML (2022a) Unintentional force drifts across the human fingers: Implications for the neural control of finger tasks. *Experimental Brain Research* 240: 751–761.

Abolins V, Latash ML (2022b) Unintentional force drifts as consequences of indirect force control with spatial referent coordinates. *Neuroscience* 481: 156–165.

Abolins V, Ormanis J, Latash ML (2023) Unintentional drifts in performance during one-hand and two-hand finger force production. *Experimental Brain Research* 241: 699–712.

Abolins V, Stremoukhov A, Walter C, Latash ML (2020) On the origin of finger enslaving: Control with referent coordinates and effects of visual feedback. *Journal of Neurophysiology* 124: 1625–1636.

Adamovich SV, Levin MF, Feldman AG (1997) Central modifications of reflex parameters may underlie the fastest arm movements. *Journal of Neurophysiology* 77: 1460–1469.

Akulin VM, Carlier F, Solnik S, Latash ML (2019) Sloppy, but acceptable, control of biological movement: Algorithm-based stabilization of subspaces in abundant spaces. *Journal of Human Kinetics* 67: 49–72.

Almeida GL, Corcos DM, Latash ML (1994) Practice and transfer effects during fast single joint elbow movements in individuals with Down syndrome. *Physical Therapy* 74: 1000–1016.

Ambike S, Mattos D, Zatsiorsky VM, Latash ML (2016a) Synergies in the space of control variables within the equilibrium-point hypothesis. *Neuroscience* 315: 150–161.

Ambike S, Mattos D, Zatsiorsky VM, Latash ML (2016b) Unsteady steady-states: Central causes of unintentional force drift. *Experimental Brain Research* 234: 3597–3611.

Ambike S, Mattos D, Zatsiorsky VM, Latash ML (2016c) The nature of constant and cyclic force production: Unintentional force-drift characteristics. *Experimental Brain Research* 234: 197–208.

Ambike S, Mattos D, Zatsiorsky VM, Latash ML (2018) Systematic, unintended drifts in the cyclic force produced with the fingertips. *Motor Control* 22: 82–99.

Ambike S, Paclet F, Zatsiorsky VM, Latash ML (2014) Factors affecting grip force: Anatomy, mechanics, and referent configurations. *Experimental Brain Research* 232: 1219–1231.

Ambike S, Zatsiorsky VM, Latash ML (2015a) Processes underlying unintentional finger force changes in the absence of visual feedback. *Experimental Brain Research* 233: 711–721.

Ambike S, Zhou T, Zatsiorsky VM, Latash ML (2015b) Moving a hand-held object: Reconstruction of referent coordinate and apparent stiffness trajectories. *Neuroscience* 298: 336–356.

Andersen RA, Snyder LH, Bradley DC, Xing J (1997) Multimodal representation of space in the posterior parietal cortex and its use in planning movements. *Annual Reviews in Neuroscience* 20: 303–330.

Anderson FC, Pandy MG (2001) Static and dynamic optimization solutions for gait are practically equivalent. *Journal of Biomechanics* 34: 153–161.

Aoki T, Latash ML, Zatsiorsky VM (2007) Adjustments to local friction in multi-finger prehension. *Journal of Motor Behavior* 39: 276–290.

Arbib MA, Iberall T, Lyons D (1985) Coordinated control programs for movements of the hand. In: Goodwin AW, Darian-Smith I (Eds.) *Hand Function and the Neocortex*, pp. 111–129. Springer Verlag: Berlin.

Arce F, Novick I, Mandelblat-Cerf Y, Vaadia E (2010) Neuronal correlates of memory formation in motor cortex after adaptation to force field. *Journal of Neuroscience* 30: 9189–9198.

Arias P, Espinosa N, Robles-García V, Cao R, Cudeiro J (2012) Antagonist muscle co-activation during straight walking and its relation to kinemaltics: Insight from young, elderly and Parkinson's disease. *Brain Research* 1455: 124–131.

Arimoto S, Nguyen PTA, Han HY, Doulgeri Z (2000) Dynamics and control of a set of dual fingers with soft tips. *Robotica* 18: 71–80.

Arimoto S, Tahara K, Yamaguchi M, Nguyen PTA, Han HY (2001) Principles of superposition for controlling pinch motions by means of robot fingers with soft tips. *Robotica* 19: 21–28.

Aruin AS, Almeida GL (1996) A coactivation strategy in anticipatory postural adjustments in persons with Down syndrome. *Motor Control* 1: 178–191.

Aruin AS, Forrest WR, Latash ML (1998) Anticipatory postural adjustments in conditions of postural instability. *Electroencephalography and Clinical Neurophysiology* 109: 350–359.

Aruin AS, Kanekar N (2013) Effect of a textured insole on balance and gait symmetry. *Experimental Brain Research* 231: 201–208.

Aruin AS, Latash ML (1995) The role of motor action in anticipatory postural adjustments studied with self-induced and externally-triggered perturbations. *Experimental Brain Research* 106: 291–300.

Aruin AS, Rao N (2018) The effect of a single textured insole in gait rehabilitation of individuals with stroke. *International Journal of Rehabilitation Research* 41: 218–223.

Asaka T, Wang Y, Fukushima J, Latash ML (2008) Learning effects on muscle modes and multi-mode synergies. *Experimental Brain Research* 184: 323–338.

Asaka T, Yahata K, Mani H, Wang Y (2011) Modulations of muscle modes in automatic postural responses induced by external surface translations. *Journal of Motor Behavior* 43: 165–172.

Asanuma H (1973) Cerebral cortical control of movements. *Physiologist* 16: 143–166.

Auyang AG, Yen JT, Chang YH (2009) Neuromechanical stabilization of leg length and orientation through interjoint compensation during human hopping. *Experimental Brain Research* 192: 253-264.

Babinski F (1899) De l'asynergie cerebelleuse. *Revue Neurologique* 7: 806-816.

Bagesteiro LB, Sainburg RL (2002) Handedness: Dominant arm advantages in control of limb dynamics. *Journal of Neurophysiology* 88: 2408-2421.

Bagesteiro LB, Sainburg RL (2003) Nondominant arm advantages in load compensation during rapid elbow joint movements. *Journal of Neurophysiology* 90: 1503-1513.

Barrea A, Delhaye BP, Lefèvre P, Thonnard JL (2018) Perception of partial slips under tangential loading of the fingertip. *Scientific Reports* 8(1): 7032.

Bassin PV, Bernstein NA, Latash LP (1966) On the problem of the relation between structure and function in the brain from a contemporary point of view. In: Grastschenkov NI (Ed.) *Physiology in Clinical Practice*, pp. 38-71. Nauka: Moscow (in Russian).

Bastian AJ, Mugnaini E, Thach WT (1999) Cerebellum. In: Zigmond MJ, Bloom FE, Landis SC, Roberts JL, Squire LR (Eds.) *Fundamental Neuroscience*, pp. 973-992. Academic Press: San Diego.

Bazalgette D, Zattara M, Bathien N, Bouisset S, Rondot P (1986) Postural adjustments associated with rapid voluntary arm movements in patients with Parkinson's disease. *Advances in Neurology* 45: 371-374.

Belen'kii VY, Gurfinkel VS, Pal'tsev YI (1967) Elements of control of voluntary movements. *Biofizika* 10: 135-141.

Benamati A, Ricotta JM, De SD, Latash ML (2024) Three levels of neural control contributing to force-stabilizing synergies in multi-finger tasks. *Neuroscience* 551: 262-275.

Bennett DJ, Hollerbach JM, Xu Y, Hunter IW (1992) Time-varying stiffness of human elbow joint during cyclic voluntary movement. *Experimental Brain Research* 88: 433-442.

Berger W, Trippel M, Discher M, Dietz V (1992) Influence of subjects' height on the stabilization of posture. *Acta Otolaryngology* 112: 22-30.

Berkinblit MB, Feldman AG, Fukson OI (1986a) Adaptability of innate motor patterns and motor control mechanisms. *Behavior and Brain Science* 9: 585-638.

Berkinblit MB, Gelfand IM, Feldman AG (1986b) A model for the control of multijoint movements. *Biofizika* 31: 128-138.

Bernstein NA (1930) A new method of mirror cyclographie and its application towards the study of labor movements during work on a workbench. *Hygiene, Safety and Pathology of Labor* 5: 3-9; 6: 3-11 (in Russian).

Bernstein NA (1935) The problem of interrelation between coordination and localization. *Archives of Biological Science* 38: 1-35 (in Russian).

Bernstein NA (1947) On the construction of movements. Medgiz: Moscow (in Russian). English translation is in: Latash ML (Ed.) (2020) *Bernstein's Construction of Movements*. Routledge: Abingdon, UK.

Bernstein NA (1966) *Essays on the Physiology of Movements and Physiology of Activity.* Meditsina: Moscow (in Russian).

Bernstein NA (1967) *The Co-ordination and Regulation of Movements.* Pergamon Press: Oxford.

Bernstein NA (1996) On dexterity and its development. In: Latash ML, Turvey MT (Eds.) *Dexterity and Its Development*, pp. 1-244. Erlbaum Publ.: Mahwah, NJ.

Bernstein NA (2003) *Contemporary Studies on the Physiology of the Neural Process.* Smysl: Moscow (in Russian).

Bernstein NA, Popova TS (1930) Studies on the biodynamics of the piano strike. *Proceedings of the Piano-Methodological Section of the State Institute of Music Science* 1: 5-47 (in Russian).

Bigland-Ritchie B, Cafarelli E, Vollestad NK (1986a) Fatigue of submaximal static contractions. *Acta Physiolpgica Scandinavica* 128(Suppl. 556): 137-148.

Bigland-Ritchie BR, Dawson NJ, Johansson RS, Lippold OC (1986b) Reflex origin for the slowing of motoneurone firing rates in fatigue of human voluntary contractions. *Journal of Physiology* 379: 451-459.

Bigland-Ritchie B, Johansson R, Lippold OCJ, Woods JJ (1983) Contractile speed and EMG changes during fatigue of sustained maximal voluntary contractions. *Journal of Neurophysiology* 50: 313-324.

Binder-Macleod SA (1995) Variable-frequency stimulation patterns for the optimization of force during muscle fatigue. Muscle wisdom and the catch-like property. *Advances in Experimental Medicine and Biology* 384: 227-240.

Birbaumer N, Piccione F, Silvoni S, Wildgruber M (2012) Ideomotor silence: The case of complete paralysis and brain-computer interfaces (BCI). *Psychological Research* 76: 183-191.

Bizzi E, Accornero N, Chapple W, Hogan N (1982) Arm trajectory formation in monkeys. *Experimental Brain Research* 46: 139-143.

Bizzi E, Giszter SF, Loeb E, Mussa-Ivaldi FA, Saltiel P (1995) Modular organization of motor behavior in the frog's spinal cord. *Trends in Neuroscience* 18: 442-446.

Blanton S, Wilsey H, Wolf SL (2008) Constraint-induced movement therapy in stroke rehabilitation: Perspectives on future clinical applications. *NeuroRehabilitation* 23: 15-28.

Bobath B (1978) *Adult Hemiplegia: Evaluation and Treatment.* William Heinemann: London.

Boehm WL, Nichols KM, Gruben KG (2019) Frequency-dependent contributions of sagittal-plane foot force to upright human standing. *Journal of Biomechanics* 83: 305-309.

Bolbecker AR, Apthorp D, Martin AS, Tahayori B, Moravec L, Gomez KL, O'Donnell BF, Newman SD, Hetrick WP (2018) Disturbances of postural sway components in cannabis users. *Drug and Alcohol Dependence* 190: 54-61.

Bongaardt R (2001) How Bernstein conquered movement. In: Latash ML, Zatsiorsky VM (Eds.) *Classics in Movement Science*, pp. 59-84. Human Kinetics: Urbana, IL.

Bosco G, Poppele RE (2002) Encoding of hindlimb kinematics by spinocerebellar circuitry. *Archives Italiennes de Biologie* 140: 185-192.

Bottasso C, Prilutsky BI, Croce A, Imberti E, Sartirana S (2006) A numerical procedure for inferring from experimental data the optimization cost functions using a multibody model of the neuro-musculoskeletal system. *Multibody System Dynamics* 16: 123-154.

Bouisset S, Do M-C (2008) Posture, dynamic stability, and voluntary movement. *Neurophysiologie clinique* 38: 345-362.

Bouisset S, Zattara M (1987) Biomechanical study of the programming of anticipatory postural adjustments associated with voluntary movement. *Journal of Biomechanics* 20: 735-742.

Bray CW (1928) Transfer of learning. *Journal of Experimental Psychology* 11: 443–469.

Brooks JX, Cullen KE (2019) Predictive sensing: The role of motor signals in sensory processing. *Biological Psychiatry, Cognitive Neuroscience and Neuroimaging* 4: 842–850.

Brueckner D, Kiss R, Muehlbauer T (2018) Associations between practice-related changes in motor performance and muscle activity in healthy individuals: A systematic review. *Sports Medicine Open* 4(1): 9.

Bunday KL, Perez MA (2012) Motor recovery after spinal cord injury enhanced by strengthening corticospinal synaptic transmission. *Current Biology* 22: 2355–2361.

Burgess PR, Clark FJ (1969) Characteristics of knee joint receptors in the cat. *Journal of Physiology* 203: 317–335.

Burke RE, Rudomin P, Zajac FE (1970) Catch property in single mammalian motor units. *Science* 168: 122–124.

Burke RE, Rudomin P, Zajac FE (1976) The effect of activation history on tension production by individual muscle units. *Brain Research* 109: 515–529.

Burstedt MK, Flanagan JR, Johansson RS (1999) Control of grasp stability in humans under different frictional conditions during multidigit manipulation. *Journal of Neurophysiology* 82: 2393–2405.

Cabel DW, Cisek P, Scott SH (2001) Neural activity in primary motor cortex related to mechanical loads applied to the shoulder and elbow during a postural task. *Journal of Neurophysiology* 86: 2102–2108.

Cafarelli E, Kostka CE (1981) Effect of vibration on static force sensation in man. *Experimental Neurology* 74: 331–340.

Calvin-Figuiere S, Romaiguere P, Gilhodes JC, Roll JP (1999) Antagonist motor responses correlate with kinesthetic illusions induced by tendon vibration. *Experimental Brain Research* 124: 342–350.

Campbell MJ, McComas AJ, Petito F (1973) Physiological changes in aging muscles. *Journal of Neurology Neurosurgery and Psychiatry* 36: 174–182.

Canales J (2015) *The Physicist and the Philosopher: Einstein, Bergson, and the Debate That Changed Our Understanding of Time*. Princeton University Press: Princeton, NJ.

Carmena JM, Lebedev MA, Crist RE, O'Doherty JE, Santucci DM, Dimitrov DF, Patil PG, Henriquez CS, Nicolelis MA (2003) Learning to control a brain-machine interface for reaching and grasping by primates. *PLoS Biology* 1(2): E42.

Carson RG (1995) The dynamics of isometric bimanual coordination. *Experimental Brain Research* 105: 465–476.

Casadio M, Morasso PG, Sanguineti V (2005) Direct measurement of ankle stiffness during quiet standing: Implications for control modelling and clinical application. *Gait and Posture* 21: 410–424.

Cauraugh JH, Summers JJ (2005) Neural plasticity and bilateral movements: A rehabilitation approach for chronic stroke. *Progress in Neurobiology* 75: 309–320.

Cavallari P, Bolzoni F, Bruttini C, Esposti R (2016) The organization and control of intralimb anticipatory postural adjustments and their role in movement performance. *Frontiers in Human Neuroscience* 10: 525.

Cesari P, Newell KM (1999) The scaling of human grip configurations. *Journal of Experimental Psychology: Human Perception and Performance* 25: 927–935.

Cesari P, Newell KM (2002) Scaling the components of prehension. *Motor Control* 6: 347–365.

Chan CWY, Kearney RE (1982) Is the functional stretch reflex servo controlled or preprogrammed? *Electroencephalography and Clinical Neurophysiology* 53: 310–324.

Chikh S, Watelain E, Faupin A, Pinti A, Jarraya M, Garnier C (2016) Adaptability and prediction of anticipatory muscular activity parameters to different movements in the sitting position. *Perceptual and Motor Skills* 123: 190–231.

Christakos CN, Papadimitriou NA, Erimaki S (2006) Parallel neuronal mechanisms underlying physiological force tremor in steady muscle contractions of humans. *Journal of Neurophysiology* 95: 53–66.

Christova P, Kossev A (1998) Motor unit activity during long-lasting intermittent muscle contractions in humans. *European Journal of Applied Physiology and Occupational Physiology* 77: 379–387.

Cirstea MC, Levin MF (2000) Compensatory strategies for reaching in stroke. *Brain* 123: 940–953.

Clark FJ, Burgess PR (1975) Slowly adapting receptors in cat knee joint: Can they signal joint angle? *Journal of Neurophysiology* 38: 1448–1463.

Coelho CJ, Studenka BE, Rosenbaum DA (2014) End-state comfort trumps handedness in object manipulation. *Journal of Experimental Psychology: Human Perception and Performance* 40: 718–730.

Cole J, Paillard J (1995) Living without touch and peripheral information about body position and movement: Studies with deafferented subjects. In: Bermudes JL, Marcel A, Eilan N (Eds.) *The Body and the Self*, pp. 245–266, MIT Press: Cambridge, MA.

Cole KJ (2006) Age-related directional bias of fingertip force. *Experimental Brain Research* 175: 285–291.

Cole KJ, Abbs JH (1987) Kinematic and electromyographic responses to perturbation of a rapid grasp. *Journal of Neurophysiology* 57: 1498–1510.

Cole KJ, Johansson RS (1993) Friction at the digit-object interface scales the sensorimotor transformation for grip responses to pulling loads. *Experimental Brain Research* 95: 523–532.

Colebatch JG, Gandevia SC (1989) The distribution of muscular weakness in upper motor neuron lesions affecting the arm. *Brain* 112: 749–763.

Collins JJ, De Luca CJ (1993) Open-loop and closed-loop control of posture: A random-walk analysis of center-of-pressure trajectories. *Experimental Brain Research* 95: 308–318.

Corbetta D, Sirtori V, Castellini G, Moja L, Gatti R (2015) Constraint-induced movement therapy for upper extremities in people with stroke. *Cochrane Database Systematic Reviews* 2015(10): CD004433.

Corcos DM, Agarwal GC, Flaherty BP, Gottlieb GL (1990) Organizing principles for single-joint movements: IV. Implications for isometric contractions. *Journal of Neurophysiology* 64: 1033–1042.

Corcos DM, Gottlieb GL, Agarwal GC (1989) Organizing principles for single joint movements. II. A speed-sensitive strategy. *Journal of Neurophysiology* 62: 358–368.

Corcos DM, Gottlieb GL, Latash ML, Almeida GL, Agarwal GC (1992) Electromechanical delay: An experimental artifact. *Journal of Electromyography and Kinesiology* 2: 59–68.

Corcos DM, Gottlieb GL, Penn RD, Myklebust B, Agarwal GC (1986) Movement deficits caused by hyperexcitable stretch reflexes in spastic humans. *Brain* 109: 1043–1058.

Corcos DM, Myklebust BM, Latash ML (2022) The legacy of Gerald L. Gottlieb in human movement neuroscience. *Journal of Neurophysiology* 128: 148–159.

Cordo PJ, Nashner LM (1982) Properties of postural adjustments associated with rapid arm movements. *Journal of Neurophysiology* 47: 1888–1905.

Costa EC, Santinelli FB, Moretto GF, Figueiredo C, von Ah Morano AE, Barela JA, Barbieri FA (2022) A multiple domain postural control assessment in people with Parkinson's disease: Traditional, non-linear, and rambling and trembling trajectories analysis. *Gait and Posture* 97: 130–136.

Côté JN, Feldman AG, Mathieu PA, Levin MF (2008) Effects of fatigue on intermuscular coordination during repetitive hammering. *Motor Control* 12: 79–92.

Côté JN, Mathieu PA, Levin MF, Feldman AG (2002) Movement reorganization to compensate for fatigue during sawing. *Experimental Brain Research* 146: 394–398.

Craske B (1977) Perception of impossible limb positions induced by tendon vibration. *Science* 196: 71–73.

Cuadra C, Bartsch A, Tiemann P, Reschechtko S, Latash ML (2018) Multi-finger synergies and the muscular apparatus of the hand. *Experimental Brain Research* 236: 1383–1393.

Cuadra C, Corey J, Latash ML (2021a) Distortions of the efferent copy during force perception: A study of force drifts and effects of muscle vibration. *Neuroscience* 457: 139–154.

Cuadra C, Falaki A, Sainburg RL, Sarlegna FR, Latash ML (2019) Case studies in neuroscience. The central and somatosensory contributions to finger inter-dependence and co-ordination: Lessons from a study of a "deafferented person." *Journal of Neurophysiology* 121: 2083–2087.

Cuadra C, Gilmore R, Latash ML (2021b) Finger force matching and verbal reports: Testing predictions of the Iso-Perceptual Manifold (IPM) concept. *Journal of Motor Behavior* 53: 598–610.

Cuadra C, Latash ML (2019) Exploring the concept of iso-perceptual manifold (IPM): A study of finger force matching. *Neuroscience* 401: 130–141.

Cuadra C, Wojnicz W, Kozinc Z, Latash ML (2020) Perceptual and motor effects of muscle co-activation in a force production task. *Neuroscience* 437: 34–44.

Cusumano JP, Cesari P (2006) Body-goal variability mapping in an aiming task. *Biological Cybernetics* 94: 367–379.

Dakin R, Segre PS, Altshuler DL (2020) Individual variation and the biomechanics of maneuvering flight in hummingbirds. *Journal of Experimental Biology* 223: jeb161828.

Danion F, Schöner G, Latash ML, Li S, Scholz JP, Zatsiorsky VM (2003) A force mode hypothesis for finger interaction during multi-finger force production tasks. *Biological Cybernetics* 88: 91–98.

Danna-Dos-Santos A, Degani AM, Latash ML (2008) Flexible muscle modes and synergies in challenging whole-body tasks. *Experimental Brain Research* 189: 171–187.

Danna-Dos-Santos A, Slomka K, Zatsiorsky VM, Latash ML (2007) Muscle modes and synergies during voluntary body sway. *Experimental Brain Research* 179: 533–550.

d'Avella A, Giese M, Ivanenko YP, Schack T, Flash T (2015) Editorial: Modularity in motor control: From muscle synergies to cognitive action representation. *Frontiers in Computational Neuroscience* 9: 126.

De SD, Ricotta JM, Benamati A, Latash ML (2024) Two classes of action-stabilizing synergies reflecting spinal and supraspinal circuitry. *Journal of Neurophysiology* 131: 152–165.

De Freitas PB, Freitas SMSF, Lewis MM, Huang X, Latash ML (2018) Stability of steady hand force production explored across spaces and methods of analysis. *Experimental Brain Research* 236: 1545–1562.

De Freitas PB, Freitas SMSF, Lewis MM, Huang X, Latash ML (2019) Individual preferences in motor coordination seen across the two hands: Relations to movement stability and optimality. *Experimental Brain Research* 237: 1–13.

De Freitas PB, Freitas SMSF, Reschechtko S, Corson T, Lewis MM, Huang X, Latash ML (2020) Synergic control of action in levodopa-naïve Parkinson's disease patients: I. Multi-finger interaction and coordination. *Experimental Brain Research* 238: 229–245.

De Havas J, Gomi H, Haggard P (2017) Experimental investigations of control principles of involuntary movement: A comprehensive review of the Kohnstamm phenomenon. *Experimental Brain Research* 235: 1953–1997.

Delafontaine A, Vialleron T, Hussein T, Yiou E, Honeine JL, Colnaghi S (2019) Anticipatory postural adjustments during gait initiation in stroke patients. *Frontiers in Neurology* 10: 352.

Delhaye BP, Jarocka E, Barrea A, Thonnard JL, Edin B, Lefèvre P (2021) High-resolution imaging of skin deformation shows that afferents from human fingertips signal slip onset. *Elife* 10: e64679.

de Lussanet MH, Smeets JB, Brenner E (2002) Relative damping improves linear mass-spring models of goal-directed movements. *Human Movement Science* 21: 85–100.

Del Vecchio A, Marconi Germer C, Kinfe TM, Nuccio S, Hug F, Eskofier B, Farina D, Enoka RM (2023) The forces generated by agonist muscles during isometric contractions arise from motor unit synergies. *Journal of Neuroscience* 43: 2860–2873.

DeWald JP, Pope PS, Given JD, Buchanan TS, Rymer WZ (1995) Abnormal muscle coactivation patterns during isometric torque generation at the elbow and shoulder in hemiparetic subjects. *Brain* 118: 495–510.

De Wolf S, Slijper H, Latash ML (1998) Anticipatory postural adjustments during self-paced and reaction-time movements. *Experimental Brain Research* 121: 7–19.

Diedrichsen J, Shadmehr R, Ivry RB (2010) The coordination of movement: Optimal feedback control and beyond. *Trends in Cognitive Science* 14: 31–39.

Dietz V, Quintern J, Berger W (1984) Corrective reactions to stumbling in man: Functional significance of spinal and transcortical reflexes. *Neuroscience Letters* 44: 131–135.

DiZio P, Lackner JR (1995) Motor adaptation to Coriolis force perturbations of reaching movements: Endpoint but not trajectory adaptation transfers to the nonexposed arm. *Journal of Neurophysiology* 74: 1787–1792.

Domen K, Zatsiorsky VM, Latash ML (1999) Reconstruction of equilibrium trajectories during whole-body movements. *Biological Cybernetics* 80: 195–204.

Dominici N, Ivanenko YP, Cappellini G, d'Avella A, Mondì V, Cicchese M, Fabiano A, Silei T, Di Paolo A, Giannini C, Poppele RE, Lacquaniti F (2011) Locomotor primitives in newborn babies and their development. *Science* 334: 997–999.

Domkin D, Laczko J, Djupsjöbacka M, Jaric S, Latash ML (2005) Joint angle variability in 3D bimanual pointing: Uncontrolled manifold analysis *Experimental Brain Research* 163: 44–57.

Domkin D, Laczko J, Jaric S, Johansson H, Latash ML (2002) Structure of joint variability in bimanual pointing tasks. *Experimental Brain Research* 143: 11–23.

Dounskaia N (2007) Kinematic invariants during cyclical arm movements. *Biological Cybernetics* 96: 147–163.

Dounskaia N (2010) Control of human limb movements: The leading joint hypothesis and its practical applications. *Exercise and Sport Science Reviews* 38: 201–208.

Dromerick AW, Lum PS, Hidler J (2006) Activity-based therapies. *NeuroRx* 3: 428–438.

Duarte M, Freitas SMSF (2010) Revision of posturography based on force plate for balance evaluation. *Brazilian Journal of Physical Therapy* 14: 183–192.

Duarte M, Zatsiorsky VM (1999) Patterns of center of pressure migration during prolonged unconstrained standing. *Motor Control* 3: 12–27.

Duchateau J, Balestra C, Carpentier A, Hainaut K (2002) Reflex regulation during sustained and intermittent submaximal contractions in humans. *Journal of Physiology* 541: 959–967.

Duchateau J, Baudry S (2014) The neural control of coactivation during fatiguing contractions revisited. *Journal of Electromyography and Kinesiology* 24: 780–788.

Dyhre-Poulsen P, Simonsen EB, Voigt M (1991) Dynamic control of muscle stiffness and H reflex modulation during hopping and jumping in man. *Journal of Physiology* 437: 287–304.

Earhart GM, Jones GM, Horak FB, Block EW, Weber KD, Fletcher WA (2002) Podokinetic after-rotation following unilateral and bilateral podokinetic stimulation. *Journal of Neurophysiology* 87: 1138–1141.

Eisen A, Entezari-Taher M, Stewart H (1996) Cortical projections to spinal motoneurons: Changes with aging and amyotrophic lateral sclerosis. *Neurology* 46: 1396–1404.

Elble RJ, Moody C, Leffler K, Sinha R (1994) The initiation of normal walking. *Movement Disorders* 2: 139–146.

Eklund G, Hagbarth KE (1966) Normal variability of tonic vibration reflexes in man. *Experimental Neurology* 16: 80–92.

Eklund G, Hagbarth KE (1967) Vibratory induced motor effects in normal man and in patients with spastic paralysis. *Electroencephalography and Clinical Neurophysiology* 23: 393.

Elias LJ, Bryden MP, Bulman-Fleming MB (1998) Footedness is a better predictor than is handedness for emotional lateralization. *Neuropsychologia* 36: 37–43.

Emge N, Prebeg G, Uygur M, Jaric S (2013) Effects of muscle fatigue on grip and load force coordination and performance of manipulation tasks. *Neuroscience Letters* 550: 46–50.

Engel AK, Fries P (2010) Beta-band oscillations: Signaling the status quo? *Current Opinions in Neurobiology* 20: 156–165.

Enoka RM (2019) Physiological validation of the decomposition of surface EMG signals. *Journal of Electromyography and Kinesiology* 46: 70–83.

Enoka RM, Duchateau J (2008) Muscle fatigue: What, why and how it influences muscle function. *Journal of Physiology* 586: 11–23.

Enoka RM, Stuart DG (1992) Neurobiology of muscle fatigue. *Journal of Applied Physiology* 72: 1631–1648.

Erimaki S, Christakos CN (2008) Coherent motor unit rhythms in the 6–10 Hz range during time-varying voluntary muscle contractions: Neural mechanism and relation to rhythmical motor control. *Journal of Neurophysiology* 99: 473–483.

Erlhagen W, Schöner G (2002) Dynamic field theory of movement preparation. *Psychological Reviews* 109: 545–572.

Evarts EV (1968) Relation of pyramidal tract activity to force exerted during voluntary movement. *Journal of Neurophysiology* 31: 14–27.

Fahn S, Jankovic J (2007) *Principle and Practice of Movement Disorders*. Churchill, Livingstone: Philadelphia, PA.

Falaki A, Cuadra C, Lewis MM, Prado-Rico JM, Huang X, Latash ML (2023) Multi-muscle synergies in preparation to gait initiation in Parkinson's disease. *Clinical Neurophysiology* 154: 12–24.

Falaki A, Huang X, Lewis MM, Latash ML (2016) Impaired synergic control of posture in Parkinson's patients without postural instability. *Gait and Posture* 44: 209–215.

Falaki A, Huang X, Lewis MM, Latash ML (2017a) Motor equivalence and structure of variance: Multi-muscle postural synergies in Parkinson's disease. *Experimental Brain Research* 235: 2243–2258.

Falaki A, Huang X, Lewis MM, Latash ML (2017b) Dopaminergic modulation of multi-muscle synergies in postural tasks performed by patients with Parkinson's disease. *Journal of Electromyography and Kinesiology* 33: 20–26.

Falaki A, Jo HJ, Lewis MM, O'Connell B, De Jesus S, McInerney J, Huang X, Latash ML (2018) Systemic effects of deep brain stimulation on synergic control in Parkinson's disease. *Clinical Neurophysiology* 129: 1320–1332.

Fallon JB, Macefield VG (2007) Vibration sensitivity of human muscle spindles and Golgi tendon organs. *Muscle and Nerve* 36: 21–29.

Farina D, Enoka RM (2023) Evolution of surface electromyography: From muscle electrophysiology towards neural recording and interfacing. *Journal of Electromyography and Kinesiology* 71: 102796.

Farina D, Merletti R, Enoka RM (2014) The extraction of neural strategies from the surface EMG: An update. *Journal of Applied Physiology* 117: 1215–1230.

Farmer SF, Bremner FD, Halliday DM, Rosenberg JR, Stephens JA (1993) The frequency content of common synaptic inputs to motoneurones studied during voluntary isometric contraction in man. *Journal of Physiology* 470: 127–155.

Farrar DJ, Locke SE, Kantrowitz FG (1995) Chronic fatigue syndrome. 1: Etiology and pathogenesis. *Behavioral Medicine* 21: 5–16.

Fautrelle L, Bonnetblanc F (2012) On-line coordination in complex goal-directed movements: A matter of interactions between several loops. *Brain Research Bulletin* 89: 57–64.

Fee MS (2014) The role of efference copy in striatal learning. *Currents Opinions in Neurobiology* 25: 194–200.

Feigenberg IM (1969) Probabilistic prognosis and its significance in normal and pathological subjects. In: Cole M, Malzman I (Eds.) *Handbook of Contemporary Soviet Psychology*, pp. 355–360. Basic Books: New York.

Feigenberg IM (1998) The model of the future in motor control. In: Latash ML (Ed.) *Progress in Motor Control: Vol. 1: Bernstein's Traditions in Movement Studies*, pp. 89–104. Human Kinetics: Urbana, IL.

Feigenberg IM (2014) *Nikolai Bernstein. From Reflexes to Model of the Future*. LIT Verlag: Münster, Germany.

Feldman AG (1966) Functional tuning of the nervous system with control of movement or maintenance of a steady posture. II. Controllable parameters of the muscle. *Biophysics* 11: 565–578.

Feldman AG (1986) Once more on the equilibrium-point hypothesis (λ-model) for motor control. *Journal of Motor Behavior* 18: 17–54.

Feldman AG (2009) New insights into action-perception coupling. *Experimental Brain Research* 194: 39–58.

Feldman AG (2015) *Referent Control of Action and Perception: Challenging Conventional Theories in Behavioral Science*. Springer: New York.

Feldman AG (2016) Active sensing without efference copy: Referent control of perception. *Journal of Neurophysiology* 116: 960–976.

Feldman AG (2019) Indirect, referent control of motor actions underlies directional tuning of neurons. *Journal of Neurophysiology* 121: 823–841.

Feldman AG, Latash ML (1982a) Interaction of afferent and efferent signals underlying joint position sense: Empirical and theoretical approaches. *Journal of Motor Behavior* 14: 174–193.

Feldman AG, Latash ML (1982b) Afferent and efferent components of joint position sense: Interpretation of kinaesthetic illusions. *Biological Cybernetics* 42: 205–214.

Feldman AG, Latash ML (1982c) Inversions of vibration-induced senso-motor events caused by supraspinal influences in man. *Neuroscience Letters* 31: 147–151.

Feldman AG, Latash ML (2005) Testing hypotheses and the advancement of science: Recent attempts to falsify the equilibrium-point hypothesis. *Experimental Brain Research* 161: 91–103.

Feldman AG, Levin MF, Garofolini A, Piscitelli D, Zhang L (2021) Central pattern generator and human locomotion in the context of referent control of motor actions. *Clinical Neurophysiology* 132: 2870–2889.

Feldman AG, Levin MF, Mitnitski AM, Archambault P (1998) 1998 ISEK Congress Keynote Lecture: Multi-muscle control in human movements. *Journal of Electromyography and Kinesiology* 8: 383–390.

Feldman AG, Orlovsky GN (1972) The influence of different descending systems on the tonic stretch reflex in the cat. *Experimental Neurology* 37: 481–494.

Feldman DE (2012) The spike-timing dependence of plasticity. *Neuron* 75: 556–571.

Ferrario C, Condoluci C, Tarabini M, Manzia CM, Galli M (2023) Anticipatory postural adjustments and kinematic analysis of step ascent and descent in adults with Down syndrome. *Journal of Intellectual Disability Research* 67: 475–487.

Fitts PM (1954) The information capacity of the human motor system in controlling the amplitude of movement. *Journal of Experimental Psychology* 47: 381–391.

Fitts PM, Peterson JR (1964) Information capacity of discrete motor responses. *Journal of Experimental Psychology* 67: 103–112.

Fitts RH, Courtright JB, Kim DH, Witzmann FA (1982) Muscle fatigue with prolonged exercise: Contractile and biochemical alterations. *American Journal of Physiology* 242: C65–C73.

Flanagan JR, Wing AM (1993) Modulation of grasp force with load force during point-to-point arm movements. *Experimental Brain Research* 95: 131–143.

Flanagan JR, Wing AM (1995) The stability of precision grasp forces during cyclic arm movements with a hand-held load. *Experimental Brain Research* 105: 455–464.

Flash T (1987) The control of hand equilibrium trajectories in multi-joint arm movements. *Biological Cybernetics* 57: 257–274.

Flash T, Hochner B (2005) Motor primitives in vertebrates and invertebrates. *Current Opinion in Neurobiology* 15: 660–666.

Flash T, Hogan N (1985) The coordination of arm movements: An experimentally confirmed mathematical model. *Journal of Neuroscience* 5: 1688–1703.

Ford KR, van den Bogert J, Myer GD, Shapiro R, Hewett TE (2008) The effects of age and skill level on knee musculature co-contraction during functional activities: A systematic review. *British Journal of Sports Medicine* 42: 561–566.

Forestier N, Nougier V (1998) The effects of muscular fatigue on the coordination of a multijoint movement in human. *Neuroscience Letters* 252: 187–190.

Fortier PA, Kalaska JF, Smith AM (1989) Cerebellar neuronal activity related to whole-arm reaching movements in the monkey. *Journal of Neurophysiology* 62: 198–211.

Freitas SMSF, de Freitas PB, Falaki A, Corson T, Lewis MM, Huang X, Latash ML (2020) Synergic control of action in levodopa-naïve Parkinson's disease patients. II. Multi-muscle synergies stabilizing vertical posture. *Experimental Brain Research* 238: 2931–2945.

Freitas SMSF, Scholz JP, Latash ML (2010) Analyses of joint variance related to voluntary whole-body movements performed in standing. *Journal of Neuroscience Methods* 188: 89–96.

Freitas SM, Scholz JP, Stehman AJ (2007) Effect of motor planning on use of motor abundance. *Neuroscience Letters* 417: 66–71.

Frenkel-Toledo S, Solomon JM, Shah A, Baniña MC, Berman S, Soroker N, Liebermann DG, Levin MF (2021) Tonic stretch reflex threshold as a measure of spasticity after stroke: Reliability, minimal detectable change and responsiveness. *Clinical Neurophysiology* 132: 1226–1233.

Friedman J, Skm V, Zatsiorsky VM, Latash ML (2009) The sources of two components of variance: An example of multifinger cyclic force production tasks at different frequencies. *Experimental Brain Research* 196: 263–277.

Fritz SL, Butts RJ, Wolf SL (2012) Constraint-induced movement therapy: From history to plasticity. *Expert Review of Neurotherapeutics* 12: 191–198.

Fu Q, Santello M (2015) Retention and interference of learned dexterous manipulation: Interaction between multiple sensorimotor processes. *Journal of Neurophysiology* 113: 144–155.

Fuglevand AJ, Zackowski KM, Huey KA, Enoka RM (1993) Impairment of neuromuscular propagation during human fatiguing contractions at submaximal forces. *Journal of Physiology* 460: 549–572.

Fujita T, Nakamura S, Ohue M, Fujii Y, Miyauchi A, Takagi Y, Tsugeno H (2005) Effect of age on body sway assessed by computerized posturography. *Journal of Bone Minerals and Metabolism* 23: 152–156.

Fukson OI, Berkinblit MB, Feldman AG (1980) The spinal frog takes into account the scheme of its body during the wiping reflex. *Science* 209: 1261–1263.

Full RJ, Kubow T, Schmitt J, Holmes P, Koditschek D (2002) Quantifying dynamic stability and maneuverability in legged locomotion. *Integrative and Comparative Biology* 42: 149–157.

Gallego JA, Perich MG, Chowdhury RH, Solla SA, Miller LE (2020) Long-term stability of cortical population dynamics underlying consistent behavior. *Nature Neuroscience* 23: 260–270.

Gandevia SC (1982) The perception of motor commands or effort during muscular paralysis. *Brain* 105: 151–159.

Gao F, Latash ML, Zatsiorsky VM (2005) Control of finger force direction in the flexion-extension plane. *Experimental Brain Research* 161: 307–315.

Garland SJ, Enoka RM, Serrano LP, Robinson GA (1994) Behavior of motor units in human biceps brachii during a submaximal fatiguing contraction. *Journal of Applied Physiology* 76: 2411–2419.

Gates DH, Dingwell JB (2008) The effects of neuromuscular fatigue on task performance during repetitive goal-directed movements. *Experimental Brain Research* 187: 573–585.

Gelfand IM, Latash ML (1998) On the problem of adequate language in movement science. *Motor Control* 2: 306–313.

Gelfand IM, Tsetlin ML (1962) On some methods of control of complex systems. *Uspekhi Matematicheskih Nauk* 17: 3–25 (in Russian), English version in *Russian Mathematical Surveys* 17: 95–117.

Gelfand IM, Tsetlin ML (1971) Some methods of controlling complex systems. In: Gelfand IM, Gurfinkel VS, Fomin SV, Tsetlin ML (Eds.) *Models of the Structural-Functional Organization of Certain Biological Systems*, pp. 329–345. MIT Press: Cambridge, MA.

Georgopoulos AP, Ashe J, Smyrnis N, Taira M (1992) The motor cortex and the coding of force. *Science* 256: 1692–1695.

Georgopoulos AP, Carpenter AF (2015) Coding of movements in the motor cortex. *Current Opinions in Neurobiology* 33: 34–39.

Georgopoulos AP, Kalaska JF, Caminiti R, Massey JT (1982) On the relations between the direction of two-dimensional arm movements and cell discharge in primate motor cortex. *Journal of Neuroscience* 2: 1527–1537.

Georgopoulos AP, Kalaska JF, Caminiti R, Massey JT (1983) Interruption of motor cortical discharge subserving aimed arm movements. *Experimental Brain Research* 49: 327–340.

Georgopoulos AP, Lurito JT, Petrides M, Schwartz AB, Massey JT (1989) Mental rotation of the neuronal population vector. *Science* 243: 234–236.

Georgopoulos AP, Schwartz AB, Kettner RE (1986) Neural population coding of movement direction. *Science* 233: 1416–1419.

Gera G, Freitas SM, Scholz JP (2016a) Relationship of diminished interjoint coordination after stroke to hand path consistency. *Experimental Brain Research* 234: 741–751.

Gera G, McGlade KE, Reisman DS, Scholz JP (2016b) Trunk muscle coordination during upward and downward reaching in stroke survivors. *Motor Control* 20: 50–69.

Gerasimenko Y, Roy RR, Edgerton VR (2008) Epidural stimulation: Comparison of the spinal circuits that generate and control locomotion in rats, cats and humans. *Experimental Neurology* 209: 417–425.

Ghez C, Gordon J (1987) Trajectory control in targeted force impulses. I. Role of opposing muscles. *Experimental Brain Research* 67: 225–240.

Gibson JJ (1979) *The Ecological Approach to Visual Perception*. Houghton Mifflin: Boston, MA.

Gielen CC, Houk JC, Marcus SL, Miller LE (1984) Viscoelastic properties of the wrist motor servo in man. *Annals of Biomedical Engineering* 12: 599–620.

Gilbertson T, Lalo E, Doyle L, Di Lazzaro V, Cioni B, Brown P (2005) Existing motor state is favored at the expense of new movement during 13–35 Hz oscillatory synchrony in the human corticospinal system. *Journal of Neuroscience* 25: 7771–7779.

Gilles MA, Wing AM (2003) Age-related changes in grip force and dynamics of hand movement. *Journal of Motor Behavior* 35: 79–85.

Giszter SF (2015) Motor primitives—new data and future questions. *Current Opinions in Neurobiology* 33: 156–165.

Giszter SF, Mussa-Ivaldi FA, Bizzi E (1993) Convergent force fields organized in the frog's spinal cord. *Journal of Neuroscience* 13: 467–491.

Golenia L, Schoemaker MM, Otten E, Mouton LJ, Bongers RM (2018) Development of reaching during mid-childhood from a developmental systems perspective. *PLoS One* 13(2): e0193463.

Goodale MA, Milner AD (1992) Separate visual pathways for perception and action. *Trends in Neuroscience* 15: 20–25.

Goodale MA, Milner AD, Jakobson LS, Carey DP (1991) A neurological dissociation between perceiving objects and grasping them. *Nature* 349: 154–156.

Goodman SR, Shim JK, Zatsiorsky VM, Latash ML (2005) Motor variability within a multi-effector system: Experimental and analytical studies of multi-finger production of quick force pulses. *Experimental Brain Research* 163: 75–85.

Goodwin GM, McCloskey DI, Matthews PB (1972) The contribution of muscle afferents to kinaesthesia shown by vibration induced illusions of movement and by the effects of paralysing joint afferents. *Brain* 95: 705–748.

Gordon AM (2001) Development of hand motor control. In: Kalverboer AF, Gramsbergen A (Eds.) *Handbook of Brain and Behaviour in Human Development*, pp. 513–537. Kluwer Academic Press: Dordrecht, The Netherlands.

Gordon AM, Ingvarsson PE, Forssberg H (1997) Anticipatory control of manipulative forces in Parkinson's disease. *Experimental Neurology* 145: 477–488.

Gordon CR, Fletcher WA, Melvill Jones G, Block EW (1995) Adaptive plasticity in the control of locomotor trajectory. *Experimental Brain Research* 102: 540–545.

Gordon J, Ghez C (1987) Trajectory control in targeted force impulses. II. Pulse height control. *Experimental Brain Research* 67: 241–252.

Gorniak SL, Alberts JL (2013) Effects of aging on force coordination in bimanual task performance. *Experimental Brain Research* 229: 273–284.

Gorniak S, Zatsiorsky VM, Latash ML (2007) Hierarchies of synergies: An example of the two-hand, multi-finger tasks. *Experimental Brain Research* 179: 167–180.

Gorniak S, Zatsiorsky VM, Latash ML (2009) Hierarchical control of static prehension: II. Multi-digit synergies. *Experimental Brain Research* 194: 1–15.

Gorniak SL, Zatsiorsky VM, Latash ML (2011) Manipulation of a fragile object by elderly individuals. *Experimental Brain Research* 212: 505–516.

Gottlieb GL (1996) On the voluntary movement of compliant (inertial-viscoelastic) loads by parcellated control mechanisms. *Journal of Neurophysiology* 76: 3207–3229.

Gottlieb GL, Agarwal GC (1986) The invariant characteristic isn't. *Behavioral and Brian Sciences* 9: 608–609.

Gottlieb GL, Agarwal GC (1988) Compliance of single joints: Elastic and plastic characteristics. *Journal of Neurophysiology* 59: 937–951.

Gottlieb GL, Corcos DM, Agarwal GC (1989a) Strategies for the control of voluntary movements with one mechanical degree of freedom. *Behavioral and Brain Sciences* 12: 189–250.

Gottlieb GL, Corcos DM, Agarwal GC (1989b) Organizing principles for single joint movements. I: A speed-insensitive strategy. *Journal of Neurophysiology* 62: 342–357.

Gottlieb GL, Corcos DM, Agarwal GC, Latash ML (1990) Organizing principles for single joint movements. III: Speed-insensitive strategy as a default. *Journal of Neurophysiology* 63: 625–636.

Gottlieb GL, Latash ML, Corcos DM, Liubinskas TJ, Agarwal GC (1992) Organizing principles for single joint movements. V. Agonist-antagonist interactions. *Journal of Neurophysiology* 67: 1417–1427.

Gottlieb GL, Song Q, Hong DA, Almeida GL, Corcos D (1996) Coordinating movement at two joints: A principle of linear covariance. *Journal of Neurophysiology* 75: 1760–1764.

Graham Brown T (1914) On the nature of the fundamental activity of the nervous centres; together with an analysis of the conditioning of the rhythmic activity in progression and a theory of the evolution of function in the nervous system. *Journal of Physiology* 48: 18–46.

Granit R (1975) The functional role of the muscle spindles—facts and hypotheses. *Brain* 98: 531–556.

Gray R (2020) Changes in movement coordination associated with skill acquisition in baseball batting: Freezing/freeing degrees of freedom and functional variability. *Frontiers in Psychology* 11: 1295.

Grey MJ, Pierce CW, Milner TE, Sinkjaer T (2001) Soleus stretch reflex during cycling. *Motor Control* 5: 36–49.

Gribble PL, Ostry DJ (2000) Compensation for loads during arm movements using equilibrium-point control. *Experimental Brain Research* 135: 474–482.

Gribble PL, Ostry DJ, Sanguineti V, Laboissiere R (1998) Are complex control signals required for human arm movements? *Journal of Neurophysiology* 79: 1409–1424.

Grillner S (1975) Locomotion in vertebrates: Central mechanisms and reflex interaction. *Physiological Reviews* 55: 247–304.

Grillner S, Wallen P (1985) Central pattern generators for locomotion, with special reference to vertebrates. *Annual Reviews in Neuroscience* 8: 233–261.

Guimarães AN, Ugrinowitsch H, Dascal JB, Porto AB, Okazaki VHA (2020) Freezing degrees of freedom during motor learning: A systematic review. *Motor Control* 24: 457–471.

Gurfinkel VS, Latash ML (1978) Motor reversals in calf muscles. *Phyziologiya Cheloveka (Human Physiology)* 4: 30–35.

Gurfinkel VS, Levik YS, Kazennikov OV, Selionov VA (1998) Locomotor-like movements evoked by leg muscle vibration in humans. *European Journal of Neuroscience* 10: 1608–1612.

Gutman SR, Latash ML, Gottlieb GL, Almeida GL (1993) Kinematic description of variability of fast movements: Analytical and experimental approaches. *Biological Cybernetics* 69: 485–492.

Hagbarth KE (1993) Microneurography and applications to issues of motor control: Fifth Annual Stuart Reiner Memorial Lecture. *Muscle and Nerve* 16: 693–705.

Hagbarth K-E, Bongiovanni LG, Nordin M (1995) Reduced servo-control of fatigued human finger extensor and flexor muscles. *Journal of Physiology* 485: 865–872.

Halliday DM, Conway BA, Farmer SF, Rosenberg JR (1999) Load-independent contributions from motor-unit synchronization to human physiological tremor. *Journal of Neurophysiology* 82: 664–675.

Halperin I, Chapman DW, Behm DG (2015) Non-local muscle fatigue: Effects and possible mechanisms. *European Journal of Applied Physiology* 115: 2031–2048.

Hansen C, LaRue J, Do M-C, Latash ML (2016) Postural preparation to stepping: Coupled center of pressure shifts in the anterior-posterior and medio-lateral directions. *Journal of Human Kinetics* 54: 5–14.

Harris CM, Wolpert DM (1998) Signal-dependent noise determines motor planning. *Nature* 394: 780–784.

Hasanbarani F, Latash ML (2020) Performance-stabilizing synergies in a complex motor task: Analysis based on the uncontrolled manifold hypothesis. *Motor Control* 24: 238–252.

Hayashi R, Becker WJ, Lee RG (1990) Effects of unexpected perturbations on trajectories and EMG patterns of rapid wrist flexion movements in humans. *Neuroscience Research* 8: 100–113.

Hayashi R, Miyake A, Jijiwa H, Watanabe S (1981) Postural readjustment to body sway induced by vibration in man. *Experimental Brain Research* 43: 217–225.

Heald JB, Wolpert DM, Lengyel M (2023) The computational and neural bases of context-dependent learning. *Annual Reviews in Neuroscience* 46: 233–258.

Heckman CJ, Gorassini MA, Bennett DJ (2005) Persistent inward currents in motoneuron dendrites: Implications for motor output. *Muscle and Nerve* 31: 135–156.

Heckman CJ, Hyngstrom AS, Johnson MD (2008a) Active properties of motoneurone dendrites: diffuse descending neuromodulation, focused local inhibition. *Journal of Physiology* 586: 1225–1231.

Heckman CJ, Johnson M, Mottram C, Schuster J (2008b) Persistent inward currents in spinal motoneurons and their influence on human motoneuron firing patterns. *Neuroscientist* 14: 264–275.

Heckman CJ, Kuo JJ, Johnson MD (2004) Synaptic integration in motoneurons with hyperexcitable dendrites. *Canadian Journal of Physiology and Pharmacology* 82: 549–555.

Heilman KM, Mack L, Rothi LG, Watson RT (1987) Transitive movements in a deafferented man. *Cortex* 23: 525–530.

Heitmann S, Ferns N, Breakspear M (2012) Muscle co-contraction modulates damping and joint stability in a three-link biomechanical limb. *Frontiers in Neurorobotics* 5: 5.

Henneman E, Somjen G, Carpenter DO (1965) Excitability and inhibitibility of motoneurones of different sizes. *Journal of Neurophysiology* 28: 599–620.

Herzog M, Focke A, Maurus P, Thürer B, Stein T (2022) Random practice enhances retention and spatial transfer in force field adaptation. *Frontiers in Human Neuroscience* 16: 816197.

Higashiyama A, Yamazaki T (2022) Postural and visual aftereffects to a slanted floor in lying and sitting positions. *Vision Research* 199: 108077.

Hinder MR, Milner TE (2003) The case for an internal dynamics model versus equilibrium point control in human movement. *Journal of Physiology* 549: 953–963.

Hirai H, Miyazaki F, Naritomi H, Koba K, Oku T, Uno K, Uemura M, Nishi T, Kageyama M, Krebs HI (2015) On the origin of muscle synergies: Invariant balance in the co-activation of agonist and antagonist muscle pairs. *Frontiers in Bioengineering and Biotechnology* 3: 192.

Hirose J, Cuadra C, Walter C, Latash ML (2020) Finger interdependence and unintentional force drifts: Lessons from manipulations of visual feedback. *Human Movement Science* 74: 102714.

Hirschfeld H, Forssberg H (1991) Phase-dependent modulations of anticipatory postural activity during human locomotion. *Journal of Neurophysiology* 66: 12–19.

Hoehn M, Yahr M (1967) Parkinsonism: Onset, progression and mortality. *Neurology* 17: 427–442.

Hoffer JA, Andreassen S (1981) Regulation of soleus muscle stiffness in premammillary cats: Intrinsic and reflex components. *Journal of Neurophysiology* 45: 267–285.

Hogan N (1985) The mechanics of multi-joint posture and movement control. *Biological Cybernetics* 52: 315–331.

Hogan N, Sternad D (2012) Dynamic primitives of motor behavior. *Biological Cybernetics* 106: 727–739.

Hogan N, Sternad D (2013) Dynamic primitives in the control of locomotion. *Frontiers in Computational Neuroscience* 7: 71.

Holdefer RN, Miller LE (2002) Primary motor cortical neurons encode functional muscle synergies. *Experimental Brain Research* 146: 233–243.

Hollerbach JM (1982) Computers, brain, and the control of movements. *Trends in Neuroscience* 5: 189–192.

Hong DA, Corcos DM, Gottlieb GL (1994) Task dependent patterns of muscle activation at the shoulder and elbow for unconstrained arm movements. *Journal of Neurophysiology* 71: 1261–1265.

Horak FB, Nashner LM (1986) Central programming of postural movements: Adaptation to altered support-surface configurations. *Journal of Neurophysiology* 55: 1369–1381.

Horak FB, Nutt JG, Nashner LM (1992) Postural inflexibility in parkinsonian subjects. *Journal of Neurological Sciences* 111: 46–58.

Horak FB, Shupert CL, Mirka A (1989) Components of postural dyscontrol in the elderly: A review. *Neurobiology of Aging* 10: 727–738.

Houk JC (2005) Agents of the mind. *Biological Cybernetics* 92: 427–437.
Hsu WL, Scholz JP, Schöner G, Jeka JJ, Kiemel T (2007) Control and estimation of posture during quiet stance depends on multijoint coordination. *Journal of Neurophysiology* 97: 3024–3035.
Hu Z, Hao M, Xu S, Xiao Q, Lan N (2019) Evaluation of tremor interference with control of voluntary reaching movements in patients with Parkinson's disease. *Journal of Neuroengineering and Rehabilitation* 16(1): 38.
Hug F, Avrillon S, Ibanez J, Farina D (2023) Common synaptic input, synergies and size principle: Control of spinal motor neurons for movement generation. *Journal of Physiology* 601: 11–20.
Hughes OM, Abbs JH (1976) Labial-mandibular coordination in the production of speech: Implications for the operation of motor equivalence. *Phonetica* 33: 199–221.
Hughlings Jackson J (1889) On the comparative study of disease of the nervous system. *British Medical Journal* 355–362, Aug. 17, 1889.
Hultborn H, Brownstone RB, Toth TI, Gossard JP (2004) Key mechanisms for setting the input-output gain across the motoneuron pool. *Progress in Brain Research* 143: 77–95.
Hupé JM, Rubin N (2004) The oblique plaid effect. *Vision Research* 44: 489–500.
Iggo A, Muir AR (1969) The structure and function of a slowly adapting touch corpuscle in hairy skin. *Journal of Physiology* 200: 763–796.
Ilmane N, Sangani S, Feldman AG (2013) Corticospinal control strategies underlying voluntary and involuntary wrist movements. *Behavioral and Brain Research* 236: 350–358.
Imamizu H, Kuroda T, Miyauchi S, Yoshioka T, Kawato M (2003) Modular organization of internal models of tools in the human cerebellum. *Proceedings of the National Academy of Sciences USA* 100: 5461–5466.
Imamizu H, Kuroda T, Yoshioka T, Kawato M (2004) Functional magnetic resonance imaging examination of two modular architectures for switching multiple internal models. *Journal of Neuroscience* 24: 1173–1181.
Inglin B, Woollacott MH (1988) Anticipatory postural adjustments associated with reaction time arm movements: A comparison between young and old. *Journal of Gerontology* 43: M105–M113.
Ivanenko YP, Cappellini G, Poppele RE, Lacquaniti F (2008) Spatiotemporal organization of alpha-motoneuron activity in the human spinal cord during different gaits and gait transitions. *European Journal of Neuroscience* 27: 3351–3368.
Ivanenko YP, Dominici N, Cappellini G, Di Paolo A, Giannini C, Poppele RE, Lacquaniti F (2013) Changes in the spinal segmental motor output for stepping during development from infant to adult. *Journal of Neuroscience* 33: 3025–3036.
Ivanenko YP, Poppele RE, Lacquaniti F (2004) Five basic muscle activation patterns account for muscle activity during human locomotion. *Journal of Physiology* 556: 267–282.
Ivanenko YP, Poppele RE, Lacquaniti F (2006) Motor control programs and walking. *Neuroscientist* 12: 339–348.
Iyer MB, Christakos CN, Ghez C (1994) Coherent modulations of human motor unit discharges during quasi-sinusoidal isometric muscle contractions. *Neuroscience Letters* 170: 94–98.
Jankowska E (1979) New observations on neuronal organization of reflexes from tendon organ afferents and their relation to reflexes evoked from muscle spindle afferents.

In: Granit R, Pompeiano O (Eds.) *Reflex Control of Posture and Movement*, pp. 29–36. Elsevier: Amsterdam.

Jaric S, Gottlieb GL, Latash ML, Corcos DM (1998) Changes in the symmetry of rapid movements: Effects of velocity and viscosity. *Experimental Brain Research* 120: 52–60.

Jaric S, Latash ML (1998) Learning a motor task involving obstacles by a multi-joint, redundant limb: Two synergies within one movement. *Journal of Electromyography and Kinesiology* 8: 169–176.

Jaric S, Milanovic S, Blezic S, Latash ML (1999) Changes in movement kinematics during single-joint movements against expectedly and unexpectedly changed inertial loads. *Human Movement Science* 18: 49–66.

Jeannerod M (1988) *The Neural and Behavioural Organization of Goal-Directed Movements*. Clarendon Press: Oxford.

Jeka JJ, Lackner JR (1994) Fingertip contact influences human postural control. *Experimental Brain Research* 100: 495–502.

Jeka JJ, Oie K, Schoner G, Dijkstra T, Henson E (1998) Position and velocity coupling of postural sway to somatosensory drive. *Journal of Neurophysiology* 79: 1661–1674.

Jeneson JA, Taylor JS, Vigneron DB, Willard TS, Carvajal L, Nelson SJ, Murphy-Boesch J, Brown TR (1990) 1H MR imaging of anatomical compartments within the finger flexor muscles of the human forearm. *Magnetic Resonance in Medicine* 15: 491–496.

Jo HJ, Ambike S, Lewis MM, Huang X, Latash ML (2016a) Finger force changes in the absence of visual feedback in patients with Parkinson's disease. *Clinical Neurophysiology* 127: 684–692.

Jo HJ, Lucassen E, Huang X, Latash ML (2017) Changes in multi-digit synergies and their feed-forward adjustments in multiple sclerosis. *Journal of Motor Behavior* 49: 218–228.

Jo HJ, Maenza C, Good DC, Huang X, Park J, Sainburg RL, Latash ML (2016b) Effects of unilateral stroke on multi-finger synergies and their feed-forward adjustments. *Neuroscience* 319: 194–205.

Jo HJ, Park J, Lewis MM, Huang X, Latash ML (2015) Prehension synergies and hand function in early-stage Parkinson's disease. *Experimental Brain Research* 233: 425–440.

Jo HJ, Perez MA (2020) Corticospinal-motor neuronal plasticity promotes exercise-mediated recovery in humans with spinal cord injury. *Brain* 143: 1368–1382.

Jo HJ, Richardson MSA, Oudega M, Perez MA (2020) The potential of corticospinal-motoneuronal plasticity for recovery after spinal cord injury. *Current Physical Medicine and Rehabilitation Reports* 8: 293–298.

Jo HJ, Richardson MSA, Oudega M, Perez MA (2021) Paired corticospinal-motoneuronal stimulation and exercise after spinal cord injury. *Journal of Spinal Cord Medicine* 44(Suppl. 1): S23–S27.

Johansson RS, Westling G (1984) Roles of glabrous skin receptors and sensorimotor memory in automatic control of precision grip when lifting rougher or more slippery objects. *Experimental Brain Research* 56: 550–564.

Johnson SK, DeLuca J, Natelson BH (1999) Chronic fatigue syndrome: Reviewing the research findings. *Annals of Behavioral Medicine* 21: 258–271.

Jones LA (1995) The senses of effort and force during fatiguing contractions. *Advances in Experimental Medicine and Biology* 384: 305–313.

Jones RD, Donaldson IM, Parkin PJ (1989) Impairment and recovery of ipsilateral sensory-motor function following unilateral cerebral infarction. *Brain* 112: 113–132.

Jonides J, Smith EE, Koeppe RA, Awh E, Minoshima S, Mintun MA (1993) Spatial working memory in humans as revealed by PET. *Nature* 363: 623–625.

Kadefors R, Kaiser E, Petersen I (1968) Dynamic spectrum analysis of myo-potentials with special reference to muscle fatigue. *Electromyography* 8: 39–74.

Kaewmanee T, Liang H, Aruin AS (2020) Effect of predictability of the magnitude of a perturbation on anticipatory and compensatory postural adjustments. *Experimental Brain Research* 238: 2207–2219.

Kakei S, Hoffman DS, Strick PL (1999) Muscle and movement representations in the primary motor cortex. *Science*. 285: 2136–2139.

Kandel ER, Schwartz JH, Jessell TM (Eds.) (2012) *Principles of Neural Science*, Fifth edition. McGraw-Hill: New York.

Kang K, Shelley M, Sompolinsky H (2003) Mexican hats and pinwheels in visual cortex. *Proceedings of the National Academy of Sciences USA* 100: 2848–2853.

Kang N, Shinohara M, Zatsiorsky VM, Latash ML (2004) Learning multi-finger synergies: An uncontrolled manifold analysis. *Experimental Brain Research* 157: 336–350.

Kapur S, Friedman J, Zatsiorsky VM, Latash ML (2010a) Finger interaction in a three-dimensional pressing task. *Experimental Brain Research* 203: 101–118.

Kapur S, Zatsiorsky VM, Latash ML (2010b) Age-related changes in the control of finger force vectors. *Journal of Applied Physiology* 109: 1827–1841.

Karst GM, Hasan Z (1987) Antagonist muscle activity during human forearm movements under varying kinematic and loading conditions. *Experimental Brain Research* 67: 391–401.

Karst GM, Hasan Z (1990) Direction-dependent strategy for control of multi-joint arm movements. In: Winters JM, Woo SL-Y (Eds.) *Multiple Muscle Systems. Biomechanics and Movement Organization*, pp. 268–281. Springer-Verlag: New York.

Kavanagh JJ, Taylor JL (2022) Voluntary activation of muscle in humans: Does serotonergic neuromodulation matter? *Journal of Physiology* 600: 3657–3670.

Kawato M (1999) Internal models for motor control and trajectory planning. *Current Opinions in Neurobiology* 9: 718–727.

Kay BA, Turvey MT, Meijer OG (2003) An early oscillator model: Studies on the biodynamics of the piano strike (Bernstein & Popova 1930). *Motor Control* 7: 1–45.

Kelso JAS (1995) *Dynamic Patterns: The Self-Organization of Brain and Behavior*. MIT Press: Cambridge, MA.

Kelso JA, Holt KG (1980) Exploring a vibratory systems analysis of human movement production. *Journal of Neurophysiology* 43: 1183–1196.

Kelso JAS, Holt KG, Kugler PN, Turvey MT (1980) On the concept of coordinative structures as dissipative structures. II. Empirical lines of convergence. In: Stelmach GE, Requin J (Eds.) *Tutorials in Motor Behavior*, pp. 49–70. N-Holland Publ.: Amsterdam.

Kelso JA, Tuller B, Vatikiotis-Bateson E, Fowler CA (1984) Functionally specific articulatory cooperation following jaw perturbations during speech: Evidence for coordinative structures. *Journal of Experimental Psychology: Human Perception and Performance* 10: 812–832.

Keogh JW, Hume PA (2012) Evidence for biomechanics and motor learning research improving golf performance. *Sports Biomechanics* 11: 288–309.

Kim SW, Shim JK, Zatsiorsky VM, Latash ML (2006) Anticipatory adjustments of multi-finger synergies in preparation for self-triggered perturbations. *Experimental Brain Research* 174: 604–612.

Kim SW, Shim JK, Zatsiorsky VM, Latash ML (2008) Finger interdependence: Linking the kinetic and kinematic variables. *Human Movement Science* 27: 408–422.

Kinoshita H, Backstrom L, Flanagan JR, Johansson RS (1997) Tangential torque effects on the control of grasp forces when holding objects with a precision grip. *Journal of Neurophysiology* 78: 1619–1630.

Kinoshita H, Kawai S, Ikuta K (1995) Contributions and co-ordination of individual fingers in multiple finger prehension. *Ergonomics* 38: 1212–1230.

Kirk BJC, Trajano GS, Pulverenti TS, Rowe G, Blazevich AJ (2019) Neuromuscular factors contributing to reductions in muscle force after repeated, high-intensity muscular efforts. *Frontiers in Physiology* 10: 783.

Kitatani R, Ohata K, Sakuma K, Aga Y, Yamakami N, Hashiguchi Y, Yamada S (2016) Ankle muscle coactivation during gait is decreased immediately after anterior weight shift practice in adults after stroke. *Gait and Posture* 45: 35–40.

Klishko AN, Farrell BJ, Beloozerova IN, Latash ML, Prilutsky BI (2014) Stabilization of cat paw trajectory during locomotion. *Journal of Neurophysiology* 112: 1376–1391.

Klous M, Mikulic P, Latash ML (2011) Two aspects of feed-forward postural control: Anticipatory postural adjustments and anticipatory synergy adjustments. *Journal of Neurophysiology* 105: 2275–2288.

Knapen T, van Ee R (2006) Slant perception, and its voluntary control, do not govern the slant aftereffect: Multiple slant signals adapt independently. *Vision Research* 46: 3381–3392.

Körding KP, Wolpert DM (2006) Bayesian decision theory in sensorimotor control. *Trends in Cognitive Science* 10: 319–326.

Koshland GF, Gerilovsky L, Hasan Z (1991) Activity of wrist muscles elicited during imposed or voluntary movements about the elbow joint. *Journal of Motor Behavior* 23: 91–100.

Kovacs AJ, Buchanan JJ, Shea CH (2010) Impossible is nothing: 5:3 and 4:3 multi-frequency bimanual coordination. *Experimental Brain Research* 201: 249–259.

Kravitz DJ, Saleem KS, Baker CI, Mishkin M (2011) A new neural framework for visuo-spatial processing. *Nature Reviews in Neuroscience* 12: 217–230.

Kretchmar S, Latash ML (2022) Human movement: In search of borderlands between philosophy and physics. *Kinesiology Review* 11: 179–190.

Krishnamoorthy V, Goodman SR, Latash ML, Zatsiorsky VM (2003a) Muscle synergies during shifts of the center of pressure by standing persons: Identification of muscle modes. *Biological Cybernetics* 89: 152–161.

Krishnamoorthy V, Latash ML (2005) Reversals of anticipatory postural adjustments during voluntary sway. *Journal of Physiology* 565: 675–684.

Krishnamoorthy V, Latash ML, Scholz JP, Zatsiorsky VM (2003b) Muscle synergies during shifts of the center of pressure by standing persons. *Experimental Brain Research* 152: 281–292.

Krishnamoorthy V, Latash ML, Scholz JP, Zatsiorsky VM (2004) Muscle modes during shifts of the center of pressure by standing persons: Effects of instability and additional support. *Experimental Brain Research* 157: 18-31.

Krishnamoorthy V, Scholz JP, Latash ML (2007) The use of flexible arm muscle synergies to perform an isometric stabilization task. *Clinical Neurophysiology* 118: 525-537.

Krishnan V, Aruin AS (2011) Postural control in response to a perturbation: Role of vision and additional support. *Experimental Brain Research* 212: 385-397.

Krishnan V, Aruin AS, Latash ML (2011) Two stages and three components of postural preparation to action. *Experimental Brain Research* 212: 47-63.

Krishnan V, Rosenblatt NJ, Latash ML, Grabiner MD (2013) The effects of age on stabilization of the mediolateral trajectory of the swing foot. *Gait and Posture* 38: 923-928.

Kudo K, Tsutsui S, Ishikura T, Ito T, Yamamoto Y (2000) Compensatory coordination of release parameters in a throwing task. *Journal of Motor Behavior* 32: 337-345.

Kugler PN, Kelso JAS, Turvey MT (1980) On the concept of coordinative structures as dissipative structures. I. Theoretical lines of convergence. In: Stelmach GE, Requin J (Eds.) *Tutorials in Motor Behavior*, pp. 3-45. N-Holland Publ.: Amsterdam.

Kugler PN, Turvey MT (1987) *Information, Natural Law, and the Self-Assembly of Rhythmic Movement.* Erlbaum: Hillsdale, NJ.

Kukulka CG, Clamann HP (1981) Comparison of the recruitment and discharge properties of motor units in human brachial biceps and adductor pollicis during isometric contractions. *Brain Research* 219: 45-55.

Kuo AD (2002) The relative roles of feedforward and feedback in the control of rhythmic movements. *Motor Control* 6: 129-145.

Kurtzer IL (2015) Long-latency reflexes account for limb biomechanics through several supraspinal pathways. *Frontiers in Integrative Neuroscience* 8: 99.

Kwakkel G, Veerbeek JM, van Wegen EE, Wolf SL (2015) Constraint-induced movement therapy after stroke. *Lancet Neurology* 14: 224-234.

Lackner JR, DiZio P (1994) Rapid adaptation to Coriolis force perturbations of arm trajectory. *Journal of Neurophysiology* 72: 1-15.

Lackner JR, Taublieb AB (1984) Influence of vision on vibration-induced illusions of limb movement. *Experimental Neurology* 85: 97-106.

Lafe CW, Pacheco MM, Newell KM (2016) Adapting relative phase of bimanual isometric force coordination through scaling visual information intermittency. *Human Movement Science* 47: 186-196.

Laidlaw DH, Bilodeau M, Enoka RM (2000) Steadiness is reduced and motor unit discharge is more variable in old adults. *Muscle and Nerve* 23: 600-612.

Lance JW (1980) The control of muscle tone, reflexes, and movement: Robert Wartenberg Lecture. *Neurology* 30: 1303-1313.

Landau WM (1974) Spasticity: The fable of a neurological demon and the emperor's new therapy. *Archives of Neurology* 31: 217-219.

Latash LP, Latash ML, Mejier OG (1999) Thirty years later: On the problem of the relation between structure and function in the brain from a contemporary viewpoint (1966). Part I. *Motor Control* 3: 329-345.

Latash LP, Latash ML, Mejier OG (2000) Thirty years later: On the problem of the relation between structure and function in the brain from a contemporary viewpoint (1966). Part II. *Motor Control* 4: 125–149.

Latash ML (1992) Virtual trajectories, joint stiffness, and changes in natural frequency during single-joint oscillatory movements. *Neuroscience* 49: 209–220.

Latash ML (1993) *Control of Human Movement*. Human Kinetics: Urbana, IL.

Latash ML (1994) Reconstruction of equilibrium trajectories and joint stiffness patterns during single-joint voluntary movements under different instructions. *Biological Cybernetics* 71: 441–450.

Latash ML (1999) Mirror writing: Learning, transfer, and implications for internal inverse models. *Journal of Motor Behavior* 31: 107–112.

Latash ML (2007) Learning motor synergies by persons with Down syndrome. *Journal of Intellectual Disability Research* 51: 962–971.

Latash ML (2008) *Synergy*. Oxford University Press: New York.

Latash ML (2010a) Motor synergies and the equilibrium-point hypothesis. *Motor Control* 14: 294–322.

Latash ML (2010b) Stages in learning motor synergies: A view based on the equilibrium-point hypothesis. *Human Movement Science* 29: 642–654.

Latash ML (2012a) The bliss (not the problem) of motor abundance (not redundancy). *Experimental Brain Research* 217: 1–5.

Latash ML (2012b) *Fundamentals of Motor Control*. Academic Press: New York.

Latash ML (2012c) Movements that are both variable and optimal. *Journal of Human Kinetics* 34: 5–13.

Latash ML (2017) Biological movement and laws of physics. *Motor Control* 21: 327–344.

Latash ML (2018a) Stability of kinesthetic perception in efferent-afferent spaces: The concept of iso-perceptual manifold. *Neuroscience* 372: 97–113.

Latash ML (2018b) Muscle co-activation: Definitions, mechanisms, and functions. *Journal of Neurophysiology* 120: 88–104.

Latash ML (2019) *Physics of Biological Action and Perception*. Academic Press: New York.

Latash ML (Ed.) (2020a) *Bernstein's Construction of Movements*. Routledge: Abingdon, UK.

Latash ML (2020b) On primitives in motor control. *Motor Control* 24: 318–346.

Latash ML (2021a) Laws of nature that define biological action and perception. *Physics of Life Reviews* 36: 47–67.

Latash ML (2021b) One more time about motor (and non-motor) synergies. *Experimental Brain Research* 239: 2951–2967.

Latash ML (2021c) Efference copy in kinesthetic perception: A copy of what is it? *Journal of Neurophysiology* 125: 1079–1094.

Latash ML (2023) Optimality, stability, and agility of human movement: New optimality criterion and trade-offs. *Motor Control* 27: 123–159.

Latash ML, Aruin AS, Neyman I, Nicholas JJ, Shapiro MB (1995a) Feedforward postural reactions in patients with Parkinson's disease in a two-joint motor task. *Electroencephalography and Clinical Neurophysiology* 97: 77–89.

Latash ML, Aruin AS, Shapiro MB (1995b) The relation between posture and movement: A study of a simple synergy in a two-joint task. *Human Movement Science* 14: 79–107.

Latash ML, Aruin AS, Zatsiorsky VM (1999) The basis of a simple synergy: Reconstruction of joint equilibrium trajectories during unrestrained arm movements. *Human Movement Science* 18: 3–30.

Latash ML, Gottlieb GL (1990) Compliant characteristics of single joints: Preservation of equifinality with phasic reactions. *Biological Cybernetics* 62: 331–336.

Latash ML, Gottlieb GL (1991) Reconstruction of elbow joint compliant characteristics during fast and slow voluntary movements. *Neuroscience* 43: 697–712.

Latash ML, Gottlieb GL (1992) Virtual trajectories of single-joint movements performed under two basic strategies. *Neuroscience* 47: 357–365.

Latash ML, Gurfinkel VS (1976) Tonic vibration reflex and position of the body. *Physiologiya Cheloveka (Human Physiology)* 2: 593–598.

Latash ML, Gutman SR, Gottlieb GL (1991) Relativistic effects in single-joint voluntary movements. *Biological Cybernetics* 65: 401–406.

Latash ML, Huang X (2015) Neural control of movement stability: Lessons from studies of neurological patients. *Neuroscience* 301: 39–48.

Latash ML, Kalugina E, Nicholas JJ, Orpett C, Stefoski D, Davis F (1996) Myogenic and central neurogenic factors in fatigue in multiple sclerosis. *Multiple Sclerosis* 1: 236–241.

Latash ML, Kang N, Patterson D (2002a) Finger coordination in persons with Down syndrome: Atypical patterns of coordination and the effects of practice. *Experimental Brain Research* 146: 345–355.

Latash ML, Li S, Danion F, Zatsiorsky VM (2002b) Central mechanisms of finger interaction during one- and two-hand force production at distal and proximal phalanges. *Brain Research* 924: 198–208.

Latash ML, Li Z-M, Zatsiorsky VM (1998) A principle of error compensation studied within a task of force production by a redundant set of fingers. *Experimental Brain Research* 122: 131–138.

Latash ML, Madarshahian S, Ricotta J (2023) Intra-muscle synergies: Their place in the neural control hierarchy. *Motor Control* 27: 402–441.

Latash ML, Penn RD (1996) Changes in voluntary motor control induced by intrathecal baclofen. *Physiotherapy Research International* 1: 229–246.

Latash ML, Penn RD, Corcos DM, Gottlieb GL (1989) Short-term effects of intrathecal baclofen in spasticity. *Experimental Neurology* 103: 165–172.

Latash ML, Penn RD, Corcos DM, Gottlieb GL (1990) Effects of intrathecal baclofen on voluntary motor control in spastic paresis. *Journal of Neurosurgery* 72: 388–392.

Latash ML, Scholz JF, Danion F, Schöner G (2001) Structure of motor variability in marginally redundant multi-finger force production tasks. *Experimental Brain Research* 141: 153–165.

Latash ML, Scholz JF, Danion F, Schöner G (2002c) Finger coordination during discrete and oscillatory force production tasks. *Experimental Brain Research* 146: 412–432.

Latash ML, Scholz JP, Schöner G (2002d) Motor control strategies revealed in the structure of motor variability. *Exercise and Sport Science Reviews* 30: 26–31.

Latash ML, Scholz JP, Schöner G (2007) Toward a new theory of motor synergies. *Motor Control* 11: 276–308.

Latash ML, Shim JK, Smilga AV, Zatsiorsky V (2005) A central back-coupling hypothesis on the organization of motor synergies: A physical metaphor and a neural model. *Biological Cybernetics* 92: 186–191.

Latash ML, Shim JK, Zatsiorsky VM (2004) Is there a timing synergy during multi-finger production of quick force pulses? *Experimental Brain Research* 159: 65–71.

Latash ML, Talis VL (2021) Bernstein's philosophy of time: An unknown manuscript by Nikolai Bernstein (1949). *Motor Control* 25: 315–336.

Latash ML, Yarrow K, Rothwell JC (2003) Changes in finger coordination and responses to single pulse TMS of motor cortex during practice of a multi-finger force production task. *Experimental Brain Research* 151: 60–71.

Latash ML, Yee M, Orpett C, Slingo A, Nicholas JJ (1994) Combining electrical muscle stimulation with voluntary contraction for studying muscle fatigue. *Archives of Physical Medicine and Rehabilitation* 75: 29–35.

Latash ML, Zatsiorsky VM (1993) Joint stiffness: Myth or reality? *Human Movement Science* 12: 653–692.

Latash ML, Zatsiorsky VM (2016) *Biomechanics and Motor Control: Defining Central Concepts*. Academic Press: New York.

Lee WA, Buchanan TS, Rogers MW (1987) Effects of arm acceleration and behavioral conditions on the organization of postural adjustments during arm flexion. *Experimental Brain Research* 66: 257–270.

Lee Y, Ashton-Miller JA (2011) The effects of gender, level of co-contraction, and initial angle on elbow extensor muscle stiffness and damping under a step increase in elbow flexion moment. *Annals of Biomedical Engineering* 39: 2542–2549.

Lee YJ, Chen B, Aruin AS (2015) Older adults utilize less efficient postural control when performing pushing task. *Journal of Electromyography and Kinesiology* 25: 966–972.

Lemon RN, Morecraft RJ (2023) The evidence against somatotopic organization of function in the primate corticospinal tract. *Brain* 146: 1791–1803.

Leone FC, Nottingham RB, Nelson LS (1961) The folded normal distribution. *Technometrics* 3: 543–550.

Lepers R, Brenière Y (1995) The role of anticipatory postural adjustments and gravity in gait initiation. *Experimental Brain Research* 107: 118–124.

Lévénez M, Garland SJ, Klass M, Duchateau J (2008) Cortical and spinal modulation of antagonist coactivation during a submaximal fatiguing contraction in humans. *Journal of Neurophysiology* 99: 554–563.

Levin MF, Dimov M (1997) Spatial zones for muscle coactivation and the control of postural stability. *Brain Research* 757: 43–59.

Levin MF, Feldman AG (1994) The role of stretch reflex threshold regulation in normal and impaired motor control. *Brain Research* 657: 23–30.

Levin MF, Michaelsen SM, Cirstea CM, Roby-Brami A (2002) Use of the trunk for reaching targets placed within and beyond the reach in adult hemiparesis. *Experimental Brain Research* 143: 171–180.

Levin O, Wenderoth N, Steyvers M, Swinnen SP (2003) Directional invariance during loading-related modulations of muscle activity: Evidence for motor equivalence. *Experimental Brain Research* 148: 62–76.

Levy G, Flash T, Hochner B (2015) Arm coordination in octopus crawling involves unique motor control strategies. *Current Biology* 25: 1195–1200.

Lewis MM, Lee E-Y, Jo HJ, Park J, Latash ML, Huang X (2016) Synergy as a new and sensitive marker of basal ganglia dysfunction: A study of asymptomatic welders. *Neurotoxicology* 56: 76–85.

Lewis MM, Slagle CG, Smith AB, Truong Y, Bai P, McKeown MJ, Mailman RB, Belger A, Huang X (2007) Task specific influences of Parkinson's disease on the striato-thalamo-cortical and cerebello-thalamo-cortical motor circuitries. *Neuroscience* 147: 224–235.

Li ZM, Latash, ML, Zatsiorsky VM (1998) Force sharing among fingers as a model of the redundancy problem. *Experimental Brain Research* 119: 276–286.

Li ZM, Zatsiorsky VM, Latash ML (2000) Contribution of the extrinsic and intrinsic hand muscles to the moments in finger joints. *Clinical Biomechanics* 15: 203–211.

Liddell EGT, Sherrington CS (1924) Reflexes in response to stretch (myotatic reflexes). *Proceedings of the Royal Society of London, Series B* 96: 212–242.

Lin JK, Pawelzik K, Ernst U, Sejnowski TJ (1998) Irregular synchronous activity in stochastically-coupled networks of integrate-and-fire neurons. *Network* 9: 333–344.

Loeb GE (1999) What might the brain know about muscles, limbs and spinal circuits? *Progress in Brain Research* 123: 405–409.

Loeb GE (2012) Optimal isn't good enough. *Biological Cybernetics* 106: 757–765.

Loram ID, Maganaris CN, Lakie M (2005) Active, non-spring-like muscle movements in human postural sway: How might paradoxical changes in muscle length be produced? *Journal of Physiology* 564: 281–293.

Lowery MM, Myers LJ, Erim Z (2007) Coherence between motor unit discharges in response to shared neural inputs. *Journal of Neuroscience Methods* 163: 384–391.

Lukos J, Ansuini C, Santello M (2007) Choice of contact points during multidigit grasping: Effect of predictability of object center of mass location. *Journal of Neuroscience* 27: 3894–3903.

Lund S (1980) Postural effects of neck muscle vibration in man. *Experientia* 36: 1398.

Lundberg A (1966) Integration in reflex pathway. In: Granit R (Ed.) *Muscular Afferents and Motor Control. Nobel Symposium, I*, pp. 275–305. Almqvist & Wiksell: Stockholm.

Luo J, Xue N, Chen J (2022) A review: Research progress of neural probes for brain research and brain-computer interface. *Biosensors (Basel)* 12(12): 1167.

Luu BL, Day BL, Cole JD, Fitzpatrick RC (2011) The fusimotor and reafferent origin of the sense of force and weight. *Journal of Physiology* 589: 3135–3147.

Ma C, Ma X, Fan J, He J (2017) Neurons in primary motor cortex encode hand orientation in a reach-to-grasp task. *Neuroscience Bulletin* 33: 383–395.

Madarshahian S, Latash ML (2022a) Reciprocal and coactivation commands at the level of individual motor units in an extrinsic finger flexor-extensor muscle pair. *Experimental Brain Research* 240: 321–340.

Madarshahian S, Latash ML (2022b) Effects of hand muscle function and dominance on intra-muscle synergies. *Human Movement Science* 82: 102936.

Madarshahian S, Letizi J, Latash ML (2021) Synergic control of a single muscle: The example of flexor digitorum superficialis. *Journal of Physiology* 599: 1261–1279.

Madarshahian S, Ricotta J, Latash ML (2022) Intra-muscle synergies stabilizing reflex-mediated force changes. *Neuroscience* 505: 59–77.

Madeleine P, Lundager B, Voigt M, Arendt-Nielsen L (2003) Standardized low-load repetitive work: Evidence of different motor control strategies between experienced workers and a reference group. *Applied Ergonomics* 34: 533–542.

Madeleine P, Voigt M, Mathiassen SE (2008) The size of cycle-to-cycle variability in biomechanical exposure among butchers performing a standardised cutting task. *Ergonomics* 51: 1078–1095.

Maenza C, Good DC, Winstein CJ, Wagstaff DA, Sainburg RL (2020) Functional deficits in the less-impaired arm of stroke survivors depend on hemisphere of damage and extent of paretic arm impairment. *Neurorehabilitation and Neural Repair* 34: 39–50.

Maki BE, Holliday PJ, Fernie GR (1990) Aging and postural control. A comparison of spontaneous- and induced-sway balance tests. *Journal of American Geriatric Society* 38: 1–9.

Malfait N, Gribble PL, Ostry DJ (2005) Generalization of motor learning based on multiple field exposures and local adaptation. *Journal of Neurophysiology* 93: 3327–3338.

Mangalam M, Rein R, Fragaszy DM (2018) Bearded capuchin monkeys use joint synergies to stabilize the hammer trajectory while cracking nuts in bipedal stance. *Proceedings of the Royal Society Biological Sciences* 285(1889): 20181797.

Mari S, Serrao M, Casali C, Conte C, Martino G, Ranavolo A, Coppola G, Draicchio F, Padua L, Sandrini G, Pierelli F (2014) Lower limb antagonist muscle co-activation and its relationship with gait parameters in cerebellar ataxia. *Cerebellum* 13: 226–236.

Mariappan YK, Manduca A, Glaser KJ, Chen J, Amrami KK, Ehman RL (2010) Vibration imaging for localization of functional compartments of the extrinsic flexor muscles of the hand. *Journal of Magnetic Resonance Imaging* 31: 1395–1401.

Mark VW, Taub E (2004) Constraint-induced movement therapy for chronic stroke hemiparesis and other disabilities. *Restorative Neurology and Neuroscience* 22: 317–336.

Marsden CD, Merton RA, Morton HB (1976) Stretch reflex and servo action in a variety of human muscles. *Journal of Physiology* 259: 531–560.

Marsden J, Stevenson V, Jarrett L (2023) Treatment of spasticity. *Handbook on Clinical Neurology* 196: 497–521.

Martin JR, Budgeon MK, Zatsiorsky VM, Latash ML (2011) Stabilization of the total force in multi-finger pressing tasks studied with the 'inverse piano' technique. *Human Movement Science* 30: 446–458.

Martin V, Reimann H, Schöner G (2019) A process account of the uncontrolled manifold structure of joint space variance in pointing movements. *Biological Cybernetics* 113: 293–307.

Martin V, Scholz JP, Schöner G (2009) Redundancy, self-motion, and motor control. *Neural Computations* 21: 1371–1414.

Marzke MW (1992) Evolutionary development of the human thumb. *Hand Clinics* 8: 1–8.

Massion J (1992) Movement, posture and equilibrium: Interaction and coordination. *Progress in Neurobiology* 38: 35–56.

Matthews PBC (1959) The dependence of tension upon extension in the stretch reflex of the soleus of the decerebrate cat. *Journal of Physiology* 47: 521–546.

Matthews PBC (1966) The reflex excitation of the soleus muscle of the decerebrate cat caused by vibration applied to its tendon. *Journal of Physiology* 184: 450–472.

Matthews PBC (1972) *Mammalian Muscle Receptors and Their Central Actions*. Williams & Wilkins: Baltimore.

Matthews PB, Stein RB (1969) The sensitivity of muscle spindle afferents to small sinusoidal changes of length. *Journal of Physiology* 200: 723–743.

Mattos D, Kuhl J, Scholz JP, Latash ML (2013) Motor equivalence (ME) during reaching: Is ME observable at the muscle level? *Motor Control* 17: 145–175.

Mattos D, Latash ML, Park E, Kuhl J, Scholz JP (2011) Unpredictable elbow joint perturbation during reaching results in multijoint motor equivalence. *Journal of Neurophysiology* 106: 1424–1436.

Mattos D, Schöner G, Zatsiorsky VM, Latash ML (2015a) Motor equivalence during accurate multi-finger force production. *Experimental Brain Research* 233: 487–502.

Mattos D, Schöner G, Zatsiorsky VM, Latash ML (2015b) Task-specific stability of abundant systems: Structure of variance and motor equivalence. *Neuroscience* 310: 600–615.

Meijer OG (2001) Making things happen: An introduction to the history of movement science. In: Latash ML, Zatsiorsky VM (Eds.) *Classics in Movement Science*, pp. 1–58. Human Kinetics: Champaign, IL.

Meijer OG, Kots YM, Edgerton VR (2001) Low-dimensional control: Tonus (1963). *Motor Control* 5: 1–22.

Melzer I, Benjuya N, Kaplanski J (2004) Postural stability in the elderly: A comparison between fallers and non-fallers. *Age and Ageing* 33: 602–607.

Merleau-Ponty M (1942/1963) *The Structure of Behavior* (Trans. by A Fisher). Beacon Press: Boston.

Merletti R, Holobar A, Farina D (2008) Analysis of motor units with high-density surface electromyography. *Journal of Electromyography and Kinesiology* 18: 879–890.

Merton PA (1953) Speculations on the servo-control of movements. In: Malcolm JL, Gray JAB, Wolstenholm GEW (Eds.) *The Spinal Cord*, pp. 183–198. Little, Brown: Boston.

Miall RC, Kitchen NM, Nam S-H, Lefumat H, Renault AG, Orstavik K, Cole JD, Sarlegna FR. Proprioceptive loss and the perception, control and learning of arm movements in humans: Evidence from sensory neuronopathy. *Experimental Brain Research* 236: 2137–2155.

Milner TE (2002) Adaptation to destabilizing dynamics by means of muscle cocontraction. *Experimental Brain Research* 143: 406–416.

Milner TE, Cloutier C (1993) Compensation for mechanically unstable loading in voluntary wrist movement. *Experimental Brain Research* 94: 522–532.

Miyake T, Okabe M (2022) Roles of mono- and bi-articular muscles in human limbs: Two-joint link model and applications. *Integrative Organismal Biology* 4(1): obac042.

Mochizuki L, Duarte M, Amadio AC, Zatsiorsky VM, Latash ML (2006) Changes in postural sway and its fractions in conditions of postural instability. *Journal of Applied Biomechanics* 22: 51–66.

Monks J (1989) Experiencing symptoms in chronic illness: Fatigue in multiple sclerosis. *International Disability Studies* 11: 78–83.

Morasso P (1983) Three-dimensional arm trajectories. *Biological Cybernetics* 48: 187–194.

Morasso P (1981) Spatial control of arm movements. *Experimental Brain Research* 42: 223–227.

Morasso PG, Sanguineti V (2002) Ankle muscle stiffness alone cannot stabilize balance during quiet standing. *Journal of Neurophysiology* 88: 2157–2162.

Morrison A, McGrath D, Wallace ES (2016) Motor abundance and control structure in the golf swing. *Human Movement Science* 46: 129–147.

Müller H, Sternad D (2003) A randomization method for the calculation of covariation in multiple nonlinear relations: Illustrated with the example of goal-directed movements. *Biological Cybernetics* 89: 22–33.

Mullick AA, Musampa NK, Feldman AG, Levin MF (2013) Stretch reflex spatial threshold measure discriminates between spasticity and rigidity. *Clinical Neurophysiology* 124: 740–751.

Mullick AA, Turpin NA, Hsu SC, Subramanian SK, Feldman AG, Levin MF (2018) Referent control of the orientation of posture and movement in the gravitational field. *Experimental Brain Research* 236: 381–398.

Muratori LM, McIsaac TL, Gordon AM, Santello M (2008) Impaired anticipatory control of force sharing patterns during whole-hand grasping in Parkinson's disease. *Experimental Brain Research* 185: 41–52.

Mutha PK, Haaland KY, Sainburg RL (2012) The effects of brain lateralization on motor control and adaptation. *Journal of Motor Behavior* 44: 455–469.

Naber M, Gruenhage G, Einhäuser W (2010) Tri-stable stimuli reveal interactions among subsequent percepts: Rivalry is biased by perceptual history. *Vision Research* 50: 818–828.

Nadin M (Ed.) (2015) *Anticipation: Learning from the Past. The Russian/Soviet Contributions to the Science of Anticipation.* Springer: New York.

Nagai K, Yamada M, Uemura K, Yamada Y, Ichihashi N, Tsuboyama T (2011) Differences in muscle coactivation during postural control between healthy older and young adults. *Archives of Gerontology and Geriatrics* 53: 338–343.

Nakao M, Inoue Y, Murkami H (1989) Aging process of leg muscle endurance in males and females. *European Journal of Applied Physiology* 59: 209–214.

Nardini AG, Freitas SMSF, Falaki A, Latash ML (2019) Preparation to a quick whole-body action: Control with referent body orientation and multi-muscle synergies. *Experimental Brain Research* 237: 1361–1374.

Nardon M, Pascucci F, Cesari P, Bertucco M, Latash ML (2022) Synergies stabilizing vertical posture in spaces of control variables. *Neuroscience* 500: 79–94.

Nardone A, Schieppati M (1988) Postural adjustments associated with voluntary contractions of leg muscles in standing man. *Experimental Brain Research* 69: 469–480.

Nashner LM, Cordo PJ (1981) Relation of automatic postural responses and reaction-time voluntary movements of human leg muscles. *Experimental Brain Research* 43: 395–405.

Nelson W (1983) Physical principles for economies of skilled movements. *Biological Cybernetics* 46: 135–147.

Newell KM (1991) Motor skill acquisition. *Annual Reviews in Psychology* 42: 213–237.

Nichols TR (1994) A biomechanical perspective on spinal mechanisms of coordinated muscular action: An architecture principle. *Acta Anatomica* 151: 1–13.

Nichols TR (2002) Musculoskeletal mechanics: A foundation of motor physiology. *Advances in Experimental and Medical Biology* 508: 473–479.

Nichols TR (2018) Distributed force feedback in the spinal cord and the regulation of limb mechanics. *Journal of Neurophysiology* 119: 1186–1200.

Nichols TR, Houk JC (1976) Improvement in linearity and regulation of stiffness that results from actions of stretch reflex. *Journal of Neurophysiology* 39: 119–142.

Nichols TR, Steeves JD (1986) Resetting of resultant stiffness in ankle flexor and extensor muscles in the decerebrate cat. *Experimental Brain Research* 62: 401–410.

Nicol C, Kuitunen S, Kyrolainen H, Avela J, Komi PV (2003) Effects of long- and short-term fatiguing stretch-shortening cycle exercises on reflex EMG and force of the tendon-muscle complex. *European Journal of Applied Physiology* 90: 470–479.

Nicolelis MA, Lebedev MA (2009) Principles of neural ensemble physiology underlying the operation of brain-machine interfaces. *Nature Reviews in Neuroscience* 10: 530–540.

Nijs J, Aelbrecht S, Meeus M, Van Oosterwijck J, Zinzen E, Clarys P (2011) Tired of being inactive: A systematic literature review of physical activity, physiological exercise capacity and muscle strength in patients with chronic fatigue syndrome. *Disability and Rehabilitation* 33: 1493–1500.

Niu X, Latash ML, Zatsiorsky VM (2012) Reproducibility and variability of the cost functions reconstructed from experimental recordings in multi-finger prehension. *Journal of Motor Behavior* 44: 69–85.

Nouillot P, Bouisset S, Do MC (1992) Do fast voluntary movements necessitate anticipatory postural adjustments even if equilibrium is unstable? *Neuroscience Letters* 147: 1–4.

Novak KE, Miller LE, Houk JC (2000) Kinematic properties of rapid hand movements in a knob turning task. *Experimental Brain Research* 132: 419–433.

Olafsdottir H, Kim SW, Zatsiorsky VM, Latash ML (2008) Anticipatory synergy adjustments in preparation to self-triggered perturbations in elderly individuals. *Journal of Applied Biomechanics* 24: 175–179.

Olafsdottir H, Yoshida N, Zatsiorsky VM, Latash ML (2005a) Anticipatory covariation of finger forces during self-paced and reaction time force production. *Neuroscience Letters* 381: 92–96.

Olafsdottir H, Yoshida N, Zatsiorsky VM, Latash ML (2007a) Elderly show decreased adjustments of motor synergies in preparation to action. *Clinical Biomechanics* 22: 44–51.

Olafsdottir H, Zatsiorsky VM, Latash ML (2005b) Is the thumb a fifth finger? A study of digit interaction during force production tasks. *Experimental Brain Research* 160: 203–213.

Olafsdottir H, Zhang W, Zatsiorsky VM, Latash ML (2007b) Age related changes in multi-finger synergies in accurate moment of force production tasks. *Journal of Applied Physiology* 102: 1490–1501.

Oliveira MA, Hsu J, Park J, Clark JE, Shim JK (2008) Age-related changes in multi-finger interactions in adults during maximum voluntary finger force production tasks. *Human Movement Science* 27: 714–727.

Oludare SO, Ma CC, Aruin AS (2017) Unilateral discomfort increases the use of contralateral side during sit-to-stand transfer. *Rehabilitation Research and Practice* 2017: 4853840.

Orlovsky GN, Deliagina TG, Grillner S (1999) *Neuronal Control of Locomotion. From Mollusc to Man*. Oxford University Press: New York.

Ostry DJ, Feldman AG (2003) A critical evaluation of the force control hypothesis in motor control. *Experimental Brain Research* 153: 275–288.

Osu R, Ota K, Fujiwara T, Otaka Y, Kawato M, Liu M (2011) Quantifying the quality of hand movement in stroke patients through three-dimensional curvature. *Journal of Neuroengineering and Rehabilitation* 8: 62.

Overduin SA, d'Avella A, Roh J, Carmena JM, Bizzi E (2015) Representation of muscle synergies in the primate brain. *Journal of Neuroscience* 35: 12615–12624.

Owen AM, McMillan KM, Laird AR, Bullmore E (2005) N-back working memory paradigm: A meta-analysis of normative functional neuroimaging studies. *Human Brain Mapping* 25: 46–59.

Owings TM, Grabiner MD (1998) Normally aging older adults demonstrate the bilateral deficit during ramp and hold contractions. *Journal of Gerontology Series A, Biological Sciences and Medical Sciences* 53: B425–B429.

Park J, Jo HJ, Lewis MM, Huang X, Latash ML (2013a) Effects of Parkinson's disease on optimization and structure of variance in multi-finger tasks. *Experimental Brain Research* 231: 51–63.

Park J, Lewis MM, Huang X, Latash ML (2013b) Effects of olivo-ponto-cerebellar atrophy (OPCA) on finger interaction and coordination. *Clinical Neurophysiology* 124: 991–998.

Park J, Lewis MM, Huang X, Latash ML (2014) Dopaminergic modulation of motor coordination in Parkinson's disease. *Parkinsonism and Related Disorders* 20: 64–68.

Park J, Sun Y, Zatsiorsky VM, Latash ML (2011) Age-related changes in optimality and motor variability: An example of multi-finger redundant tasks. *Experimental Brain Research* 212: 1–18.

Park J, Wu Y-H, Lewis MM, Huang X, Latash ML (2012) Changes in multi-finger interaction and coordination in Parkinson's disease. *Journal of Neurophysiology* 108: 915–924.

Parsa B, O'Shea DJ, Zatsiorsky VM, Latash ML (2016) On the nature of unintentional action: A study of force/moment drifts during multi-finger tasks. *Journal of Neurophysiology* 116: 698–708.

Parsa B, Terekhov AV, Zatsiorsky VM, Latash ML (2017) Optimality and stability of intentional and unintentional actions: I. Origins of drifts in performance. *Experimental Brain Research* 235: 481–496.

Partridge LD (1965) Modifications of neural output signals by muscles: A frequency response study. *Journal of Applied Physiology* 20: 150–156.

Pataky TC, Latash ML, Zatsiorsky VM (2007) Finger interaction during maximal radial and ulnar deviation efforts: Experimental data and linear neural network modeling. *Experimental Brain Research* 179: 301–312.

Patejdl R, Zettl UK (2022) The pathophysiology of motor fatigue and fatigability in multiple sclerosis. *Frontiers in Neurology* 13: 891415.

Patil AC, Thakor NV (2016) Implantable neurotechnologies: A review of micro- and nanoelectrodes for neural recording. *Medical and Biological Engineering and Computing* 54: 23–44.

Pawlowski M, Furmanek MP, Sobota G, Marszalek W, Slomka KJ, Bacik B, Juras G (2021) Number of trials necessary to apply analysis within the framework of the uncontrolled manifold analysis at different levels of hierarchical synergy control. *Journal of Human Kinetics* 76: 131–143.

Penfield W, Rasmussen T (1950) *The Cerebral Cortex of Man. A Clinical Study of Localization of Function.* MacMillan: New York.

Penn RD, Kroin JS (1987) Long-term intrathecal baclofen infusion for treatment of spasticity. *Journal of Neurosurgery* 66: 181–185.

Penn RD, Savoy SM, Corcos D, Latash M, Gottlieb G, Parke B, Kroin JS (1989) Intrathecal baclofen for severe spinal spasticity. *New England Journal of Medicine* 320: 1517–1521.

Perreault EJ, Kirsch RF, Crago PE (2004) Multijoint dynamics and postural stability of the human arm. *Experimental Brain Research* 157: 507–517.

Pichler P, Lagnado L (2020) Motor behavior selectively inhibits hair cells activated by forward motion in the lateral line of zebrafish. *Current Biology* 30: 150–157.

Pilon J-F, De Serres SJ, Feldman AG (2007) Threshold position control of arm movement with anticipatory increase in grip force. *Experimental Brain Research* 181: 49–67.

Piscitelli D, Falaki A, Solnik S, Latash ML (2017) Anticipatory postural adjustments and anticipatory synergy adjustments: Preparing to a postural perturbation with predictable and unpredictable direction. *Experimental Brain Research* 235: 713–730.

Piscitelli D, Turpin NA, Subramanian SK, Feldman AG, Levin MF (2020) Deficits in corticospinal control of stretch reflex thresholds in stroke: Implications for motor impairment. *Clinical Neurophysiology* 131: 2067–2078.

Polit A, Bizzi E (1978) Processes controlling arm movements in monkey. *Science* 201: 1235–1237.

Polit A, Bizzi E (1979) Characteristics of motor programs underlying arm movement in monkey. *Journal of Neurophysiology* 42: 183–194.

Poon C, Chin-Cottongim LG, Coombes SA, Corcos DM, Vaillancourt DE (2012) Spatiotemporal dynamics of brain activity during the transition from visually guided to memory-guided force control. *Journal of Neurophysiology* 108: 1335–1348.

Prilutsky BI (2000) Coordination of two- and one-joint muscles: Functional consequences and implications for motor control. *Motor Control* 4: 1–44.

Prilutsky BI, Zatsiorsky VM (2002) Optimization-based models of muscle coordination. *Exercise and Sport Science Reviews* 30: 32–38.

Prochazka A, Clarac F, Loeb GE, Rothwell JC, Wolpaw JR (2000) What do reflex and voluntary mean? Modern views on an ancient debate. *Experimental Brain Research* 130: 417–432.

Proske U, Allen T (2019) The neural basis of the senses of effort, force and heaviness. *Experimental Brain Research* 237: 589–599.

Proske U, Gandevia SC (2012) The proprioceptive senses: Their roles in signaling body shape, body position and movement, and muscle force. *Physiological Reviews* 92: 1651–1697.

Pruszynski JA (2014) Primary motor cortex and fast feedback responses to mechanical perturbations: A primer on what we know now and some suggestions on what we should find out next. *Frontiers in Integrative Neuroscience* 8: 72.

Pruszynski JA, Johansson RS, Flanagan JR (2016) A rapid tactile-motor reflex automatically guides reaching toward handheld objects. *Current Biology* 26: 788–792.

Raibert MH (1977) *Motor Control and Learning by the State Space Model.* Doct. Diss., MIT, Cambridge, MA.

Ralston HJ, Inman VT, Strait LA, Shaffrath MD (1947) Mechanics of human isolated voluntary muscle. *American Journal of Physiology* 151: 612–620.

Ramadan R, Hummert C, Jokeit JS, Schöner G (2022) Estimating the time structure of descending activation that generates movements at different speeds. *Journal of Neurophysiology* 128: 1091–1105.

Rannama I, Zusa A, Latash ML (2023) Unintentional force drifts in the lower extremities. *Experimental Brain Research* 241: 1309–1318.

Raptis H, Burtet L, Forget R, Feldman AG (2010) Control of wrist position and muscle relaxation by shifting spatial frames of reference for motoneuronal recruitment: Possible involvement of corticospinal pathways. *Journal of Physiology* 588: 1551–1570.

Rasouli O, Solnik S, Furmanek MP, Piscitelli D, Falaki A, Latash ML (2017) Unintentional drifts during quiet stance and voluntary body sway. *Experimental Brain Research* 235: 2301–2316.

Reisman D, Scholz JP (2003) Aspects of joint coordination are preserved during pointing in persons with post-stroke hemiparesis. *Brain* 126: 2510–2527.

Rektor I, Sochurkova D, Bockova M (2006) Intracerebral ERD/ERS in voluntary movement and in cognitive visuomotor task. *Progress in Brain Research* 159: 311–330.

Reschechtko S, Cuadra C, Latash ML (2018) Force illusions and drifts observed during muscle vibration. *Journal of Neurophysiology* 119: 326–336.

Reschechtko S, Hasanbarani F, Akulin VM, Latash ML (2017) Unintentional force changes in cyclical tasks performed by an abundant system: Empirical observations and a dynamical model. *Neuroscience* 350: 94–109.

Reschechtko S, Latash ML (2017) Stability of hand force production: I. Hand level control variables and multi-finger synergies. *Journal of Neurophysiology* 118: 3152–3164.

Reschechtko S, Latash ML (2018) Stability of hand force production: II. Ascending and descending synergies. *Journal of Neurophysiology* 120: 1045–1060.

Reschechtko S, Pruszynski JA (2020) Stretch reflexes. *Current Biology* 30: R1025–R1030.

Reschechtko S, Zatsiorsky VM, Latash ML (2014) Stability of multi-finger action in different spaces. *Journal of Neurophysiology* 112: 3209–3218.

Reschechtko S, Zatsiorsky VM, Latash ML (2015) Task-specific stability of multi-finger steady-state action. *Journal of Motor Behavior* 47: 365–377.

Richards CL, Malouin F (2013) Cerebral palsy: Definition, assessment and rehabilitation. *Handbook of Clinical Neurology* 111: 183–195.

Ricotta JM, De SD, Nardon M, Benamati A, Latash ML (2023a) Effects of fatigue on intramuscle force-stabilizing synergies. *Journal of Applied Physiology* 135: 1023–1035.

Ricotta JM, Nardon M, De SD, Jiang J, Graziani W, Latash ML (2023b) Motor unit based synergies in a non-compartmentalized muscle. *Experimental Brain Research* 241: 1367–1379.

Rijntjes M, Buechel C, Kiebel S, Weiller C (1999) Multiple somatotopic representations in the human cerebellum. *Neuroreport* 10: 3653–3658.

Riley MA, Wong S, Mitra S, Turvey MT (1997) Common effects of touch and vision on postural parameters. *Experimental Brain Research* 117: 165–170.

Rinaldi M, Ranavolo A, Conforto S, Martino G, Draicchio F, Conte C, Varrecchia T, Bini F, Casali C, Pierelli F, Serrao M (2017) Increased lower limb muscle coactivation reduces gait performance and increases metabolic cost in patients with hereditary spastic paraparesis. *Clinical Biomechanics* 48: 63–72.

Robert T, Zatsiorsky VM, Latash ML (2008) Multi-muscle synergies in an unusual postural task: Quick shear force production. *Experimental Brain Research* 187: 237–253.

Rogers MW, Kukulka CG, Soderberg GL (1992) Age-related changes in postural responses preceding rapid self-paced and reaction time arm movements. *Journal of Gerontology* 47: M159–M165.

Roll JP, Vedel JP (1982) Kinaesthetic role of muscle afferents in man, studied by tendon vibration and microneurography. *Experimental Brain Research* 47: 177–190.

Roll JP, Vedel JP, Roll R (1989) Eye, head and skeletal muscle spindle feedback in the elaboration of body references. *Progress in Brain Research* 80: 113–123.

Roos MR, Rice CL, Vandervoort AA (1997) Age-related changes in motor unit function. *Muscle and Nerve* 20: 679–690.

Rose DK, Winstein CJ (2004) Bimanual training after stroke: Are two hands better than one? *Topics in Stroke Rehabilitation* 11: 20–30.

Rosenblatt NJ, Grabiner MD (2012) Relationship between obesity and falls by middle-aged and older women. *Archives of Physical Medicine and Rehabilitation* 93: 718–722.

Rosenblatt NJ, Hurt CP, Latash ML, Grabiner MD (2014) An apparent contradiction: Increasing variability to achieve greater precision? *Experimental Brain Research* 232: 403–413.

Rothwell JC (1994) *Control of Human Voluntary Movement*, Second edition. Chapman & Hall: London.

Rothwell JC, Traub MM, Day BL, Obeso JA, Thomas PK, Marsden CD (1982a) Manual motor performance in a deafferented man. *Brain* 105: 515–542.

Rothwell JC, Traub MM, Marsden CD (1982b) Automatic and "voluntary" responses compensating for disturbances of human thumb movements. *Brain Research* 248: 33–41.

Royer N, Coates K, Aboodarda SJ, Camdessanché JP, Millet GY (2022) How is neuromuscular fatigability affected by perceived fatigue and disability in people with multiple sclerosis? *Frontiers in Neurology* 13: 983643.

Rozand V, Senefeld JW, Hassanlouei H, Hunter SK (2017) Voluntary activation and variability during maximal dynamic contractions with aging. *European Journal of Applied Physiology* 117: 2493–2507.

Safavynia SA, Ting LH (2013) Long-latency muscle activity reflects continuous, delayed sensorimotor feedback of task-level and not joint-level error. *Journal of Neurophysiology* 110: 1278–1290.

Sahrmann SA, Norton BJ (1977) The relationship of voluntary movement of spasticity in the upper motor neuron syndrome. *Annals of Neurology* 2: 460–465.

Sainburg RL (2002) Evidence for a dynamic-dominance hypothesis of handedness. *Experimental Brain Research* 142: 241–258.

Sainburg RL (2005) Handedness: Differential specializations for control of trajectory and position. *Exercise and Sport Science Reviews* 33: 206–213.

Sainburg RL (2014) Convergent models of handedness and brain lateralization. *Frontiers in Psychology* 5: 1092.

Sainburg RL, Ghilardi MF, Poizner H, Ghez C (1995) Control of limb dynamics in normal subjects and patients without proprioception. *Journal of Neurophysiology* 73: 820–835.

Sainburg RL, Kalakanis D (2000) Differences in control of limb dynamics during dominant and nondominant arm reaching. *Journal of Neurophysiology* 83: 2661–2675.

Sainburg RL, Poizner H, Ghez C (1993) Loss of proprioception produces deficits in interjoint coordination. *Journal of Neurophysiology* 70: 2136–2147.

Sainburg RL, Schaefer SY, Yadav V (2016) Lateralized motor control processes determine asymmetry of interlimb transfer. *Neuroscience* 334: 26–38.Sanei K, Keir PJ (2013) Independence and control of the fingers depend on direction and contraction mode. *Human Movement Science* 32: 457–471.

Sangari S, Kirshblum S, Guest JD, Oudega M, Perez MA (2021) Distinct patterns of spasticity and corticospinal connectivity following complete spinal cord injury. *Journal of Physiology* 599: 4441–4454.

Sangari S, Perez MA (2022) Prevalence of spasticity in humans with spinal cord injury with different injury severity. *Journal of Neurophysiology* 128: 470–479.

Santello M, Bianchi M, Gabiccini M, Ricciardi E, Salvietti G, Prattichizzo D, Ernst M, Moscatelli A, Jorntell H, Kappers AM, Kyriakopoulos K, Schaeffer AA, Castellini C, Bicchi A (2016) Hand synergies: Integration of robotics and neuroscience for understanding the control of biological and artificial hands. *Physics of Life Reviews* 17: 1–23.

Santello M, Soechting JF (2000) Force synergies for multifingered grasping. *Experimental Brain Research* 133: 457–467.

Sarabon N, Panjan A, Latash ML (2013) The effects of aging on the rambling and trembling components of postural sway: Effects of motor and sensory challenges. *Gait and Posture* 38: 637–642.

Sarna S, Sahi T, Koskenvuo M, Kaprio J (1993) Increased life expectancy of world class male athletes. *Medicine and Science in Sports and Exercise* 25: 237–244.

Sato H (1982) Functional characteristics of human skeletal muscle revealed by spectral analysis of the surface electromyogram. *Electromyography and Clinical Neurophysiology* 22: 459–516.

Schaefer SY, Haaland KY, Sainburg RL (2007) Ipsilesional motor deficits following stroke reflect hemispheric specializations for movement control. *Brain* 130: 2146–2158.

Schaefer SY, Mutha PK, Haaland KY, Sainburg RL (2012) Hemispheric specialization for movement control produces dissociable differences in online corrections after stroke. *Cerebral Cortex* 22: 1407–1419.

Schaffer JE, Sarlegna FR, Sainburg RL (2021) A rare case of deafferentation reveals an essential role of proprioception in bilateral coordination. *Neuropsychologia* 160: 107969.

Scheidt RA, Ghez C (2007) Separate adaptive mechanisms for controlling trajectory and final position in reaching. *Journal of Neurophysiology* 98: 3600–3613.

Schieber MH (2001) Constraints on somatotopic organization in the primary motor cortex. *Journal of Neurophysiology* 86: 2125–2143.

Schieber MH, Santello M (2004) Hand function: Peripheral and central constraints on performance. *Journal of Applied Physiology* 96: 2293–2300.

Schmidt RA (1975) A schema theory of discrete motor skill learning. *Psychological Reviews* 82: 225–260.

Schmidt RA (1980) Past and future issues in motor programming. *Research Quarterly of Exercise and Sport* 51: 122–140.

Schmidt RA, McGown C (1980) Terminal accuracy of unexpected loaded rapid movements: Evidence for a mass-spring mechanism in programming. *Journal of Motor Behavior* 12: 149–161.

Schmidt RC, Carello C, Turvey MT (1990) Phase transitions and critical fluctuations in the visual coordination of rhythmic movements between people. *Journal of Experimental Psychology: Human Perception and Performance* 16: 227–247.

Scholz JP, Danion F, Latash ML, Schöner G (2002) Understanding finger coordination through analysis of the structure of force variability. *Biological Cybernetics* 86: 29–39.

Scholz JP, Latash ML (1998) A study of a bimanual synergy associated with holding an object. *Human Movement Science* 17: 753–779.

Scholz JP, Schöner G (1999) The uncontrolled manifold concept: Identifying control variables for a functional task. *Experimental Brain Research* 126: 289–306.

Scholz JP, Schöner G (2014) Use of the uncontrolled manifold (UCM) approach to understand motor variability, motor equivalence, and self-motion. *Advances in Experimental and Medical Biology* 826: 91–100.

Scholz JP, Schöner G, Hsu WL, Jeka JJ, Horak F, Martin V (2007) Motor equivalent control of the center of mass in response to support surface perturbations. *Experimental Brain Research* 180: 163–179.

Scholz JP, Schöner G, Latash ML (2000) Identifying the control structure of multijoint coordination during pistol shooting. *Experimental Brain Research* 135: 382–404.

Schöner G (1995) Recent developments and problems in human movement science and their conceptual implications. *Ecological Psychology* 8: 291–314.

Schöner G (2002) Timing, clocks, and dynamical systems. *Brain and Cognition* 48: 31–51.

Schoner G, Kelso JAS (1988) Dynamic pattern generation in behavioral and neural systems. *Science* 239: 1513–1520.

Schöner G, Scholz JP (2007) Analyzing variance in multi-degree-of-freedom movements: Uncovering structure versus extracting correlations. *Motor Control* 11: 259–275.

Schöner G, Thelen E (2006) Using dynamic field theory to rethink infant habituation. *Psychological Reviews* 113: 273–299.

Schotland JL, Lee WA, Rymer WZ (1989) Wiping reflex and flexion withdrawal reflexes display different EMG patterns prior to movement onset in the spinalized frog. *Experimental Brain Research* 78: 649–653.

Schrödinger E (1944/2012) *What Is Life?* Canto Classics. Cambridge University Press: Cambridge, UK.

Schrödinger E (1949) "Do electrons think?" BBC Lecture. https://www.youtube.com/watch?v=hCwR1ztUXtU.

Scott SH, Kalaska JF (1995) Changes in motor cortex activity during reaching movements with similar hand paths but different arm postures. *Journal of Neurophysiology* 73: 2563–2567.

Sefati S, Nevelin ID, Roth E, Mitchell TRT, Snyder JB, Maciver MA, Fortune ES, Cowan NJ (2013) Mutually opposing forces during locomotion can eliminate the tradeoff between maneuverability and stability. *Proceedings of the National Academy of Sciences USA* 110: 18798–18803.

Seif-Naraghi AH, Winters JM (1990) Optimized strategies for scaling goal-directed dynamic limb movements. In: Winters JM, Woo SL-Y (Eds.) *Multiple Muscle Systems. Biomechanics and Movement Organization*, pp. 312–334. Springer-Verlag: New York.

Selionov VA, Ivanenko YP, Solopova IA, Gurfinkel VS (2009) Tonic central and sensory stimuli facilitate involuntary air-stepping in humans. *Journal of Neurophysiology* 101: 2847–2858.

Semon R (1921) *The Mneme*. George Allen & Unwin: London.

Sergio LE, Kalaska JF (1997) Systematic changes in directional tuning of motor cortex cell activity with hand location in the workspace during generation of static isometric forces in constant spatial directions. *Journal of Neurophysiology* 78: 1170–1174.

Sergio LE, Kalaska JF (1998) Changes in the temporal pattern of primary motor cortex activity in a directional isometric force versus limb movement task. *Journal of Neurophysiology* 80: 1577–1583.

Serlin DM, Schieber MH (1993) Morphologic regions of the multitendoned extrinsic finger muscles in the monkey forearm. *Acta Anatomica* 146: 255–266.

Shadmehr R (2017) Distinct neural circuits for control of movement vs. holding still. *Journal of Neurophysiology* 117: 1431–1460.

Shadmehr R, Mussa-Ivaldi FA (1994) Adaptive representation of dynamics during learning of a motor task. *Journal of Neuroscience* 14: 3208–3224.

Shadmehr R, Wise SP (2005) *The Computational Neurobiology of Reaching and Pointing*. MIT Press: Cambridge, MA.

Shaklai S, Mimouni-Bloch A, Levin M, Friedman J (2017) Development of finger force coordination in children. *Experimental Brain Research* 235: 3709–3720.

Shapiro MB, Aruin AS, Latash ML (1995) Velocity dependent activation of postural muscles in a simple two-joint synergy. *Human Movement Science* 14: 351–369.

Shapkova EYu (2004) Spinal locomotor capability revealed by electrical stimulation of the lumbar enlargement in paraplegic patients. In: Latash ML, Levin MF (Eds.) *Progress in Motor Control-3*, pp. 253–290. Human Kinetics: Champaign, IL.

Sheeran WM, Ahmed OJ (2020) The neural circuitry supporting successful navigation despite variable movement speeds. *Neuroscience and Biobehavioral Reviews* 108: 821–833.

Shemmell J, Krutky MA, Perreault EJ (2010) Stretch sensitive reflexes as an adaptive mechanism for maintaining limb stability. *Clinical Neurophysiology* 121: 1680–1689.

Shenoy KV, Sahani M, Churchland MM (2013) Cortical control of arm movements: A dynamical systems perspective. *Annual Reviews in Neuroscience* 36: 337–359.

Shik ML, Orlovsky GN (1976) Neurophysiology of locomotor automatism. *Physiological Reviews* 56: 465–501.

Shim JK, Latash ML, Zatsiorsky VM (2003) Prehension synergies: Trial-to-trial variability and hierarchical organization of stable performance. *Experimental Brain Research* 152: 173–184.

Shim JK, Latash ML, Zatsiorsky VM (2005a) Prehension synergies in three dimensions. *Journal of Neurophysiology* 93: 766–776.

Shim JK, Lay B, Zatsiorsky VM, Latash ML (2004) Age-related changes in finger coordination in static prehension tasks. *Journal of Applied Physiology* 97: 213–224.

Shim JK, Olafsdottir H, Zatsiorsky VM, Latash ML (2005b) The emergence and disappearance of multi-digit synergies during force production tasks. *Experimental Brain Research* 164: 260–270.

Shim JK, Park J, Zatsiorsky VM, Latash ML (2006) Adjustments of prehension synergies in response to self-triggered and experimenter-triggered load and torque perturbations. *Experimental Brain Research* 175: 641–653.

Shinohara M, Latash ML, Zatsiorsky VM (2003a) Age effects on force production by the intrinsic and extrinsic hand muscles and finger interaction during maximal contraction tasks. *Journal of Applied Physiology* 95: 1361–1369.

Shinohara M, Li S, Kang N, Zatsiorsky VM, Latash ML (2003b) Effects of age and gender on finger coordination in MVC and sub-maximal force-matching tasks. *Journal of Applied Physiology* 94: 259–270.

Shinohara M, Scholz JP, Zatsiorsky VM, Latash ML (2004) Finger interaction during accurate multi-finger force production tasks in young and elderly persons. *Experimental Brain Research* 156: 282–292.

Shiratori T, Latash ML (2000) The roles of proximal and distal muscles in anticipatory postural adjustments under asymmetrical perturbations and during standing on roller-skates. *Clinical Neurophysiology* 111: 613–623.

Shiratori T, Latash ML (2001) Anticipatory postural adjustments during load catching by standing subjects. *Clinical Neurophysiology* 112: 1250–1265.

Siemienski A (2006) Direct solution of the inverse optimization problem of load sharing between muscles. *Journal of Biomechanics* 39: S45.

Simmons RW, Richardson C (1984) Maintenance of equilibrium point control during an unexpectedly loaded rapid limb movement. *Brain Research* 302: 239–244.

Singh T, Latash ML (2011) Effects of muscle fatigue on multi-muscle synergies. *Experimental Brain Research* 214: 335–350.

Singh T, Skm V, Zatsiorsky VM, Latash ML (2010) Fatigue and motor redundancy: Adaptive increase in force variance in multi-finger tasks. *Journal of Neurophysiology* 103: 2990–3000.

Singh T, Zatsiorsky VM, Latash ML (2012) Effects of fatigue on synergies in a hierarchical system. *Human Movement Science* 31: 1379–1398.

Singh T, Zatsiorsky VM, Latash ML (2013) Adaptations to fatigue of a single digit violate the principle of superposition in a multi-finger static prehension task. *Experimental Brain Research* 225: 589–602.

Sittig AC, Denier van der Gon JJ, Gielen CC (1985) Separate control of arm position and velocity demonstrated by vibration of muscle tendon in man. *Experimental Brain Research* 60: 445–453.

Sjogaard G, Kiens B, Jorgensen K, Saltin B (1986) Intramuscular pressure, EMG and blood flow during low-level prolonged static contraction in man. *Acta Physiologica Scandinavica* 128: 475–484.

Sjogaard G, Savard G, Juel C (1988) Muscle blood flow during isometric activity and its relation to muscle fatigue. *European Journal of Applied Physiology and Occupational Physiology* 57: 327–335.

Slifkin AB, Vaillancourt DE, Newell KM (2000) Intermittency in the control of continuous force production. *Journal of Neurophysiology* 84: 1708–1718.

Slijper H, Latash ML (2000) The effects of instability and additional hand support on anticipatory postural adjustments in leg, trunk, and arm muscles during standing. *Experimental Brain Research* 135: 81–93.

Slobounov S, Chiang H, Johnston J, Ray W (2002) Modulated cortical control of individual fingers in experienced musicians: An EEG study. Electroencephalographic study. *Clinical Neurophysiology* 113: 2013–2024.

Smeets JB, Brenner E (1999) A new view on grasping. *Motor Control* 3: 237–271.

Smeets JB, Brenner E (2001) Independent movements of the digits in grasping. *Experimental Brain Research* 139: 92–100.

Smeets JBJ, van der Kooij K, Brenner E (2019) A review of grasping as the movements of digits in space. *Journal of Neurophysiology* 122: 1578–1597.

Solnik S, Qiao M, Latash ML (2017) Effects of visual feedback and memory on unintentional drifts in performance during finger pressing tasks. *Experimental Brain Research* 235: 1149–1162.

Solnik S, Reschechtko S, Wu Y-H, Zatsiorsky VM, Latash ML (2015) Force-stabilizing synergies in motor tasks involving two actors. *Experimental Brain Research* 233: 2935–2949.

Solopova IA, Selionov VA, Zhvansky DS, Gurfinkel VS, Ivanenko Y (2016) Human cervical spinal cord circuitry activated by tonic input can generate rhythmic arm movements. *Journal of Neurophysiology* 115: 1018–1030.

Sperry RW (1950) Neural basis of the spontaneous optokinetic response produced by visual inversion. *Journal of Comparative Physiology and Psychology* 43: 482–489.

St. Gregory Palamas (1983) *The Triads*. Classics of Western Spirituality. Paulist Press: Mahwah, NJ.

St. Gregory Palamas (1988) *The One Hundred and Fifty Chapters*. Pontifical Institute of Mediaeval Studies: Toronto.

Stein PSG (1984) Central pattern generators in the spinal cord. In: Davidoff RA (Ed.) *Handbook of the Spinal Cord, Vol. 2-3: Anatomy and Physiology*, pp. 647–672. Marcel Dekker: New York, Basel.

Sternad D (2002) Wachholder and Altenberger 1927: Foundational experiments for current hypotheses on equilibrium-point control in voluntary movements. *Motor Control* 6: 299–302.

Sternad D, Park SW, Müller H, Hogan N (2010) Coordinate dependence of variability analysis. *PLoS Computational Biology* 6(4): e1000751.

Straka H, Chagnaud BP (2017) Moving or being moved: That makes a difference. *Journal of Neurology* 264(Suppl. 1): 28–33.

Subramanian SK, Feldman AG, Levin MF (2018) Spasticity may obscure motor learning ability after stroke. *Journal of Neurophysiology* 119: 5–20.

Talis VL (2022) *The Doctor Who Loved Steam Engines. Recollections about Nikolai Aleksandrovich Bernstein*. Novoe Literaturnoe Obozrenie: Moscow (in Russian).

Tan H, Wade C, Brown P (2016) Post-movement beta activity in sensorimotor cortex indexes confidence in the estimation from internal models. *Journal of Neuroscience* 36: 1516–1528.

Tang PF, Woollacott MH (1998) Inefficient postural responses to unexpected slips during walking in older adults. *Journal of Gerontology A: Biological Sciences and Medical Sciences* 53: M471–M480.

Tatton WG, Bawa P, Bruce IC, Lee RG (1978) Long loop reflexes in monkeys: An interpretive base for human reflexes. *Progress in Clinical Neurophysiology* 4: 229–245.

Taub E, Morris DM (2001) Constraint-induced movement therapy to enhance recovery after stroke. *Current Atherosclerosis Reports* 3: 279–286.

Terekhov AV, Pesin YB, Niu X, Latash ML, Zatsiorsky VM (2010) An analytical approach to the problem of inverse optimization: An application to human prehension. *Journal of Mathematic Biology* 61: 423–453.

Thach WT, Goodkin HG, Keating JG (1992) Cerebellum and the adaptive coordination of movement. *Annual Review of Neuroscience* 15: 403–442.

Thompson AK, Wolpaw JR (2015) Restoring walking after spinal cord injury: Operant conditioning of spinal reflexes can help. *Neuroscientist* 21: 203–215.

Thompson AK, Wolpaw JR (2021) H-reflex conditioning during locomotion in people with spinal cord injury. *Journal of Physiology* 599: 2453–2469.

Tillman M, Ambike S (2018) Cue-induced changes in the stability of finger force-production tasks revealed by the uncontrolled manifold analysis. *Journal of Neurophysiology* 119: 21–32.

Tillman M, Ambike S (2020) The influence of recent actions and anticipated actions on the stability of finger forces during a tracking task. *Motor Control* 24: 365–382.

Ting LH, Macpherson JM (2005) A limited set of muscle synergies for force control during a postural task. *Journal of Neurophysiology* 93: 609–613.

Ting LH, McKay JL (2007) Neuromechanics of muscle synergies for posture and movement. *Current Opinions in Neurobiology* 17: 622–628.

Todorov E (2004) Optimality principles in sensorimotor control. *Nature Neuroscience* 7: 907–915.

Todorov E, Jordan MI (2002) Optimal feedback control as a theory of motor coordination. *Nature Neuroscience* 5: 1226–1235.

Tomita Y, Feldman AG, Levin MF (2017) Referent control and motor equivalence of reaching from standing. *Journal of Neurophysiology* 117: 303–315.

Tran US, Stieger S, Voracek M (2014) Evidence for general right-, mixed-, and left-sidedness in self-reported handedness, footedness, eyedness, and earedness, and a primacy of footedness in a large-sample latent variable analysis. *Neuropsychologia* 62: 220–232.

Tremblay S, Shiller DM, Ostry DJ (2003) Somatosensory basis of speech production. *Nature* 423: 866–8669.

Tresch MC, Cheung VC, d'Avella A (2006) Matrix factorization algorithms for the identification of muscle synergies: Evaluation on simulated and experimental data sets. *Journal of Neurophysiology* 95: 2199–2212.

Turpin NA, Feldman AG, Levin MF (2017) Stretch-reflex threshold modulation during active elbow movements in post-stroke survivors with spasticity. *Clinical Neurophysiology* 128: 1891–1897.

Turvey MT (1990) Coordination. *American Psychologist* 45: 938–953.

Turvey MT (2007) Action and perception at the level of synergies. *Human Movement Science* 26: 657–697.

Uchiyama T, Johansson H, Windhorst U (2003) Static and dynamic input-output relations of the feline medial gastrocnemius motoneuron-muscle system subjected to recurrent inhibition: A model study. *Biological Cybernetics* 89: 264–273.

Uygur M, Prebeg G, Jaric S (2014) Force control in manipulation tasks: Comparison of two common methods of grip force calculation. *Motor Control* 18: 18–28.

Vaillancourt DE, Russell DM (2002) Temporal capacity of short-term visuomotor memory in continuous force production. *Experimental Brain Research* 145: 275–285.

Vaillancourt DE, Slifkin AB, Newell KM (2001) Visual control of isometric force in Parkinson's disease. *Neurophysiologia* 39: 1410–1418.

Vaillancourt DE, Thulborn KR, Corcos DM (2003) Neural basis for the processes that underlie visually guided and internally guided force control in humans. *Journal of Neurophysiology* 90: 3330–3340.

Valero-Cuevas FJ, Smaby N, Venkadesan M, Peterson M, Wright T (2003) The strength-dexterity test as a measure of dynamic pinch performance. *Journal of Biomechanics* 36: 265–270.

Vallbo AB (1971) Muscle spindle response at the onset of isometric voluntary contractions. Time difference between fusimotor and skeletomotor effects. *Journal of Physiology* 218: 405–431.

Vallbo AB (1974) Human muscle spindle discharge during isometric voluntary contractions. Amplitude relations between spindle frequency and torque. *Acta Physiologica Scandinavica* 90: 319–336.

Van Beek N, Stegeman DF, Jonkers I, de Korte CL, Veeger D, Maas H (2019) Single finger movements in the aging hand: changes in finger independence, muscle activation patterns and tendon displacement in older adults. *Experimental Brain Research* 237: 1141–1154.

Van Doren CL (1995) Pinch force matching errors predicted by an equilibrium-point model. *Experimental Brain Research* 106: 488–492.

Van Doren CL (1998) Differential effects of load stiffness on matching pinch force, finger span, and effort. *Experimental Brain Research* 120: 487–495.

van Ee R, van Dam LC, Brouwer GJ (2005) Voluntary control and the dynamics of perceptual bi-stability. *Vision Research* 45: 41–55.

van Ginneken WF, Poolton JM, Capio CM, van der Kamp J, Choi CSY, Masters RSW (2018) Conscious control is associated with freezing of mechanical degrees of freedom during motor learning. *Journal of Motor Behavior* 50: 436–456.

van Groeningen CJ, Nijhof EJ, Vermeule FM, Erkelens CJ (1999) Relation between torque history, firing frequency, decruitment levels and force balance in two flexors of the elbow. *Experimental Brain Research* 129: 592–604.

Vasiliev YM, Gelfand IM, Shik ML (1969) Interaction in biological systems. *Priroda (Nature)*, Issue 6: 13–21 (in Russian).

Vaz DV, Pinto VA, Junior RRS, Mattos DJS, Mitra S (2019) Coordination in adults with neurological impairment: A systematic review of uncontrolled manifold studies. *Gait and Posture* 69: 66–78.

Vereijken B, van Emmerick REA, Whiting HTA, Newell KM (1992) Free(z)ing degrees of freedom in skill acquisition. *Journal of Motor Behavior* 24: 133–142.

Viviani P, Terzuolo C (1980) Space-time invariance in learned motor skills. In: Stelmach GE, Requin J (Eds.) *Tutorials in Motor Behavior*, pp. 525–533. N-Holland Publ.: Amsterdam.

Vøllestad NK (1997) Measurement of human muscle fatigue. *Journal of Neuroscience Methods* 74: 219–227.

Von Holst E, Mittelstaedt H (1950/1973) Daz reafferezprincip. Wechselwirkungen zwischen Zentralnerven-system und Peripherie, Naturwiss. 37: 467–476. The reafference principle. In: *The Behavioral Physiology of Animals and Man. The Collected Papers of Erich von Holst* (Trans. R Martin), 1, pp. 139–173. University of Miami Press: Coral Gables, FL.

Voudouris D, Smeets JB, Brenner E (2012) Do obstacles affect the selection of grasping points? *Human Movement Science* 31: 1090–1102.

Voudouris D, Smeets JB, Brenner E (2013) Ultra-fast selection of grasping points. *Journal of Neurophysiology* 110: 1484–1489.

Wahnoun R, He J, Helms Tillery SI (2006) Selection and parameterization of cortical neurons for neuroprosthetic control. *Journal of Neural Engineering* 3: 162–171.

Wang Y, Asaka T, Zatsiorsky VM, Latash ML (2006) Muscle synergies during voluntary body sway: Combining across-trials and within-a-trial analyses. *Experimental Brain Research* 174: 679–693.

Wang Y, Zatsiorsky VM, Latash ML (2005) Muscle synergies involved in shifting center of pressure during making a first step. *Experimental Brain Research* 167: 196–210.

Watson JD, Colebatch JG, McCloskey D (1984) Effects of externally imposed elastic loads on the ability to estimate position and force. *Behavioral and Brain Research* 13: 267–271.

Watts RL, Koller WC (Eds.) (2004) *Movement Disorders. Neurological Principles and Practice*, Second edition. McGraw-Hill: New York.

Webb B (2004) Neural mechanisms for prediction: Do insects have forward models? *Trends in Neuroscience* 27: 278–282.

Weber KD, Fletcher WA, Gordon CR, Melvill Jones G, Block EW (1998) Motor learning in the 'podokinetic' system and its role in spatial orientation during locomotion. *Experimental Brain Research* 120: 377–385.

Welsh T, Elliott D (2000) Preparation and control of goal-directed limb movements in persons with Down syndrome. In: Weeks DJ, Chua R, Elliott D (Eds.) *Perceptual-Motor Behavior in Down Syndrome*, pp. 49–70. Human Kinetics: Urbana, IL.

Westling G, Johansson RS (1984) Factors influencing the force control during precision grip. *Experimental Brain Research* 53: 277–284.

Wilhelm L, Zatsiorsky VM, Latash ML (2013) Equifinality and its violations in a redundant system: Multi-finger accurate force production. *Journal of Neurophysiology* 110: 1965–1973.

Windhorst U, Christakos CN, Koehler W, Hamm TM, Enoka RM, Stuart DG (1986) Amplitude reduction of motor unit twitches during repetitive activation is accompanied by relative increase of hyperpolarizing membrane potential trajectories in homonymous α-motoneurones. *Brain Research* 398: 181–184.

Wing AM (2000) Motor control: Mechanisms of motor equivalence in handwriting. *Current Biology* 10: R245–R248.

Winter DA, Patla AE, Prince F, Ishac M, Gielo-Perczak K (1998) Stiffness control of balance in quiet standing, *Journal of Neurophysiology* 80: 1211–1221.

Winter DA, Prince F, Frank JS, Powell C, Zabjek KF (1996) Unified theory regarding A/P and M/L balance in quiet stance. *Journal of Neurophysiology* 75: 2334–2343.

Wolpaw JR (2007) Spinal cord plasticity in acquisition and maintenance of motor skills. *Acta Physiologica (Oxford)* 189: 155–169.

Wolpaw JR, Bedlack RS, Reda DJ, Ringer RJ, Banks PG, Vaughan TM, Heckman SM, McCane LM, Carmack CS, Winden S, McFarland DJ, Sellers EW, Shi H, Paine T, Higgins DS, Lo AC, Patwa HS, Hill KJ, Huang GD, Ruff RL (2018) Independent home use of a brain-computer interface by people with amyotrophic lateral sclerosis. *Neurology* 91: e258–e267.

Wolpaw JR, Carp JS (1993) Adaptive plasticity in spinal cord. *Advances in Neurology* 59: 163–174.

Wolpaw JR, Kamesar A (2022) Heksor: The central nervous system substrate of an adaptive behaviour. *Journal of Physiology* 600: 3423–3452.

Wolpert DM, Miall RC, Kawato M (1998) Internal models in the cerebellum. *Trends in Cognitive Science* 2: 338–347.

Woods JJ, Furbush F, Bigland-Ritchie B (1987) Evidence for a fatigue-induced reflex inhibition of motoneuron firing rates. *Journal of Neurophysiology* 58: 125–137.

Woods S, O'Mahoney C, McKiel A, Natale L, Falk B (2023) Child-adult differences in antagonist muscle coactivation: A systematic review. *Journal of Electromyography and Kinesiology* 68: 102727.

Woollacott M, Inglin B, Manchester D (1988) Response preparation and posture control. Neuromuscular changes in the older adult. *Annals of the New York Academy of Sciences* 515: 42–53.

Woollacott MH, Shumway-Cook A (1990) Changes in posture control across the life span—a systems approach. *Physical Therapy* 70: 799–807.

Wu Y-H, Latash ML (2014) The effects of practice on coordination. *Exercise and Sport Sciences Reviews* 42: 37–42.

Wu Y-H, Pazin N, Zatsiorsky VM, Latash ML (2012) Practicing elements vs. practicing coordination: Changes in the structure of variance. *Journal of Motor Behavior* 44: 471–478.

Wu Y-H, Pazin N, Zatsiorsky VM, Latash ML (2013) Improving finger coordination in young and elderly persons. *Experimental Brain Research* 226: 273–283.

Wurtz RH (2018) Corollary discharge contributions to perceptual continuity across saccades. *Annual Review in Vision Science* 4: 215–237.

Xu Y, Terekhov AV, Latash ML, Zatsiorsky VM (2012) Forces and moments generated by the human arm: Variability and control. *Experimental Brain Research* 223: 159–175.

Yadav V, Sainburg RL (2014) Limb dominance results from asymmetries in predictive and impedance control mechanisms. *PLoS One* 9(4): e93892.

Yamada K, Nagakane Y, Yoshikawa K, Kizu O, Ito H, Kubota T, Akazawa K, Oouchi H, Matsushima S, Nakagawa M, Nishimura T (2007) Somatotopic organization of thalamocortical projection fibers as assessed with MR tractography. *Radiology* 242: 840–845.

Yamagata M, Falaki A, Latash ML (2019a) Effects of voluntary agonist-antagonist coactivation on stability of vertical posture. *Motor Control* 23: 304–326.

Yamagata M, Gruben K, Falaki A, Ochs WL, Latash ML (2021) Biomechanics of vertical posture and control with referent joint configurations. *Journal of Motor Behavior* 53: 72–83.

Yamagata M, Popow M, Latash ML (2019b) Beyond rambling and trembling: Effects of visual feedback on slow postural drift. *Experimental Brain Research* 237: 865–871.

Yang J-F, Scholz JP, Latash ML (2007) The role of kinematic redundancy in adaptation of reaching. *Experimental Brain Research* 176: 54–69.

Yang YK, Lin CY, Chen PH, Jhou HJ (2023) Timing and dose of constraint-induced movement therapy after stroke: A systematic review and meta-regression. *Journal of Clinical Medicine* 12(6): 2267.

Yen JT, Auyang AG, Chang YH (2009) Joint-level kinetic redundancy is exploited to control limb-level forces during human hopping. *Experimental Brain Research* 196: 4349–4351.

Yu H, Sternad D, Corcos DM, Vaillancourt DE (2007) Role of hyperactive cerebellum and motor cortex in Parkinson's disease. *NeuroImage* 35: 222–233.

Yufik YM, Friston K (2016) Life and understanding: Origins of the understanding capacity in self-organizing nervous systems. *Frontiers in the System Neuroscience* 10: 98.

Zatsiorsky VM (1998) *Kinematics of Human Motion*. Human Kinetics: Champaign, IL.

Zatsiorsky VM (2002) *Kinetics of Human Motion*. Human Kinetics: Champaign, IL.

Zatsiorsky VM, Duarte M (1999) Instant equilibrium point and its migration in standing tasks: Rambling and trembling components of the stabilogram. *Motor Control* 3: 28–38.

Zatsiorsky VM, Duarte M (2000) Rambling and trembling in quiet standing. *Motor Control* 4: 185–200.

Zatsiorsky VM, Gao F, Latash ML (2003) Prehension synergies: Effects of object geometry and prescribed torques. *Experimental Brain Research* 148: 77–87.

Zatsiorsky VM, Gao F, Latash ML (2005) Motor control goes beyond physics: Differential effects of gravity and inertia on finger forces during manipulation of hand-held objects. *Experimental Brain Research* 162: 300–308.

Zatsiorsky VM, Gregory RW, Latash ML (2002a) Force and torque production in static multi-finger prehension: Biomechanics and control. Part I. Biomechanics. *Biological Cybernetics* 87: 50–57.

Zatsiorsky VM, Gregory RW, Latash ML (2002b) Force and torque production in static multi-finger prehension: Biomechanics and control. Part II. Control. *Biological Cybernetics* 87: 40–49.

Zatsiorsky VM, Latash ML (2008) Multi-finger prehension: An overview. *Journal of Motor Behavior* 40: 446–476.

Zatsiorsky VM, Latash ML, Gao F, Shim JK (2004) The principle of superposition in human prehension. *Robotica* 22: 231–234.

Zatsiorsky VM, Li ZM, Latash ML (1998) Coordinated force production in multi-finger tasks: Finger interaction and neural network modeling. *Biological Cybernetics* 79: 139–150.

Zatsiorsky VM, Li ZM, Latash ML (2000) Enslaving effects in multi-finger force production. *Experimental Brain Research* 131: 187–195.

Zatsiorsky VM, Prilutsky BI (2012) *Biomechanics of Skeletal Muscles*. Human Kinetics: Urbana, IL.

Zelik KE, La Scaleia V, Ivanenko YP, Lacquaniti F (2014) Can modular strategies simplify neural control of multidirectional human locomotion? *Journal of Neurophysiology* 111: 1686–1702.

Zelman I, Titon M, Yekutieli Y, Hanassy S, Hochner B, Flash T (2013) Kinematic decomposition and classification of octopus arm movements. *Frontiers in Computational Neuroscience* 7: 60.

Zhang L, Feldman AG, Levin MF (2018) Vestibular and corticospinal control of human body orientation in the gravitational field. *Journal of Neurophysiology* 120: 3026–3041.

Zhang L, Straube A, Eggert T (2016) Under threshold position control, peripheral mechanisms compensate efficiently for small perturbations of arm movements. *Motor Control* 20: 87–108.

Zhang W, Sainburg RL, Zatsiorsky VM, Latash ML (2006) Hand dominance and multi-finger synergies. *Neuroscience Letters* 409: 200–204.

Zhou T, Falaki A, Latash ML (2016) Unintentional movements induced by sequential transient perturbations in a multi-joint positional task. *Human Movement Science* 46: 1–9.

Zhou T, Solnik S, Wu Y-H, Latash ML (2014a) Equifinality and its violations in a redundant system: Control with referent configurations in a multi-joint positional task. *Motor Control* 18: 405–424.

Zhou T, Solnik S, Wu Y-H, Latash ML (2014b) Unintentional movements produced by back-coupling between the actual and referent body configurations: Violations of equifinality in multi-joint positional tasks. *Experimental Brain Research* 232: 3847–3859.

Zhou T, Zatsiorsky VM, Latash ML (2015a) Unintentional changes in the apparent stiffness of the multi-joint limb. *Experimental Brain Research* 233: 2989–3004.

Zhou T, Zhang L, Latash ML (2015b) Intentional and unintentional multi-joint movements: Their nature and structure of variance. *Neuroscience* 289: 181–193.

Zhou T, Zhang L, Latash ML (2015c) Characteristics of unintentional movements by a multi-joint effector. *Journal of Motor Behavior* 47: 352–361.

Zimnik AJ, Churchland MM (2021) Independent generation of sequence elements by motor cortex. *Nature Neuroscience* 24: 412–424.

Index

For the benefit of digital users, indexed terms that span two pages (e.g., 52–53) may, on occasion, appear on only one of those pages.

Figures are indicated by an italic *f* following the page number.

adaptive changes, 321–22, 326–27
adequate language, 112, 297, 340, 343–44
Agarwal GC, 39–40
agility, 123, 126–38, 199
   definition, 129–30, 137–38
aging, 111, 133, 145, 206–7, 278, 289, 310, 317–20, 321–22, 382*f*
   changes
      adaptive, 321–22
      in anticipatory postural adjustments, 114, 123, 319
      in anticipatory synergy adjustments, 123, 130, 320–21
      in grip adjustments, 319–20
      in motor units, 317–19, 318*f*
      in synergies, 320–21
   death of neurons, 317–19, 318*f*
   denervation, 317–19, 318*f*
   neural plasticity, 320–21
   reinnervation, 317–19, 318*f*
Altenburger, H, 27–28
amyotrophic lateral sclerosis, 228
anticipation, 111–25
   anticipatory grip adjustments, 117–20, 177, 335
   anticipatory postural adjustments (APA), 112–15, 116–17, 117*f*, 123–24, 125, 335
   anticipatory synergy adjustments (ASA), 120–24, 125, 130, 136, 138, 145, 146*f*, 149–50, 185, 333–35, 333*f*, 334*f*, 336–38, 356
      in cerebellar disorders, 336
      in multiple sclerosis, 130, 336–37
      in Parkinson's disease, 130, 333–35, 333*f*, 334*f*
      in stroke, 130, 339, 339*f*
   desired future, 111
Archimedes, 343–44

Aristoteles, 44
artificial intelligence (AI), 371–72
Asanuma, H, 140, 142

Babinski, F, 79, 95
back-coupling, 69*f*, 95–96, 101, 146–47, 147*f*, 148, 204–5, 303
basal ganglia, 20, 21, 85–86, 95, 102, 110, 145, 151, 165, 166, 167, 249, 331–32, 332*f*, 335–36
   disorders of the basal ganglia, 95, 332*f*
   Parkinson's disease (PD), 20, 133, 150–51, 169, 216, 233, 305, 331–35, 340
      cardinal features, 332, 333
      Hoehn and Yahr stages, 333
      postural control in PD, 123, 202–3, 216, 276, 332–33
      synergies in PD, 145, 146*f*, 276, 305, 333–34, 333*f*, 335
      anticipatory synergy adjustments in PD, 123, 145, 146*f*, 305, 333*f*, 334*f*, 334–35
      treatment of PD, 334–35, 336–38
Bergson, H, 5, 12–13
Bernstein NA, 9, 15–25, 26, 28, 43, 44–45, 46, 48, 58, 59–60, 79–83, 95, 96–97, 111, 140–41, 144–45, 155, 191–92, 249, 297–99, 307, 328, 335–36, 355–56, 358
   levels of movement construction, 15–25
      leading level, 21
      level of tone (level A), 18, 155
      level of synergies (level B), 19, 19*f*, 20, 22, 80, 355–56
      level of spatial field (levels, C1 and, C2), 20, 22, 23, 25
      level of actions (level D), 21, 25
      levels of symbolic actions (level E), 21, 25
   paleokinetic and neokinetic mechanisms, 17

Brage, T, 344
brain-computer interface, 144

cerebellum, 21, 52, 145, 249, 335–36, 336*f*, 346*f*
  cerebellar disorders, 20, 21, 79, 95, 133, 336
  cerebellum and synergies, 20, 79, 85–86, 95, 102, 110, 151, 165, 166, 167, 335, 336*f*, 337–38
Churchill, W, 323
coactivation, 22–23, 38, 50, 58–59, 81, 105, 123–24, 204, 211–12, 222, 231, 254*f*, 254–56, 277, 278–93, 351–52, 363
  C-command, 36*f*, 36–37, 37*f*, 38*f*, 93, 94*f*, 99–100, 99*f*, 100*f*, 101–3, 105–9, 106*f*–7*f*, 115, 119–20, 120*f*, 124*f*, 125, 140–41, 144, 146–47, 149, 159–63, 168, 175, 177, 178*f*, 180–81, 184–85, 186, 204–5, 205*f*, 206–7, 211–12, 211*f*, 221, 221*f*, 235–39, 242, 248, 253, 255–59, 255*f*, 257*f*, 259*f*, 260, 262, 277, 278, 280–81, 282, 283*f*, 284–87, 291, 292–93, 299, 309, 314, 322, 363, 363*f*–66*f*, 364–67, 370, 373–74, 374*f*
  effects on perception, 224–26, 289–91, 292
  increased coactivation, 291–92, 298, 314, 315*f*, 319–20, 322
  mechanical effects, 22–23, 278, 291
  negative coactivation, 287–89
  relation to stability, 278, 282–83, 291–92
  role in EMG patterns, 280–82, 281*f*, 285–86, 286*f*

damping, 40, 50, 279, 350, 354–55, 358, 359–61, 366–67
  effects of coactivation, 279–80
  fractional power damping, 40, 360–61
  reflex-mediated damping, 40, 50, 354–55, 360–62, 366–67
desired future, 111
Down syndrome, 118–19, 304, 322, 323*f*
drifts, 231–48
  and the EP-hypothesis, 235–39
  of control variables, 194, 235–39, 241–42, 262–63, 290–92, 290*f*
  of enslaving, 66–67, 243–47, 248
  of force, 137, 137*f*, 150–51, 150*f*, 231, 232–34, 239–40, 257–59, 259*f*, 290–91, 351–52

  Kohnstamm phenomenon, 231, 251–52
  podokinetic effects, 231
  of position, 241–42
  of posture, 242–43
  unintentional movements, 231–48, 291–92
  visual feedback, 232–33, 237*f*, 239, 239*f*, 242, 244–45, 245*f*, 246, 260–61
  working memory, 233
dynamic field theory, 12, 146
dynamical systems approach, 46, 346–47
  relative timing, 46, 346–47

efference copy, 219–20, 249–63
  definition, 249
  exafference, 250, 251*f*
  reafference, 250, 251*f*, 251–52
  role in kinesthetic perception, 219–20, 249–63
Einstein, A, 5
electromyography, electromyogram (EMG), 27–28, 44–45, 141*f*, 157, 202–3, 215, 310, 311, 314*f*, 363, 368, 375
  control with EMG patterns, 47–52, 202–3, 340
    dual-strategy hypothesis, 48–50, 49*f*, 350, 351*f*, 353
    excitation pulse, 49, 49*f*, 350, 353
    pulse-step model, 48–49, 50–51, 51*f*, 353
  EMG-based modes, 51, 79–80, 81, 121, 209–10, 215, 272–73, 316–17, 378
  EMG-based primitives, 51, 79–80, 306
  tri-phasic pattern, 48*f*, 50–51, 208, 209*f*, 280–82, 281*f*, 285–86, 322, 348, 350, 351*f*, 374*f*
equifinality, 40, 42, 188–89, 194–95, 264–77, 300–1, 350–51
  violations of equifinality, 40, 194–95, 264, 269–71, 353–54
    under Coriolis force, 40, 41, 41*f*, 267–68, 268*f*, 353–54
    within the EP-hypothesis, 41, 267–69, 353–54
    under force perturbations, 269, 270, 270*f*, 271*f*, 273–74, 274*f*
    under negative damping, 40, 268, 353–54
    spontaneous, 277, 353–54
equilibrium-point hypothesis, 26–43, 56, 98, 194, 264, 267–69, 278, 284–85, 300*f*,

325–26, 345, 348, 350–52, 353–54, 356–57, 362–67
  alpha-model, 18, 39–42, 300–2, 300f, 301f, 325–26, 326f
  equilibrium point, 7f, 8–9, 45–46, 362–67
  single-joint control, 35–39
  single-muscle control, 30–34, 325–26
equilibrium trajectory, 35f, 42, 190, 283f, 284f, 349, 354–55, 354f, 355f
error compensation, 20–21, 60–61, 73, 356
Euclid, 343–44
Evarts EV, 140, 142
evolution, 3, 14, 16, 17–21

fatigue, 145, 166, 231, 233, 261–62, 266–67, 291–92, 303, 310, 311–15, 327
  adaptive effects, 266–67, 313–14, 317, 321–22
  effects on synergies, 166, 315–17
  peripheral and central effects, 311–12, 312f, 313f, 314–15
Feigenberg IM, 111
Feldman AG, 18–19, 26, 30, 39–40, 43
finger, 71–72, 72f, 73–74, 87, 93–94, 136–37, 158–59, 233–34, 315–16, 364, 376f, 379f
  cortical control, 172–73, 245–46, 248, 320–21, 321f
  enslaving, 65–67, 67f, 87, 172–75, 182, 243–47, 315–16, 319, 320, 326–27
  force deficit, 172–73
  force stabilization, 88–89, 89f, 90f, 101, 108–9, 315–16, 320
  mode, 66, 67f, 87, 173–75, 182, 243
  moment minimization, 24, 320
  moment stabilization, 88–89, 89f, 90f, 101, 320
  sharing, 24, 24f, 73–74, 73f, 133f, 134f, 134–35
  virtual finger, 118f, 177–79
Fitts' law, 13
friction, 117, 118f, 175–76

Gelfand IM, 16, 57, 343–44, 345
Georgopoulos AP, 142
Gibson JJ, 14, 20
Gottlieb GL, 39–40
grip (Grasp), 117–20, 163, 175–77, 180–81, 231, 265, 266
  force, 89f, 117–19, 322

safety margin, 117, 118–19, 175–76, 176f, 240
Gurfinkel VS, 112–13
hand, 170–86
  dominance, 59–60, 135–37, 138, 148–51, 162f, 165–66, 170, 196, 197–99, 233, 356
    dynamic dominance hypothesis, 135–36, 148–51, 195f, 233
  muscles, 106f, 170–71, 319
    compartments, 155, 158–59, 171, 172
    extrinsic, 155, 158–59, 162–63, 170–71, 284–85, 319
    intrinsic, 170–71, 319
  prehension, 136–37, 146–47, 177–82, 233–34
  synergies, 177–82

Heisenberg, W, 358
Helmholtz von, H, 219–20
hierarchical control, 15, 177–79, 178f, 180, 242, 246f
Houk JC, 144–45
Hughlings Jackson, J, 55, 79, 95, 98, 327–28, 329
hypothesis, 343–57, 371, 372, 378
  development, 353–56
  falsification, 356–57, 374–75, 381
  formulation, 345–47, 356, 374f, 374–75
  testing, 345–49, 356–57, 371, 376–77, 380

impedance, 149, 359
internal model, 3, 48, 52, 53f, 54–57, 139–40, 187, 345–46, 346f
  direct model, 53–54, 54f
  inverse model, 52–53, 53f, 54f
inverse dynamics, 16, 52
inverse kinematics, 52

Jaric, S, 344

Kelso JAS, 46
Kepler, J, 344
Kugler PN, 46

Lance JW, 327, 329, 367
large-fiber peripheral neuropathy, 33, 41–42, 217, 300, 301–2, 325–27
  changes in joint coordination, 217, 301–2, 325, 326–27

law of nature, 3-5, 10, 12-13, 14, 16, 52, 54-55, 231-32
  biologically-specific, 6, 10, 13
  definition, 4
  examples, 4, 6
  variables and parameters, 4-5, 6
leading-joint hypothesis, 187-88
living system, 6, 16
locomotion, 148, 189-90, 201, 212-15, 303-4, 306
  in animal preparations, 148, 213-14
  central pattern generator (CPG), 6, 27, 141, 147-48, 189-90, 212-15, 216
  stimulation-induced locomotion, 148, 213, 214

Matthews PBC, 30, 39-40
Merleau-Ponty, M, 10, 10f, 11f, 12, 14, 231-32
Merton PA, 29
methods of brain study, 139
Mittelstaedt, H, 28-29, 250
motoneurons
  alpha, 29-30, 29f, 32, 33, 44, 47, 48, 83, 101, 139, 155, 168, 172, 202-3, 220, 228, 251-52, 300, 306, 311, 313, 327-28, 346, 352-53, 370
  alpha-gamma coactivation, 29-30, 125, 220, 222, 352-53
  gamma, 29-30, 29f, 114-15, 125, 218, 220, 352-53, 352f
  threshold for activation, 32, 33, 33f, 37-38, 83, 84f, 370
motor development, 215, 297, 305-6
motor equivalence, 71-74, 87, 88f, 90, 105, 127-28, 193-94, 194f, 241, 264-77, 356
  as an index of stability, 72-73, 74, 83, 87, 127-28, 127f, 212, 271-72, 275-76
  in handwriting, 45, 71, 186, 264
motor learning, 297
  effects on synergies, 302-4
  stages, 15, 298-99, 302-4
motor programming, 44-57
  engram, 45-46, 71, 347
    topology vs. metrics, 45-46, 71
  generalized motor program, 45-46, 347
motor rehabilitation, 297, 306-9, 337-38
  after stroke, 307-9, 308f, 338-39
    within the EP-hypothesis, 308-9, 308f, 339

motor units, 68, 83, 105, 106, 120, 155, 156-59, 284-85, 311, 351-52, 368, 369-70
motor unit modes, 68, 83-86, 84f, 98, 105-8, 106f, 125, 156-59, 160f, 160-61, 161f, 284-85, 285f, 292
  synergies among motor units, 85-86, 105-8, 125, 156-59
multiple sclerosis (MS), 305, 310, 312-13, 329
  anticipatory synergy adjustments in MS, 305, 336-37, 337f
  synergies in MS, 305, 336-37, 337f
muscle, 6-9, 155-69, 311
  activation patterns, 38
  catch property, 40-41, 268-69
  muscle mode (M-mode), 67-68, 209f, 209-10
  muscle tone, 15, 18, 306, 307, 307f, 327, 328-29, 346, 369, 373-74
  spindle, 7-8
  tri-phasic EMG pattern, 48f, 50-51, 208, 209f

neuronal, 139-52
  homunculus, 140
  plasticity, 147-48, 166, 167, 324-25, 329-30
  population vector, 142-44
  representations, 141
Newton, I, 344

optimization, 23-25, 61, 71-72, 87, 126-38, 187-88, 199, 264, 303, 305, 320
  cost function, 24-25, 128-29, 130-32
  inverse optimization, 25, 128-29, 132-33, 320
  optimal feedback control, 130-32

parametric control, 6, 8-9, 10-11, 12, 14, 37-38, 39, 98, 139-40, 301-2, 353-54
Penfield, W, 140
perception, 39, 151, 217-28, 249-63, 351-52
  direct perception, 217
  effects of coactivation, 224-26, 289-91
  force perception, 218, 219-22, 223-24, 225-26, 253, 256-59, 279, 289-90
    force matching, 254f, 255f, 256, 257-58, 258f, 259-60, 261-62, 289-90, 291, 351-52, 377f
  iso-perceptual manifold, 222-24

kinesthetic illusions, 39, 253
  perception-to-act, 217, 253–55, 257
  Perception-to-report, 217, 253–55, 257
  perceptual equivalence, 223, 277
  proprioceptors, 218–19, 253
  role of referent coordinate, 252–56
  sense of effort, 261–62
persistent inward current (PIC), 33f, 34, 262–63, 314
piano playing, 15, 297–99
posture, 93, 94f, 112–13, 141, 201–16, 242–43, 264–65, 343, 345, 374f, 378
  effects of coactivation, 204, 206–7, 211–12, 282–83, 319, 381
  postural adjustments, 123–24, 205–7
    anticipatory postural adjustments (APA), 112–15, 116–17, 117f, 123–24, 205–6, 212, 319
    anticipatory synergy adjustments (ASA), 212–13, 213f, 332–33
    compensatory postural adjustments, 113, 206–7, 332–33
    early postural adjustments, 115–17
    postural adjustment to step initiation, 115–16, 116f, 212–13, 213f
  posture-movement paradox, 28–29, 28f, 38, 207–8, 207f, 277, 345, 347–48
  referent body orientation, 204–6, 205f, 207f, 208, 209f, 242, 243, 243f, 349f, 349, 373–74, 374f
  sway, 38, 105, 114–15, 122, 202–5, 215, 242, 319, 332–33, 351–52, 373–74, 382f
    rambling, 203–4, 204f, 215, 319
    trembling, 203–4, 204f, 215
  voluntary, 208, 209f, 276, 277, 316–17, 348, 349f, 374f
prediction, 53–54
preflexes, 206
pre-programmed reaction, 28–29, 177, 206–7, 265, 314–15, 332–33
principle
  of abundance, 58–75, 297
  of activity, 9, 15–16
  of sensory corrections, 19–20, 19f
  of superposition, 181–82
proprioceptors, 48, 218–19
  articular receptors, 218, 219, 222
  Golgi tendon organs, 48, 218–19, 220, 222, 227, 253

muscle spindles, 48, 218–19, 220, 222, 225, 227, 251–52, 253, 361

reaching, 52, 55–56, 58, 72–73, 187–200
  adaptation, 55–56, 56f
  equifinality, 188–89, 190, 265–66
    violations of equifinality, 40, 41, 41f, 268
  multi-joint synergy, 63–64, 69, 86–93, 191–94
  reach to grasp, 187, 195–97
reciprocal activation, 123–24
  R-command, 36f, 36, 37f, 37, 38f, 93, 94f, 99–100, 101–3, 106, 106f, 107–9, 107f, 115, 120, 124f, 125, 140–41, 144, 146–47, 149, 159–63, 168, 175, 177, 178f, 180–81, 184–85, 186, 204–5, 205f, 206–7, 211–12, 211f, 221, 221f, 235–39, 242, 248, 253, 255–59, 255f, 257f, 259f, 260, 262, 277, 280–81, 282, 283f, 284–87, 291, 292, 299, 309, 314, 322, 363, 363f–66f, 364–67, 370, 373–74, 374f
recurrent inhibition, 101, 101f
redundancy, 56, 58, 126, 182, 187, 222
  problem of motor redundancy, 15, 19, 22–23, 22f, 58–59, 59f, 155
  state redundancy, 22, 22f, 126
  trajectory redundancy, 22, 22f, 126
referent coordinate (RC), 36, 42, 140–41, 144, 166, 215, 220, 344, 359, 370
  as a control variable, 98–110, 119–20, 120f, 122–23, 135, 146, 175, 177, 187, 190–91, 196–97, 231–32, 235–39, 247–48, 250, 256–57, 257f, 261–62, 270, 319, 351–52, 367
  definition, 36
  hierarchical control with RC, 95, 96f, 101–2, 107, 109, 166, 214, 269, 270–71, 292–93, 297, 298–99, 299f, 303, 305, 306, 309, 320, 324–25, 343
  synergies in the RC space, 93–95, 102–8, 184–85, 185f
reflex, 11–12, 32, 33, 39, 43, 47, 48, 52–53, 92–93, 102, 163–65, 172, 202–3, 214, 217, 311, 314–15, 361–62, 361f, 364
  interjoint reflexes, 64, 326, 363
  long-loop reflex, 11–12, 28–29, 177, 206–7
  magnus-De Kleijn reflex, 11–12
  monosynaptic reflex, 13, 163–64, 265, 361
  pupillary reflex, 11–12

reflex (*cont.*)
  stretch reflex, 6–9, 13, 27, 28, 30, 34, 73–74, 93, 98, 156, 156*f*, 163–65, 219–22, 265, 267, 299*f*, 308–9, 313, 325, 352–54, 352*f*, 359–60, 362–63, 362*f*, 367
    threshold, 6, 30, 32, 37–38, 39, 45–46, 93, 98, 156, 156*f*, 219–22, 267, 267*f*, 299*f*, 307, 307*f*, 308–9, 308*f*, 325, 328–29, 328*f*, 331*f*, 348*f*, 353–54, 362–63, 362*f*, 367, 368*f*, 370
  tonic vibration reflex, 224
  unloading reflex, 47
  wiping reflex, 95, 188–89, 266–67

Schmidt RA, 46
Schöner, G, 61, 95–96, 146
Schrödinger, E, 249
servo-model (servo hypothesis), 29–30, 29*f*, 352–53, 352*f*
sharing, 23–25, 43
Sherrington CS, 27
skill acquisition, 15, 19, 45, 80, 167–68, 167*f*
skill transfer, 46, 186
spasticity, 13, 166, 327–31, 369
  definition, 327–28, 328*f*, 329
  following spinal cord injury, 327–31
  following stroke, 329, 331*f*, 338
  interpretation within the EP-hypothesis, 39, 328–29, 328*f*, 351–52, 367–68, 368*f*
  treatment, 329–30, 330*f*
speech, 265
Sperry RW, 250
spinal cord, 188–90
  injury, 169, 327–31
St Gregory Palamas, 79
stability, 19, 20–21, 61, 62*f*, 81–83, 82*f*, 113, 115, 120–21, 126–38, 150–51, 170, 199, 201, 209–10, 214–15, 222–24, 233–34, 265, 288–89, 299, 320, 355–56
stereotypy, 63, 303, 304–5
stiffness, 4, 190, 193–94, 358, 359–60
  apparent stiffness, 7, 36–37, 93, 94*f*, 99–100, 99*f*, 100*f*, 102–3, 109, 119–20, 120*f*, 135, 175, 177, 184–85, 185*f*, 202, 211–12, 211*f*, 236, 237*f*, 256–57, 257*f*, 279–80, 279*f*, 284–85, 290–91, 290*f*, 354–55, 359–60, 360*f*, 361–62, 363, 373–74, 374*f*, 377*f*
  ellipsoid of stiffness, 190
stroke, 79, 95, 145, 146*f*, 149, 151, 199, 253–54, 266, 338–39

synergy, 15, 38, 51, 79–97, 145, 201, 214, 215, 227–28, 236–38, 271–72, 297, 315–17, 344, 370, 373–74, 383*f*
  in animal studies, 96–97, 189–90
  in cerebellar disorders, 20, 79, 85–86, 95
  changes with practice, 167–68, 167*f*, 302–4
  among control variables, 93–95, 102–5, 108–9, 180–81, 184–85, 185*f*, 211–12, 211*f*, 237*f*, 345, 351–52, 355–56, 364–65, 365*f*
  definition, 63, 79–80
  goal-equivalent manifold (GEM), 70
  index, 63, 85*f*, 86–87, 120–22, 121*f*, 145, 238*f*, 238, 333–35, 333*f*, 334*f*, 339, 339*f*, 378
  intra-muscle synergies, 83–86, 102, 125, 148, 149–50, 155–69, 238*f*, 317, 330–31, 356
  Jacobian, 63, 64, 65–66, 68, 84, 86, 87, 109
  kinematic synergies, 63–64, 69, 86–93, 189–90, 210, 214–15, 303–4
  kinetic synergies, 64–67, 86–93, 182–85, 210–11, 316
  level of synergies, 19, 19*f*, 20
  matrix factorization methods, 20, 51, 67–68, 81, 209–10
  multifinger synergy, 63, 65–66, 66*f*, 120–22, 121*f*, 148, 149–50, 182–86, 238*f*, 238
  multijoint synergy, 60, 64–65, 187, 191–94
  multimuscle synergy, 51, 67–68, 83–86, 109, 157, 316–17
  in multiple sclerosis, 305, 336–37, 337*f*
  in non-linear systems, 69–70
  in Parkinson's disease, 333–35, 333*f*, 334*f*
  pathological synergy, 79
  postural synergies, 94, 105, 209–12, 211*f*, 316–17
  prehension synergies, 177–81, 316
  randomization method, 69–70, 94, 104–5, 236–38
  spinal cord and synergies, 85–86, 95–96, 102, 146–48, 165–68, 185–86, 189–90, 317, 330–31
  in stroke, 79, 339, 339*f*
  subcortical circuits and synergies, 85–86, 95–96, 102, 110, 146–47, 151, 165–68, 185–86, 317
  trade-offs in a hierarchical system, 74–75, 90–92, 92*f*, 102, 162–63, 163*f*, 164–65, 165*f*, 166, 180, 299, 303

Thalamus, 139, 145, 249, 335–36
timing, 46, 68, 69*f*, 89–90, 95–96, 113, 114, 123, 146, 147*f*, 183–84
trade-offs
  agility-stability, 135–37, 305
    relation to handedness, 135–36
  optimality-stability, 133–35, 309, 320, 356
  speed-accuracy (Fitts' law), 38, 196
  synergy indices in a hierarchy, 74–75, 90–92, 92*f*, 102, 162–63, 163*f*, 164–65, 165*f*, 166, 180
transcranial magnetic stimulation (TMS), 139, 141, 329–30, 351–52
Turvey MT, 14, 20, 46

uncontrolled manifold (UCM), 58–75, 79–97, 102–9, 126–38, 157, 179, 192–93, 194–95, 195*f*, 239–40, 241, 247*f*, 247–48, 271–75, 345, 356–57, 380–81, 380*f*
  analysis of motor equivalence, 64–67, 127–28, 271–76
  stability and the UCM hypothesis, 79–97, 127, 128, 239–40, 265, 275–76, 355–56

synergies and the UCM hypothesis, 58–75, 79–97, 127, 157, 182–84, 183*f*, 184*f*, 209–10, 355–56
variance analysis, 63, 79–97, 127, 192–94, 302–4, 315–17, 336–37, 337*f*

Vallbo, A, 29–30
variability, 38, 60, 61, 62*f*, 68, 69–70, 191–92, 304–5, 314, 315
  structure of variability, 62*f*, 63, 74, 83, 126, 305
vibration, 259–61, 262–63, 291
  effects on monosynaptic reflexes, 224
  effects on sensory endings, 224, 225, 226, 253, 259–60, 263
  vibration-induced fallings, 224–26
  vibration-induced illusions, 224–26, 253, 259–61, 263
  vibration-induced stepping, 214, 224, 225–26
Von Holst, E, 28–29, 250–52

Wachhölder, K, 27–28

Zatsiorsky VM, 358